FUNGI WITHOUT GILLS
(HYMENOMYCETES and GASTEROMYCETES)

FUNGI WITHOUT GILLS
(HYMENOMYCETES and GASTEROMYCETES)

An Identification Handbook

MARTIN B. ELLIS BSc., PhD., (Lond.)
and
J. PAMELA ELLIS BSc., Dip. syst. Mycol.

CHAPMAN AND HALL

LONDON • NEW YORK • TOKYO • MELBOURNE • MADRAS

UK   Chapman and Hall, 11 New Fetter Lane, London EC4P 4EE

USA   Chapman and Hall, 29 West 35th Street, New York NY10001

JAPAN   Chapman and Hall Japan, Thomson Publishing Japan, Hirakawacho
            Nemoto Building, 7F, 1-7-11 Hirakawa-cho, Chiyoda-ku, Tokyo 102

AUSTRALIA   Chapman and Hall Australia, Thomas Nelson Australia, 480 La Trobe
            Street, PO Box 4725, Melbourne 3000

INDIA   Chapman and Hall India, R. Sheshadri, 32 Second Main Road, CIT East,
            Madras 600 035

First edition 1990

© 1990 Martin B. Ellis and J. Pamela Ellis

Typeset in 10 on 12pt Galliard by Leaper & Gard Ltd, Bristol
Printed in Great Britain by St Edmundsbury Press,
Bury St Edmonds, Suffolk

ISBN 0 412 36970 2

British Library Cataloguing in Publication Data

Ellis, M. B. (Martin Beazor), 1911-
    Fungi without gills.
    1. Lichens. Fungi
    I. Title   II. Ellis, J. Pamela
    589.2

ISBN 0–412–36970–2

Library of Congress Cataloging-in-Publication Data

Ellis, Martin B. (Martin Beazor)
    Fungi without gills (hymenomycetes and gasteromycetes) : and
identification handbook / Martin B. Ellis and J. Pamela Ellis.   p.   cm.
    Includes bibliographical references.
    ISBN 0–412–36970–2
    1. Hymenomycetes–Identification.   2. Gasteromycetes–Identification.
    I. Ellis, J. Pamela. II. Title.
QK626.E45 1990
589.2′–dc20   89-70861
                                                              CIP

CONTENTS

INTRODUCTION

Many amateur mycologists nowadays have good microscopes and interest in fungi which can be named with certainty only when they have been examined microscopically is increasing. Some fungi without gills, such as Dryad's Saddle (*Polyporus squamosus*), Jew's Ear (*Auricularia auricula-judae*), Common Bird's-nest Fungus (*Crucibulum laeve*), Beefsteak Fungus (*Fistulina hepatica*), Stinkhorn (*Phallus impudicus*) and Sulphur Polypore (*Laetiporus sulphureus*) are recognized easily macroscopically, but a microscope is needed for the determination of most of them.

During the past two years this book, in manuscript form, was used without reference to any other work to identify several thousand fresh collections of non-gilled fungi. We worked together as a team, every microscopic preparation was seen by both of us and we were able to sort out taxonomic problems through discussion. The numerous keys to all genera and species proved invaluable. Inaccuracies found we were able to put right when we found them. In previous years we used many books, monographs and other scientific papers to help us identify fungi without gills and it was often a long and laborious task. Each year from 1973 onwards as successive volumes of *The Corticiaceae of North Europe*, beautifully illustrated by John Eriksson, were published by Fungiflora, Oslo, we received fresh stimulus and help in this work. Leif Ryvarden's two volumes *The Polyporaceae of North Europe* (1976, 1978), also published by Fungiflora, proved most useful, as did Roy Watling's *Boletaceae* (British Fungus Flora 1, 1970). It was not until 1984, however, that keys to many groups became available with the publication by Gustav Fischer Verlag, Stuttgart, of Walter Jülich's excellent book *Die Nichtblätterpilze, Gallertpilze und Bauchpilze*.

In 1986 J. Breitenbach and F. Kränzlin described 528 species of non-gilled fungi in Vol. 2 of *Fungi of Switzerland*. Their beautiful colour photographs, which portray these fungi so accurately, make it possible for even quite inexperienced amateur mycologists to start identifying many of the species straightaway. There is no doubt that this book will stimulate interest in groups of fungi which have hitherto attracted relatively few workers. With a little practice resupinate fungi, club fungi, jelly fungi, brackets, earth-stars, puffballs and so on are not at all difficult to identify. Many of them flourish and sporulate freely during the winter months when there are few other basidiomycetes to be seen. January, February and March are particularly good collecting months for resupinate and jelly fungi and there are few mosquitoes around then. It is possible to study mycology the whole year round if you are prepared to look at all groups of fungi.

In this book measurements, except where stated otherwise, are in thousandths of a millimetre (microns) and much space has been saved by not constantly repeating the μm sign. Terms which may be unfamiliar are explained in the Glossary. We have not included a bibliography which would have run into many pages. The books mentioned above have bibliographies which may be consulted if needed.

Nine hundred and seventy four species in 277 genera are described in this book and 543 of them are figured. There are two major keys to genera and there is a key to the species in each genus that contains more than one species. Where a genus has only one species the generic and specific descriptions are combined. The fungi described and illustrated can nearly all be found in Great Britain and many of them are quite common. A few that have been recorded in the past such as *Myriostoma coliforme* can still be found in Europe although they have not been seen in Britain for 100 years.

For quick and easy reference genera in the two main sections are arranged in alphabetical order and the systematic position of each genus is given below.

NON-GILLED HYMENOMYCETES

AGARICALES
   CREPIDOTACEAE Chromocyphella, Episphaeria, Merismodes, Pellidiscus
   TRICHOLOMATACEAE Asterophora, Calyptella, Cellypha, Cyphellostereum, Delicatula, Flagelloscypha, Geopetalum, Glolocephala, Lachnella, Leptoglossum. Stigmatolemma
APHYLLOPHORALES
   AURISCALPIACEAE Auriscalpium
   BANKERACEAE Bankera, Hydnellum, Phellodon, Sarcodon
   BOLETOPSIDIACEAE Boletopsis
   CANTHARELLACEAE Cantharellus, Craterellus, Pseudocraterellus
   CLAVARIACEAE Ceratellopsis, Clavaria, Clavariadelphus, Clavulinopsis, Macrotyphula, Mucronella, Pistillaria, Pterula, Ramariopsis, Typhula
   CLAVULINACEAE Clavulina
   CONIOPHORACEAE Coniophora, Hypochniciellum, Jaapia, Leucogyrophana, Pseudomerulius, Serpula
   CORTICIACEAE Acanthobasidium, Aleurodiscus, Amphinema, Aphanobasidium, Athelia, Athelopsis, Auriculariopsis, Boidinia, Botryobasidium, Botryohypochnus, Brevicellicium, Butlerelfia, Byssocorticium, Ceraceomyces, Cerocorticium, Confertosbasidium, Corticium, Cristinia, Crustoderma, Cylindrobasidium, Cytidia, Dacryobasidium, Dacryobolus, Dendrothele, Digitatispora,

Echinotrema, Epithele, Erythricium, Fibrodontia, Fibulomyces,
Gloeocystidiellum, Grandinia, Granulobasidium, Hyphoderma,
Hyphodermella, Hypochnella, Hypochnicium, Hypochnopsis,
Irpicodon, Laetisaria, Lagarobasidium, Laxitextum, Lazulinospora,
Leptosporomyces, Limonomyces, Lindtneria, Litschauerella, Luellia,
Lyomyces, Megalocystidium, Meruliopsis, Merulius, Mycoacia,
Parvobasidium, Paullicorticium, Peniophora, Phanerochaete, Phlebia,
Phlebiopsis, Piloderma, Pteridomyces, Pulcherricium, Resinicium,
Sarcodontia, Scopuloides, Sistotrema, Subulicystidium, Trechispora,
Tubulicrinis, Tylospora, Vesiculomyces, Vuilleminia, Xenasma,
Xenasmatella
'CYPHELLACEAE' a convenience title for cup-, saucer- and
shell-shaped aphyllophorales and agaricales which belong to several
different families. Here are included Calyptella, Cellypha,
Cyphellostereum, Flagelloscypha, Henningsomyces, Lachnella,
Leptoglossum, Merismodes, Pellidiscus, Stigmatolemma and
Stromatoscypha
FISTULINACEAE Fistulina
GANODERMATACEAE Ganoderma
GOMPHACEAE Gomphus, Lentaria, Ramaria
HERICIACEAE Artomyces, Clavicorona, Creolophus, Hericium
HYDNACEAE Hydnum
HYMENOCHAETACEAE Coltricia, Hymenochaete, Inonotus, Onnia,
Phellinus
LACHNOCLADIACEAE Asterostroma, Scytinostroma, Vararia
POLYPORACEAE Abortiporus, Albatrellus, Antrodia, Antrodiella,
Aurantioporus, Bjerkandera, Buglossoporus, Ceriporia,
Ceriporiopsis, Cerrena, Chaetoporellus, Cinereomyces, Daedalea,
Daedalcopsis, Datronia, Dendropolyporus, Dichomitus,
Donkioporia, Flaviporus, Fomes, Fomitopsis, Funalia,
Gloeophyllum, Gloeoporus, Grifola, Hapalopilus, Heterobasidion,
Ischnoderma, Laetiporus, Lenzites, Leptoporus, Loweomyces,
Meripilus, Oligoporus, Oxyporus, Perenniporia, Phaeolus,
Physisporinus, Piptoporus, Polyporus, Porpomyces, Postia,
Pycnoporus, Rigidoporus, Schizopora, Skeletocutis, Spongipellis,
Trametes, Trichaptum, Tyromyces
SCHIZOPHYLLACEAE Henningsomyces, Plicaturopsis,
Stromatoscypha
SPARASSIDACEAE Sparassis
STECCHERINACEAE Irpex, Steccherinum
STEREACEAE Amylostereum, Chondrostereum, Columnocystis,
Cotylidia, Podoscypha, Stereopsis, Stereum
THELEPHORACEAE Pseudotomentella, Thelephora, Tomentella,
Tomentellastrum, Tomentellina, Tomentellopsis

AURICULARIALES
  Achroomyces, Auricularia, Eocronartium, Helicobasidium, Helicogloea, Herpobasidium, Mycogloea, Phleogena, Pilacrella, Stilbum
BOLETALES
  Aureoboletus, Boletinus, Boletus, Gyroporus, Leccinum, Porphyrellus, Strobilomyces, Suillus, Tylopilus, Uloporus
DACRYMYCETALES
  Calocera, Dacrymyces, Dicellomyces, Ditiola, Femsjonia, Guepiniopsis
EXOBASIDIALES
  Exobasidium, Microstroma
TREMELLALES
  Basidiodendron, Bourdotia, Efibulobasidium, Eichleriella, Exidia, Exidiopsis, Heterochaetella, Myxarium, Protodontia, Pseudohydnum, Sebacina, Stypella, Tremella, Tremellodendropsis, Tremiscus, Xenolachne
TULASNELLALES
  CERATOBASIDIACEAE Cejpomyces, Ceratobasidium, Scotomyces, Thanatephorus, Uthatobasidium
  TULASNELLACEAE Tulasnella

GASTEROMYCETES

GAUTERIALES
  Gauteria
HYMENOGASTRALES
  Gastrosporium, Gymnomyces, Hydnangium, Hymenogaster, Octaviania, Rhizopogon, Sclerogaster, Stephanospora, Wakefieldia, Zelleromyces
LYCOPERDALES
  GEASTRACEAE Geastrum, Myriostoma
  LYCOPERDACEAE Bovista, Bovistella, Calvatia, Langermannia, Lycoperdon, Vascellum
  MYCENASTRACEAE Mycenastrum
MELANOGASTRALES
  Leucogaster, Melanogaster
NIDULARIALES
  NIDULARIACEAE Crucibulum, Cyathus, Mycocalia, Nidularia
  SPHAEROBOLACEAE Sphaerobolus
PHALLALES
  CLATHRACEAE Anthurus, Aseroe, Clathrus, Ileodictyon, Lysurus
  HYSTERANGIACEAE Hysterangium

PHALLACEAE Dictyophora, Mutinus, Phallus
SCLERODERMATALES
    Astraeus, Pisolithus, Scleroderma
TULOSTOMATALES
    Battarraea, Queletia, Tulostoma

In the plates at the end of this book the habit sketches of fruitbodies have been made at different magnifications and reference to the text should be made for their dimensions: camera-lucida drawings of microscopic features such as spores, basidia and cystidia are ×600. For use of their libraries and for access to herbarium specimens we are very grateful to the directors and staff of the C.A.B. International Mycological Institute and the Royal Botanic Gardens, Kew.

NON-GILLED HYMENOMYCETES

In this section are included all groups of basidiomycetes without gills, other than rusts and smuts, where the basidia form a sort of palisade covering surfaces which are smooth, ridged, folded or have projecting teeth, or line open-ended tubes. In many of them the fruitbodies are resupinate, lying flat on the substrate with the hymenium on the outside; in others they are bracket- or shelf-like, hemispherical, pulvinate, lobed or cerebriform, cup-, ear-, saucer- or shell-shaped. They may be erect, simple or branched, club-, fan-, funnel- or trumpet-shaped, resemble antlers or coral or have a cap and stalk. The majority have non-septate, often narrowly clavate basidia with 2–8, most commonly 4, slender sterigmata at the apex. Where fruitbodies have a jelly-like consistency, however, the basidia are usually either transversely or longitudinally septate or are deeply divided and resemble tuning-forks. Common examples of species with transversely 1–3-septate basidia are Jew's Ear (*Auricularia auricula-judae*) and Tripe Fungus (*A. mesenterica*). Golden Jelly Fungus (*Tremella mesenterica*) and the well-known black Witches' Butter (*Exidia glandulosa*) have basidia with longitudinal septa. Tuning-fork types are seen in the Yellow Antler Fungus (*Calocera viscosa*) and in the widespread and very common *Dacrymyces stillatus* which forms little yellow or orange cushions or blobs on wood of both conifers and deciduous trees and frequently also on rails and fence-posts.

Many of the fungi described here play a very active part in breaking down cellulose in wood and leaves so that humus is formed. Some cause white- or brown-rots in standing trees, others wet- or dry-rots in worked timber.

When examining specimens mounts should be made initially in both water and Melzer's iodine solution. In water, spore colour and the presence or absence of oil drops or vacuoles can be established, but resolution in the iodine solution is often much better and it tells you whether or not spores are amyloid or dextrinoid. If necessary further mounts can be made in lacto-phenol cotton blue or potassium hydroxide (KOH) solution. Thin sections can be cut quite easily with a sharp safety-razor blade. Cuts made should be quite vertical but need not be quite parallel to one another; if they converge slightly the very thin parts where they meet are often excellent for observing basidia and embedded cystidia. Sometimes non-sporulating material is collected, it is either too young or too old. Occasionally sporulation can be induced by keeping specimens for a short time in tubes or petri-dishes. They soon become mouldy and are best discarded.

1

KEY TO GENERA

No visible fruitbodies other than basidia and spores 1

Visible fruitbodies formed 3

1. Parasitic fungi causing leaf-spots, distortion and galls in *Camellia*, *Rhododendron* and *Vaccinium* .... *Exobasidium*

   Parasitic fungi on leaves of *Juglans* and *Quercus*, with clustered basidia erumpent through stomata .... *Microstroma*

   Parasitic in other fungi 2

2. In the hymenium of *Dacrymyces*, basidia transversely septate .... *Achroomyces*

   In the hymenium of *Dacrymyces* and *Ditiola*, basidia longitudinally septate .... *Tremella*

   In the hymenium of Coticiaceae, basidia not transversely or longitudinally septate .... *Tulasnella*

3. Fruitbodies resupinate 4

   Fruitbodies not resupinate 178

4. Hymenial surface smooth, flat, undulating, granular or tuberculate .. 5

   Hymenial surface otherwise, ridged, folded, phlebioid, merulioid or with teeth or pores 106

5. Basidia with 1–3 transverse septa 6

   Basidia with longitudinal septa 9

   Basidia without septa 16

6. Parasitic on *Lonicera* and on fern fronds .... *Herpobasidium*

   Not so 7

7. Fruitbodies small, never more than 5 cm diam., basidia cylindrical to narrowly clavate or fusiform .... *Achroomyces*

   Fruitbodies larger and/or with differently shaped basidia 8

8. Basidia cylindrical, resupinate part of fruitbody almost always accompanied by narrow, imbricate brackets .... *Auricularia*

   Basidia curved, often helically .... *Helicobasidium*

   Basidia greatly elongated, only terminal part transversely septate and each with a lateral saccate probasidium at its base .... *Helicogloea*

9. Fruitbody replacing hymenium of its discomycete host, sterigmata two, very long .... *Xenolachne*

   Fruitbodies not so 10

10. Basidia on short side-branches of an erect main hypha, remaining as empty cases after spores have been liberated .... *Basidiodendron*

    Basidia not so, usually solitary 11

11. Basidia stalked, with clamp at base of stalk but not of basidium .... *Myxarium*

    Basidia not so 12

12. Cystidia present 13

    Cystidia absent 14

13. Cystidia hyaline, thick-walled, protruding a long way above surface of hymenium .... *Heterochaetella*
    Cystidia (g) thin-walled, with yellowish or brownish contents *Bourdotia*
14. No clamps on basidia or hyphae .... *Sebacina*
    Clamps on hyphae and/or basidia 15
15. Basidia clavate, without clamps .... *Eichleriella*
    Basidia ellipsoid, mostly with clamps .... *Exidiopsis*
16. A fungus which causes disease of apples in storage .... *Butlerelfia*
    A fungus with tetraradiate spores found on wood in sea-water
    .... *Digitatispora*
    Not so 17
17. Spores irregular, mostly 3-lobed .... *Tylospora*
    Spores very large, allantoid, 15–24 × 5–7.5 .... *Vuilleminia*
    Not so 18
18. Fruitbodies with immersed or projecting brown, simple, thick-walled, pointed setae .... *Hymenochaete*
    Not so 19
19. Fruitbodies contain many radiately-branched, brown setae (asterosetae)
    .... *Asterostroma*
    Fruitbodies contain pale yellowish, repeatedly dichotomously-branched hyphae with pointed tips .... *Vararia*
    Not so 20
20. Spores coloured 21
    Spores hyaline 42
21. Spore walls rough, spiny or warted 22
    Spore walls smooth 31
22. Spores blue–green, fruitbodies pale to dark blue .... *Lazulinospora*
    Spores yellowish to brown 23
23. Cystidia or cystidioles present and seen easily 24
    Cystidia or cystidioles none or rarely seen 25
24. Cystidioles hyaline .... *Xenasma*
    Cystidia brown, septate, large .... *Tomentellina*
25. Spores and basidia stain deeply with cotton blue 26
    Not so 27
26. Hyphae wide, short-celled, without clamps .... *Botryohypochnus*
    Hyphae longer-celled, with clamps .... *Lindtneria*
27. Hyphae with clamps 28
    Hyphae without clamps 29
28. Hyphae with clamps at almost every septum; basal hyphae often uniting to form ropes or cord-like strands .... *Tomentella*
    Not so, pale hyphae often swollen at septa .... *Trechispora*
29. Spores pale, not more than 7 diam., basidia often suburniform
    .... *Tomentellopsis*
    Spores darker, mostly 7–14 diam., basidia not suburniform 30

4   Non-gilled hymenomycetes

30. Spore walls mostly echinulate, basidia clavate .... *Tomentellastrum*
    Spore walls with warts which are often notched, basidia stalked
    .... *Pseudotomentella*
31. Spores violet, fruitbodies dark violet .... *Hypochnella*
    Spores yellowish olive, turning deep violet in KOH solution
    .... *Hypochnopsis*
    Spores bluish or greenish 32
    Spores yellowish brown or brown 33
    Spores rather pale yellowish, often thick-walled 34
32. Dendrohyphidia formed, fruitbodies dark blue with whitish margin
    .... *Pulcherricium*
    No dendrohyphidia, fruitbodies greyish blue or greenish blue
    .... *Byssocorticium*
33. Basidia with subulate sterigmata up to 16 long .... *Scotomyces*
    Basidia with relatively short sterigmata .... *Coniophora*
34. Spores forming secondary spores, basidia about same width as
    subtending hyphae, sterigmata long; often parasitic on herbaceous
    plants .... *Thanatephorus*
    Not so 35
35. Spores fusiform, 16–25 long .... *Jaapia*
    Spores never fusiform, smaller 36
36. Hyphae without clamps, spores 2–4.5 diam. .... *Piloderma*
    Hyphae with clamps, spores mostly larger 37
37. Cystidia present 38
    Cystidia absent 39
38. Cystidia thick-walled except at apex .... *Crustoderma*
    Cystidia thin-walled .... *Hypochnicium*
39. Spores 6–11 × 5–8 .... *Hypochnicium*
    Spores smaller, often spherical or subspherical 40
40. Basidia without cyanophilous drops or granules .... *Hypochniciellum*
    Basidia with cyanophilous drops or granules 41
41. Hyphae with few clamps; on dung and debris .... *Dacryobasidium*
    Hyphae with clamps at all septa; on rotten wood .... *Cristinia*
42. Spores distinctly amyloid 43
    Spores not distinctly amyloid 51
43. Spore walls rough, spiny or warted 44
    Spore walls smooth 47
44. Spores spherical or subspherical, 6–7 diam. .... *Boidinia*
    Spores ellipsoid 45
45. Spores not more than 6 long .... *Gloeocystidiellum*
    Spores not less than 12 long 46
46. Spores 12–15 × 6–8, on *Cladium mariscus* .... *Acanthobasidium*
    Spores 18–22 × 12–16, mostly on Rosaceae .... *Aleurodiscus*

Fruitbodies dark blue with pale margin .... *Pulcherricium*

Not so 66

66. Spores 16–25 × 6–7, fusiform .... *Jaapia*

Spores smaller and other shapes 67

67. Basidia mostly with 6 or more sterigmata 68

Basidia with not more than 4 sterigmata 70

68. Spores stain deeply with cotton blue, hyphae often wide and at right-angles to one another .... *Botryobasidium*

Spores not staining deeply with cotton blue 69

69. Hyphae often with oily contents and swellings .... *Sistotrema*

Hyphae narrow, without oil drops or swellings .... *Paullicorticium*

70. Fruitbodies rather loosely attached to substrate 71

Fruitbodies quite firmly attached to substrate 83

71. Fruitbodies thin, pellicular, cobwebby or membranous, often made up of loosely intertwined hyphae, easily separated from the substrate

.... 72

Fruitbodies fairly easily separated from the substrate, at first often thin but increasing in thickness with age and sometimes forming basidia at different levels 81

72. Cystidia always present 73

Cystidia absent 75

73. Cystidia septate, with clamps at septa .... *Amphinema*

Cystidia not septate 74

74. Cystidia subulate, encrusted, with encrustations arranged spirally in rows .... *Subulicystidium*

Cystidia clavate or spathulate, smooth .... *Lagarobasidium*

75. Spores thick-walled, stain strongly with cotton blue .... *Cristinia*

Spores thin-walled, not staining with cotton blue 76

76. Fruitbodies   with   brown   basal   hyphae   and   rhizomorphs

.... *Confertobasidium*

Fruitbodies not so, hyphae hyaline 77

77. Hyphae commonly, but not always, swollen near septa ...... *Trechispora*

Hyphae not so 78

78. Fruitbodies   with   narrow   rhizomorphs,   hyphae   branched   and anastomosing   freely,   with   clamps   at   nearly   every   septum

.... *Fibulomyces*

Not so 79

79. Fruitbodies white or whitish, rarely cream, hyphae often coated with crystals .... *Athelia*

Fruitbodies mostly at least tinged with greenish yellow and sometimes more deeply coloured 80

80. Basidia abruptly constricted below into a stalk .... *Athelopsis*

Basidia not so .... *Leptosporomyces*

81. Subicular hyphae mostly straight, parallel to each other and to the

surface of the substrate .... *Phanerochaete*
Subicular hyphae not so 82
82. Hyphae full of oil drops, spores pyriform or lachrymoid, sticking
together in clumps of 2 or 4 .... *Cylindrobasidium*
Hyphae not full of oil drops, spores ellipsoid or subspherical
.... *Ceraceomyces*
83. Fruitbodies membranous to waxy, soft or tough 84
Fruitbodies when fresh softly leathery or firmly gelatinous–waxy,
elastic, when dry often horny 99
84. Basidial state accompanied or replaced by spherical white bulbils of
its *Aegerita* state .... *Bulbillomyces*
Not so 85
85. Spores thick-walled .... *Hypochnicium*
Spores thin-walled 86
86. Cystidia present 87
Cystidia absent 94
87. On rotting fern stems, cystidia clavate or flask-shaped ... *Parvobasidium*
Not so 88
88. Subicular hyphae mostly straight, parallel to each other and to the
surface of the substrate .... *Phanerochaete*
Not so 89
89. Cystidia with 2 or more root-like processes at the base ...... *Tubulicrinis*
Cystidia not so 90
90. Cystidia capitate, septate with clamps, encrusted with scattered
compound crystals; common on *Sambucus* .... *Lyomyces*
Not with this combination of characters 91
91. Spores stain deeply with cotton blue, hyphae yellowish, much
branched, closely packed together .... *Crustoderma*
Not so 92
92. Hyphae loosely intertwined, often branching from or opposite clamps,
stain deeply with cotton blue .... *Grandinia*
Not so 93
93. Basidia fairly large, often constricted, usually with oil drops, sterigmata
often long and curved .... *Hyphoderma*
Not so, often with brown basal hyphae and curved spores .. *Peniophora*
94. Subicular hyphae mostly straight, parallel to each other and to the
substrate .... *Phanerochaete*
Not so 95
95. Hyphae mostly 2–4 thick but with some cells much swollen, spores
4–5 diam. .... *Brevicellicium*
Not so 96
96. Basidia large, containing oil drops .... *Cerocorticium*
Not so 97
97. Spores navicular or fusiform .... *Luellia*

116. Basidia with longitudinal septa 117
     Basidia without septa 119
117. Large cystidia in main axis of each tooth, spores 4–6 × 3.5–4.5
     .... *Stypella*
     No cystidia, spores larger 118
118. Teeth up to 0.5 mm long, spores up to 8 long .... *Protodontia*
     Teeth up to 1 mm long, spores 12–22 × 6–10 .... *Eichleriella*
119. On large sedges, *Typha* and *Scirpus*, teeth sterile, spores 18–30 × 6–8
     .... *Epithele*
     Not so 120
120. Spores coloured 121
     Spores hyaline 123
121. Spores yellowish brown, walls spiny or warted .... *Tomentella*
     Spores yellowish, with rather thick, smooth walls 122
122. Basidia with cyanophilous drops .... *Cristinia*
     Basidia not so .... *Leucogyrophana*
123. Spore walls spiny 124
     Spore walls smooth 125
124. Spores 3–4.5 × 2.5–3.5, on rotten wood and bark .... *Trechispora*
     Spores 4–5 diam., on soil in rabbit burrows .... *Echinotrema*
125. Fruitbody consisting only of translucent teeth, 0.5–3 mm long
     attached directly to the substrate .... *Mucronella*
     Not so, always some other part to fruitbody 126
126. Pores as well as teeth present .... *Schizopora*
     Teeth only formed 127
127. Teeth very long, 0.5–2 cm 128
     Teeth often more than 1 mm but not more than 5 mm long 129
     Teeth short or very short, not more than 1 mm long 134
128. Teeth white to ochraceous, spores spherical or subspherical,
     5–7 diam. .... *Spongipellis*
     Teeth sulphur yellow, sometimes turning reddish, spores
     4.5–6 × 3.5–4 .... *Sarcodontia*
129. Spores and droplets in basidia stain deeply with cotton blue, spores
     spherical or subspherical 5–7 diam. .... *Cristinia*
     Not so 130
130. Hyphae usually stain with cotton blue, are loosely intertwined and
     often branch from or opposite clamps, spores mostly 4–6 long
     .... *Grandinia*
     Not so 131
131. Spores 6–9 long, teeth violet or purplish brown .... *Trichaptum*
     Spores 9–12 long 132
     Spores 3–6 long 133
132. Spores ellipsoid, 6–8 wide .... *Cerocorticium*
     Spores allantoid, 3–3.5 wide .... *Hyphoderma*

133. Fruitbodies tough, with thick-walled skeletal hyphae ..... *Steccherinum*
     Fruitbodies waxy .... *Mycoacia*
134. Basidia with 6–8 sterigmata .... *Sistotrema*
     Basidia with not more than 4 sterigmata 135
135. Cystidia absent 136
     Cystidia present 137
136. On rotting fern fronds .... *Pteridomyces*
     On wood and bark, spores apiculate, 4–5 diam. .... *Brevicellicium*
137. Hyphae usually stain with cotton blue, are loosely intertwined and
     often branched from or opposite clamps .... *Grandinia*
     Not so 138
138. Hyphae without clamps, cystidia thick-walled, encrusted, mostly
     conical, spores 3.5–4 × 1.5–2 .... *Scopuloides*
     Hyphae without clamps, cystidia tufts of encrusted hyphae, spores
     7–11 × 4–7 .... *Hyphodermella*
     Hyphae with clamps, cystidia otherwise 139
139. Spores thick-walled, stain deeply with cotton blue 140
     Spores thin-walled, not staining with cotton blue 141
140. Cystidia cylindrical .... *Hypochnicium*
     Cystidia clavate or spathulate .... *Lagarobasidium*
141. Spores allantoid, 5–7 × 1.5 .... *Dacryobolus*
     Spores other shapes, wider 142
142. Spores 3.5–4.5 × 2.5–3.5 .... *Fibrodontia*
     Spores 5–8 × 2.5–3.5 .... *Resinicium*
143. Spores with spiny walls 144
     Spores with smooth walls 146
144. Found only in rabbit burrows .... *Echinotrema*
     Not found in rabbit burrows 145
145. Spores 6–8 diam., spines 1–2 long .... *Lindtneria*
     Spores 3–5 × 2.5–4, fruitbodies soft .... *Trechispora*
     Spores 4.5–6 × 4–5, fruitbodies tough .... *Heterobasidion*
146. Brown, thick-walled, pointed setae present in the hymenium and
     sometimes also in other parts 147
     No such setae present 148
147. Flesh soft when young, tubes not stratified .... *Inonotus*
     Flesh tough, corky to woody, tubes often stratified .... *Phellinus*
148. Spores coloured yellow to brown or bluish 149
     Spores hyaline 153
149. Hymenial surface partly merulioid, phlebioid or with teeth 150
     Hymenial surface entirely poroid 152
150. Hymenium at first reticulate poroid but soon becoming toothed,
     spores 5–7 × 3.5–5 .... *Leucogyrophana*
     Hymenium in part merulioid or phlebioid 151
151. Spores 4–5 × 1.5–2.5 .... *Pseudomerulius*

Spores 8–13 × 5–9 .... *Serpula*
152. Fruitbodies tough, corky to woody, tubes stratified, pores 4–5 per
    mm, spores 4.5–6.5 × 3–5 .... *Perenniporia*
    Fruitbodies soft, spongy, easily separated from substrate, pores 2–3
    per mm, spores 3.5–4.5 diam. .... *Byssocorticium*
153. Spores thick-walled, truncate at one end and with a pore at the other
    .... *Perenniporia*
    Spores thin-walled 154
154. Hymenial surface partly merulioid or with teeth 155
    Hymenial surface entirely poroid 159
155. Hymenium in part merulioid 156
    Hymenium in part with flat teeth 158
156. Cystidia large, clavate, with resinous, reddish brown sap .... *Phlebia*
    Cystidia none or hyaline and small 157
157. Hyphae with clamps, spores allantoid, 3.5–4.5 × 1–1.5, fruitbodies
    firmly gelatinous to elastic .... *Merulius*
    Hyphae without clamps, spores up to 6–7 long, fruitbodies softly
    leathery to waxy .... *Meruliopsis*
158. Hymenial surface white to cream or ochraceous .... *Schizopora*
    Hymenial surface violet to cocoa brownish .... *Trichaptum*
159. Basidia with 6–8 sterigmata .... *Sistotrema*
    Basidia with not more than 4 sterigmata 160
160. Fruitbodies produce masses of thick-walled, yellowish to brown
    chlamydospores .... *Oligoporus*
    No such chlamydospores formed 161
161. Spores commonly more than 12 long (11–14), skeletal hyphae
    dendroid .... *Dichomitus*
    Spores never more than 12 long, skeletal hyphae, when present, not
    dendroid 162
162. Pore surface grey except for white margin, skeletal hyphae with
    strongly gelatinized walls .... *Cinereomyces*
    Not so 163
163. Fruitbodies in section seen to have a black line separating zones; pore
    surface bruising brown or black 164
    Not so 165
164. Black line between tube-layer and flesh, spores not more than 8 long
    .... *Bjerkandera*
    Black line above flesh, spores more than 8 long .... *Datronia*
165. Cystidia abundant and obvious in hymenium 166
    Cystidia, if present, small and not seen easily 167
166. Cystidia up to 200 × 8–12, with encrusted walls, pore surface pinkish
    ochraceous .... *Junghuhnia*
    Cystidia smaller, some or all apically encrusted, pore surface
    white, cream or brownish .... *Oxyporus*

167. Spores spherical or almost so, apiculate, each with one or more oil drops or vacuoles, pore surface often bruising red or reddish-brown .... *Physisporinus*
     Not so 168
168. Pore surface when young orange or pinkish, when old orange–brown, turned violet by KOH .... *Hapalopilus*
     Not so 169
169. Flesh of fruitbody with two distinct colour zones, white next to the pore surface, the rest ochraceous to brown .... *Loweomyces*
     Not so 170
170. Tubes in several layers, thick-walled skeletal hyphae olivaceous brown .... *Donkioporia*
     Tubes in one layer, skeletal hyphae, when present, hyaline 171
171. Pore surface pinkish cream to salmon pink, basidia often two-spored .... *Tyromyces*
     Not so 172
172. Fruitbodies soft, waxy or cottony 173
     Fruitbodies if soft at first not waxy or cottony, less fragile and often becoming hard when dry 175
173. Fruitbodies easily separated from substrate, spores 2.5–3.5 × 2–2.5 .... *Porpomyces*
     Not so, spores larger 174
174. Hyphae without clamps .... *Ceriporia*
     Hyphae with clamps .... *Ceriporiopsis*
     (If spores spherical, pale yellowish, 3.5–4.5 diam. .... *Byssocorticium*)
175. Fruitbodies gelatinous, brittle when dry. Pore surface white bruising ochraceous brown, spores 3.5–5 × 1–1.5 .... *Skeletocutis*
     Not so 176
176. Spores allantoid, 3–4 × 0.5–1 .... *Chaetoporellus*
     Spores sometimes curved but always larger 177
177. No true skeletal hyphae although some are thicker-walled than others, they stain readily with cresyl blue .... *Postia*
     Hyaline, thick-walled to solid skeletal hyphae present .... *Antrodia*
178. Fruitbodies compound 179
     Fruitbodies simple although sometimes concrescent 182
179. Hymenial surface smooth, without pores, compound fruiting bodies rosette-like, about 20 cm diam. .... *Podoscypha*
     Hymenial surface with pores 180
180. Fruitbodies hemispherical, up to 50 cm across, made up of a number of umbrella-like heads, spores 8–10 × 2–3.5 .... *Dendropolyporus*
     Fruitbodies made up of a number of overlapping fan-shaped caps 181
181. Fruitbodies ochraceous brown, very large, up to 80 cm diam., spores subspherical, 6–7 × 5–6 .... *Meripilus*
     Fruitbodies rather greyish brown or grey, 20–50 cm diam., spores

195. Basidia with longitudinal septa .... *Pseudohydnum*
     Basidia without septa 196
196. Fruitbodies effuso-reflexed, brackets not well-defined, often just
     fertile nodules, teeth always mixed with the pores and developed
     from them .... *Schizopora*
     Effuso-reflexed, brackets narrow, shelf-like, not more than 1–2 cm
     wide, cystidia thick-walled, encrusted 197
     Brackets wider, fan- or shell-shaped, or branched and coralloid, no
     encrusted cystidia 198
197. Hyphae without clamps, spores 5–6.5 long .... *Irpex*
     Hyphae with clamps, spores 3–4.5 long .... *Steccherinum*
198. Spores not amyloid .... *Spongipellis*
     Spores strongly amyloid 199
199. Fruitbodies much branched, flesh strongly amyloid, teeth up to 3 cm
     long .... *Hericium*
     Fruitbodies not branched, teeth shorter 200
200 Teeth not more than 2 mm long, no cystidia .... *Irpicodon*
     Teeth 5–15 mm long, smooth-walled cystidia up to 150 × 5–8
     .... *Creolophus*
201. Hymenial surface only in part poroid 202
     Hymenial surface all poroid 206
202. Hymenial surface partly with teeth 203
     Hymenial surface partly or mostly merulioid or phlebioid 204
203. Hymenial surface violet or purplish brown .... *Trichaptum*
     Hymenial surface white to cream or ochraceous .... *Schizopora*
204. Spores hyaline, allantoid, 1–1.5 wide .... *Merulius*
     Spores yellowish to brown or rust in mass 205
205. Spores 4–5 long .... *Pseudomerulius*
     Spores 9–12 long .... *Serpula*
206. Thick-walled, brown, pointed setae present in hymenium and
     sometimes also in trama or upper surface of cap 207
     No such setae formed 208
207. Perennial, woody hard, with yellowish brown skeletal hyphae, tubes
     often stratified .... *Phellinus*
     Annual, flesh softer, no skeletal hyphae, tubes not stratified .. *Inonotus*
208. Spores diagnostic, truncate at apex, brown, with two walls, the inner
     one thick and ornamented with transverse striations .... *Ganoderma*
     Spores not so, always hyaline or very pale yellow 209
209. Spore walls minutely but clearly spiny .... *Heterobasidion*
     Spore walls quite smooth 210
210. Fruitbodies fleshy-spongy, with blood-red juice when fresh *Fistulina*
     Fruitbodies not so 211
211. Spores thick-walled 212
     Spores thin-walled 213

212. Fruitbodies annual, flesh 2-layered, upper part soft, lower part firmer, fibrous, tough .... *Spongipellis*
     Fruitbodies perennial, tough and corky to hard and woody, tubes stratified .... *Perenniporia*
213. Fruitbodies narrow, shelf-like, mostly projecting less than 2 cm from substrate 214
     Fruitbodies other shapes, usually projecting further 218
214. Pores often slot-like and up to 5 × 0.1–1 mm .... *Datronia*
     Pores mostly isodiametric, round or angular 215
215. Upper surface grey or greyish, fruitbodies in section seen to have a very dark, narrow zone between the pore layer and the flesh .... *Bjerkandera*
     Upper surface white at first, turning cream, ochraceous or brownish 216
216. Lower surface greyish pink to dark purplish or reddish brown, with white margin, tubes shallow, gelatinous, easily separated from white, cottony or felt-like flesh .... *Gloeoporus*
     Not so 217
217. Spores curved 4.5–6 × 1–1.5 .... *Postia*
     Spores straight, 6 × 2 or larger .... *Antrodia*
218. Fruitbodies thin, seldom more than 1 cm thick 219
     Fruitbodies thick, 2–25 cm 225
219. Pores labyrinthine, much contorted, spores 5–7 long .... *Cerrena*
     Pores labyrinthine or with branched and anastomosing plates, spores 7–13 long .... *Gloeophyllum*
     Pores more or less isodiametric, round or angular 220
220. Fruitbodies saffron yellow to yellowish orange .... *Loweomyces*
     Fruitbodies cream to yellowish, waxy translucent .... *Antrodiella*
     Fruitbodies other colours, never yellowish 221
221. Upper surface of fruitbody usually velvety and zonate, sometimes markedly so, spores 5–7 × 1.5–3 .... *Trametes*
     Upper surface of fruitbody dark brown, reddish brown or greyish brown, covered with stiff hairs, zonate or not .... *Funalia*
     Upper surface of fruitbody seldom if ever clearly zonate 222
222. Pore surface yellow to cinnamon, pores very small, 8–11 per mm .... *Flaviporus*
     Pore surface not these colours, pores larger 223
223. Fruitbodies soft-fleshed, especially when young, white sometimes changing to blue or greyish .... *Postia*
     Fruitbodies leathery to corky, tough, or other colours 224
224. Upper surface white at first, sometimes becoming brown or dark brown with white edge, fruitbodies leathery or corky, tough, spores 3–5 × 0.5–1.5 .... *Skeletocutis*
     Upper surface milky-coffee coloured, spores 5–8 × 2.5–3 . *Bjerkandera*

225. Pores elongated radially, slot-like or labyrinthine 226
     Pores more or less isodiametric, round or angular 228
226. Pore surface whitish or greyish, bruising pink or red, brown when
     very old .... *Daedaleopsis*
     Pore surface not bruising pink or red 227
227. Pores slot-like, 1–5 × 0.5–1 mm .... *Trametes*
     Pores labyrinthine .... *Daedalea*
228. Fruitbodies hoof-shaped, very thick 229
     Fruitbodies other shapes, somewhat thinner 230
229. Fruitbodies up to 50 × 25 cm and 25 cm thick, spores 15–22
     long .... *Fomes*
     Fruitbodies smaller and not more than 15 cm thick, spores 6–9 long
     .... *Fomitopsis*
230. Fruitbodies, when mature, with black, resinous crust ...... *Ischnoderma*
     Fruitbodies brightly coloured 231
     Fruitbodies dull colours or whitish 232
231. Fruitbodies pale pink to reddish or purplish brown, pore surface
     white or pinkish, bruising pink or violet .... *Leptoporus*
     Fruitbodies sulphur yellow to yellowish orange, very large, up to
     50 cm across .... *Laetiporus*
     Fruitbodies cinnabar or reddish orange, turned black by KOH
     .... *Pycnoporus*
     Fruitbodies cinnamon to bright ochraceous, turned violet by KOH
     .... *Hapalopilus*
     Fruitbodies orange or brownish orange, turned red or carmine by
     KOH .... *Aurantioporus*
232. Spores spherical or subspherical 233
     Spores other shapes 234
233. Spores 4–5.5 diam., apically encrusted cystidia abundant, fruitbodies
     not more than 4 cm thick .... *Oxyporus*
     Spores 6–7.5 diam., cystidia subulate, not encrusted, fruitbodies up to
     8 cm thick .... *Rigidoporus*
234. Pores 1–2 per mm, spores 7–10 × 3–4 .... *Trametes*
     Pores 2–3 per mm 235
     Pores 3–4 per mm 236
235. Spores 6–9 × 2.5–4, usually on *Quercus* .... *Buglossoporus*
     Spores 4–6 × 3–4.5, mostly on *Malus* .... *Aurantioporus*
236. Fruitbodies large, 5–30 × 5–20 cm, parasitic on *Betula* .... *Piptoporus*
     Fruitbodies smaller, not parasitic on *Betula* 237
237. Hyphae not metachromatic with cresyl blue .... *Tyromyces*
     Hyphae metachromatic with cresyl blue .... *Postia*
238. Spores 4.5–6 long .... *Lenzites*
     Spores 7–13 long .... *Gloeophyllum*
239. Basidia with 3 transverse septa, forms pink pustules on stromata of

Diatrype stigma .... *Mycogloea*

Basidia without septa, fruitbodies large or very large, natural sponge-
or cauliflower-like .... *Sparassis*

Basidia with longitudinal septa 240

Basidia deeply divided forming 2 sterigmata and resembling tuning-
forks 241

240. Spores fusiform to falcate, 15–25 × 6–7.5 .... *Efibulobasidium*

Spores cylindrical, curved, with rounded ends, 9–23 × 4–7, basidia
mostly with basal clamps .... *Exidia*

Spores spherical or broadly ellipsoid, 5–15 × 2–15, basidia mostly
with basal clamps .... *Tremella*

Spores of different shapes and sizes, basidia stalked, with clamp at
base of stalk but not of basidium .... *Myxarium*

241. On leaves of *Scirpus sylvaticus*, spores 1-septate .... *Dicellomyces*

On wood and bark, spores with 0–10 transverse septa and
occasionally 1 or 2 longitudinal septa, fruitbodies gelatinous, often
orange .... *Dacrymyces*

242. Basidia with 1–3 transverse septa and long sterigmata, embedded
in stiff jelly and often difficult to see .... *Auricularia*

Basidia without septa, not embedded in jelly 243

243. On stromata of Pyrenomycetes, especially *Diatrype stigma* *Episphaeria*

On dead grasses .... *Cellypha*

On mosses 244

On other substrata 246

244. Spores reddish brown in mass, rough-walled, 7–10 diam.
.... *Chromocyphella*

Spores hyaline, smooth 245

245. Spores 3.5–4.5 long, cystidia present .... *Cyphellostereum*

Spores 6 or more long, no cystidia .... *Leptoglossum*

246. Fruitbodies usually very closely crowded together 247

Fruitbodies solitary or gregarious but not crowded together 250

247. Hairs on fruitbodies long, brown, thick-walled and encrusted with
granules except at their thinner walled, usually pale, swollen tips
.... *Merismodes*

Hairs, when present, not so 248

248. Fruitbodies on a greyish stroma, spores spherical, 5–7 diam.
.... *Stigmatolemma*

No greyish stroma, spores usually other shapes or slightly smaller
.... 249

249. Fruitbodies embedded in a soft, white or cream, cottony subiculum
and resembling a poroid resupinate .... *Stromatoscypha*

Fruitbodies without subiculum or stroma, white, tapered towards
their bases, looking rather like batches of insect eggs
.... *Henningsomyces*

250. Spores ochraceous in mass, honey-coloured by transmitted light .... *Pellidiscus*
Spores hyaline 251
251. Spores amyloid, with spiny walls, 20–30 × 20–25 .... *Aleurodiscus*
Spores amyloid, smooth, allantoid, small .... *Plicaturopsis*
Spores not amyloid, smooth 252
252. Hymenium bright red or wine colour, spores 12–15 × 4–5, on *Salix* .... *Cytidia*
Not so, fruitbodies white or other colours 253
253. Spores allantoid, 8–12 × 2–3 .... *Auriculariopsis*
Spores other shapes and sizes 254
254. Outer layer of fruitbody containing erect, coral-like hyphae which sometimes protrude, but with no distinct long hairs ...... *Calyptella*
Surface of fruitbody distinctly hairy 255
255. Hairs thick-walled, encrusted with crystals of calcium oxalate except towards the apex where they are smooth and often whip-like .... *Flagelloscypha*
Hairs often adpressed, without smooth whip-like ends, walls thick, granulate .... *Lachnella*
256. Fruitbodies cylindrical, spathulate, capitate, thread-, club-, top-, nail-, antler- or coral-shaped 257
Fruitbodies cyathiform, fan-, ear-, funnel-, or trumpet-shaped 280
Fruitbodies with cap and stalk, mushroom-shaped 298
257. Basidia with 1–3 transverse septa, fruitbodies always unbranched 258
Basidia with longitudinal septa 261
Basidia deeply divided to form 2 prongs or sterigmata, resembling tuning-forks, fruitbodies mostly yellow or orange, spores often septate 262
Basidia neither septate nor deeply divided 266
258. Basidia pear-shaped with one transverse or oblique septum, fruit-bodies coremium-like, white or yellowish .... *Stilbum*
Basidia elongated, with 3 transverse septa 259
259. On mosses, fruitbodies thread-like or narrowly clavate, spores fusiform or sickle-shaped, 18–25 × 4–5 .... *Eocronartium*
Not on mosses, fruitbodies capitate 260
260. On rotten potatoes, fruitbodies white or whitish, with discoid tops, spores 14–17 × 7–9 .... *Pilacrella*
On dead branches, fruitbodies pale brown or snuff colour, with subspherical tops, spores 8–10 × 7–9 .... *Phleogena*
261. Fruitbodies top-shaped, black or blackish brown, spores large, allantoid, on branches of deciduous trees .... *Exidia.*
Fruitbodies branched, grey or greyish, Clavaria-like, on the ground,

spores 10–22 × 5–8 .... *Tremellodendropsis*

262. Fruitbodies branched or simple, when simple lanceolate, cylindrical, spathulate or clavate, not top-shaped, capitate or discoid . *Calocera*

Fruitbodies either capitate, top-shaped or discoid, not branched 263

263. Only disc fertile 264

Disc and other parts fertile 265

264. Sterile surface with palisade of thick-walled elements ...... *Guepiniopsis*

Stalk and sterile part of head covered with slender, white hairs .... *Femsjonia*

265. Stalk whitish or cream, covered with tomentum of thick-walled, rough, septate hairs; stalk rooted in substrate .... *Ditiola*

Not so .... *Dacrymyces*

266. Spores coloured, yellowish or brown, walls spiny or warted 267

Spores hyaline, walls smooth or rough 269

267. Fruitbodies unbranched, broadly top-shaped, lilac, rose or flesh colour .... *Gomphus*

Fruitbodies branched, other colours 268

268. Fruitbodies soft-leathery, spores often subspherical, somewhat angular or irregularly lobed .... *Thelephora*

Fruitbodies brittle or tough, spores ellipsoid, oblong-ellipsoid or fusiform .... *Ramaria*

269. Terrestrial, often growing amongst grass or mosses but in many different habitats 270

On living or dead parts of plants 274

270. Basidia with 2 inwardly curved sterigmata, spores smooth-walled usually with 1 large guttule .... *Clavulina*

Basidia all or mostly with 4 sterigmata, spores with smooth or rough walls, guttules large, small or absent 271

271. Hyphae without clamps, usually broad, normally without large guttules in the spores .... *Clavaria*

Hyphae with clamps 272

272. Fruitbodies normally unbranched, 3–30 cm tall, spores mostly 10 or more long .... *Clavariadelphus*

Spores always less than 10 long except in *Clavulinopsis vernalis* which has fruitbodies not more than 1.5 cm tall 273

273. Spores never more than 5.5 long, walls minutely spiny or warted .... *Ramariopsis*

Spores often more than 5.5 long, walls mostly smooth, occasionally coarsely tuberculate or nodulose .... *Clavulinopsis*

274. Fruitbodies repeatedly and often dichotomously or verticillately branched, bushy 275

Fruitbodies either unbranched or branched simply from the base 276

275. Spores amyloid, walls finely warted, cystidia present .... *Artomyces*

Spores only weakly amyloid, walls smooth, no cystidia .... *Lentaria*

276. Fruitbodies unbranched or branched from the base, very slender, needle- or bristle-like, sometimes forming dense tufts .... *Pterula*

Fruitbodies unbranched, thread-like or slenderly or broadly club-shaped 277

277. Hyphae without clamps, some containing large oil drops, spores 3–4 × 2–3, subspherical, smooth to verruculose .... *Clavicorona*

Hyphae with clamps, spores ellipsoid, longer, walls always smooth 278

278. Fruitbodies 0.5–2 mm tall, spores 4–6 × 2–3 .... *Ceratellopsis*

Fruitbodies and/or spores larger 279

279. Fruitbodies very tall, up to 12–30 cm, never arising from sclerotia .... *Macrotyphula*

Fruitbodies mostly shorter, often arising from sclerotia .... *Typhula*

280. Basidia transversely septate, fruitbodies almost ear-shaped, reddish brown, flabby gelatinous, on branches, especially those of *Sambucus* .... *Auricularia*

Basidia longitudinally septate, fruitbodies large, salmon pink sometimes flushed orange, on ground, mostly on buried wood or sawdust piles .... *Tremiscus*

Basidia deeply divided to form 2 sterigmata, resembling tuning-forks, on wood and bark .... *Guepiniopsis*

Basidia neither septate nor deeply divided 281

281. Hymenial surface smooth, radially wrinkled or with shallow, often forked veins or ribs 282

Hymenial surface with projecting teeth 290

Hymenial surface with pores 293

282. Fruitbodies lilac, rose or flesh colour .... *Gomphus*

Fruitbodies other colours 283

283. Spores brown and echinulate .... *Thelephora*

Spores not so 284

284. Hymenium with cystidia 285

Hymenium without cystidia 286

285. Cystidia thick-walled, fusiform, sharply pointed, fruitbodies growing on bonfire sites .... *Geopetalum*

Cystidia cylindrical to narrowly clavate, fruitbodies growing on the ground, only rarely on bonfire sites .... *Cotylidia*

286. Spores not more than 4 long 287

Spores not less than 7 long 288

287. Fruitbodies spathulate to fan-shaped, spores quite smooth .. *Stereopsis*

Fruitbodies clavate to trumpet-shaped, spores often slightly verruculose .... *Clavicorona*

288. Basidia with 2 spores .... *Craterellus*

Basidia with 2–8, mostly 4 spores 289

289. Hyphae with clamps, not markedly constricted at septa . *Cantharellus*
Hyphae without clamps, markedly constricted at septa
.... *Pseudocraterellus*
290. Spores hyaline 291
Spores brown or brownish, tuberculate 292
291. Spores with spiny walls, basidia with 4 spores .... *Phellodon*
Spores with smooth walls, basidia 6–8-spored .... *Sistotrema*
292. Flesh tough, fibrous, zoned .... *Hydnellum*
Flesh brittle, not zoned .... *Sarcodon*
293. Dark brown, thick-walled setae in hymenium .... *Onnia*
No such setae in hymenium 294
294. Cystidia present in hymenium 295
No cystidia 296
295. Cystidia and basidia contain oil drops .... *Abortiporus*
Cystidia and basidia without oil drops but cystidia often have brown,
resinous contents .... *Phaeolus*
296. Spores pale golden brown, upper surface of cap zonate .... *Coltricia*
Spores hyaline, cap not zonate 297
297. Fruitbodies growing on the ground .... *Albatrellus*
Fruitbodies almost always on wood .... *Polyporus*
298. Hymenial surface smooth, radially wrinkled, or with shallow, often
forked veins or ribs 299
Hymenial surface with projecting teeth 301
Hymenial surface with pores 306
299. Piggyback toadstools which form chlamydospores and grow on old,
rotting specimens of *Russula* and *Lactarius* .... *Asterophora*
Not so 300
300. Spores J+, 7.5–9 long .... *Delicatula*
Spores J–, 15–25 long .... *Gloiocephala*
301. On pine cones, stalk usually excentric .... *Auriscalpium*
Not on pine cones, stalk commonly although not always central 302
302. Spores brown or brownish, tuberculate 303
Spores hyaline 304
303. Flesh tough, fibrous, zoned .... *Hydnellum*
Flesh brittle, not zoned .... *Sarcodon*
304. Spores with smooth walls .... *Hydnum*
Spores with echinulate or verruculose walls 305
305. Flesh not zoned, hyphae with swellings .... *Bankera*
Flesh zoned, hyphae without swellings .... *Phellodon*
306. Fruitbodies with flesh often tough, leathery or corky, if soft not soon
decaying 307
Fruitbodies with flesh soft, often spongy, soon decaying 312
307. Dark brown, thick-walled setae in hymenium .... *Onnia*
No such setae present in hymenium 308

308. Spore walls tuberculate .... *Boletopsis*
     Spore walls smooth 309
309. Spores pale golden brown .... *Coltricia*
     Spores hyaline 310
310. Large clavate or subulate cystidia, often with brown resinous
     contents, present in hymenium .... *Phaeolus*
     No such cystidia in hymenium 311
311. Fruitbodies growing on the ground .... *Albatrellus*
     Fruitbodies almost always on wood .... *Polyporus*
312. Cap grey to olivaceous black, covered with thick, woolly-floccose
     scales; spores purplish black, almost spherical, with reticulate walls
     .... *Strobilomyces*
     Not so, spores always longer than broad 313
313. Stalk with ring 314
     Stalk without ring 315
314. Cap tomentose to fibrillose scaly, tubes short, distinctly decurrent,
     pores large and radially orientated .... *Boletinus*
     Cap distinctly glutinous to viscid .... *Suillus*
315. Spores 4.5–6 long, always growing with *Alnus* .... *Uloporus*
     Stem covered with scales which darken with age, tubes and pore
     surface mostly white or off-white, if lemon yellow always bruising
     black .... *Leccinum*
     Not so 316
316. Cap glutinous to viscid 317
     Cap sometimes slightly viscid when quite young but soon becoming
     dry, velvety, tomentose or smooth 318
317. Cystidia on tube margin large and mostly filled with bright yellow sap,
     spores 11–17 long .... *Aureoboletus*
     Cystidia in fascicles, becoming encrusted, spores mostly less than 12
     long, stem often with glands at apex or throughout .... *Suillus*
318. Spores in mass lemon yellow, pores very small, round, white then
     lemon yellow .... *Gyroporus*
     Spores in mass pink or pinkish, pores broad and angular, becoming
     pink or pinkish .... *Tylopilus*
     Spores in mass purplish brown, pores vinaceous buff bruising bluish
     green .... *Porphyrellus*
     Spores in mass usually olive to olivaceous or snuff brown,
     occasionally ochraceous or rust, pore surface often brightly
     coloured, yellow or red and frequently bruising blue, green or
     violet .... *Boletus*

DESCRIPTIONS OF GENERA AND SPECIES

Abortiporus *biennis* (Bull. ex Fr.) Singer (Fig. 1)
(*Heteroporus*)
Fruitbodies erect, pileate, 3–12 cm diam., usually centrally or excentrically stalked, often overlapping and sometimes fused together forming rosette-like clusters, cap frequently fan- or funnel-shaped, irregularly lobed and thin at margin; upper surface white at first then becoming ochraceous to brown, smooth or velvety; lower surface white or whitish, bruising pinkish or reddish brown, pores 1–3 per mm or sometimes larger, angular to labyrinthine, thin-walled. Flesh white, 2-layered, turns red when cut, softer upper layer contains chlamydospores. Cystidia (g) up to 60 × 12, irregularly cylindrical or subulate. Basidia 20–35 × 6–8, contain oil drops. Spores ellipsoid, hyaline, smooth, J–, 4–7 × 3.5–4.5. On the ground, usually attached to buried stumps or pieces of wood.

Acanthobasidium *delicatum* (Wakef.) Oberw. (Fig. 2)
Fruitbodies resupinate, forming small oval patches, very thin, delicate, whitish to pale peach. Hyphae with clamps. Hyphidia rounded or pyriform, bearing numerous short spines. Basidia (pleuro-) urn-shaped, constricted in middle, swollen below, 25–35 × 10–14, often bearing short spines, sterigmata curved, subulate, 8–9 × 2. Spores ellipsoid, hyaline, closely echinulate, strongly J+, 12–15 × 6–8. On dead leaves of *Cladium mariscus*.

Achroomyces
(*Platygloea*)
Fruitbodies not more than 5 cm wide, often much smaller, resupinate and smooth, flat or pustulate, gelatinous to waxy. Hyphae hyaline, with or without clamps. Probasidia cylindrical to clavate or ellipsoid, hyaline, thin-walled. Basidia cylindrical to narrowly clavate or fusiform, sometimes bent, divided by transverse septa into 2–4 cells; sterigmata 2–4. Spores thin-walled, hyaline, smooth, J–. Some species form conidiophores and conidia.

KEY

On dung .... *fimetarius*
On other fungi 1
On wood and bark or stems 3
1. Fruitbody absent, growing in hymenium of *Dacrymyces* .... *arrhytidiae*
    Fruitbody present on surface of host fungus 2
2. On *Botryosphaeria* on *Fraxinus* and on other pyrenomycetes and
        coelomycetes .... *sebaceus*
    On basidiomycetes, mainly Corticiaceae .... *peniophorae*

3. Spores 15–30 long .... *vestitus*
    Spores not more than 10 long 4
4. Basidia 50–90 × 4–6, mainly on *Alnus* .... *effusus*
    Basidia 18–33 × 2.5–4, on *Fraxinus* and other deciduous trees
    .... *microsporus*

Achroomyces arrhytidiae (Olive) Wojewoda
Fruitbodies absent, all growth taking place inside the hymenium of
Dacrymyces. Hyphae without clamps. Probasidia clavate. Basidia cylindrical,
20–35 × 3.5–4.5, with 2–4 cells; sterigmata up to 15 long, cylindrical or
subulate. Spores ellipsoid, 7.5–10.5 × 3.5; forming secondary spores.
Conidiophores present. Conidia spherical or broadly ellipsoid, thin-walled,
smooth, 3.5–5 × 2–3.5.

Achroomyces effusus (Schroet.) Migula
Fruitbodies effused up to 5 cm, when fresh somewhat pulvinate, smooth,
slightly wrinkled or warted, gelatinous, colourless or slightly greyish, when
dry thin, horny and scarcely visible. Basidia stalked, 50–90 × 4–6, cylindrical, 4-celled; sterigmata up to 45 × 2, cylindrical. Spores broadly ellipsoid, 6–10 × 4–7. On dead branches, mostly of *Alnus* but also occasionally
on other trees.

Achroomyces fimetarius (Schum. ex Pers.) Wojewoda (Fig. 3)
Fruitbodies 2–10 mm diam., up to 1 mm thick when fresh, waxy, flesh-
coloured or pale lilac. Basidia cylindrical, 4-celled, 35–40 × 5–6; sterigmata
up to 10 long, cylindrical. Spores ellipsoid, 10–11 × 4–6. On dung of herbi-
vores, e.g. cow, horse and rabbit.

Achroomyces microsporus (McNabb) Wojewoda
Fruitbodies thin, effused up to 5 cm, greyish lilac when fresh, almost
invisible when dry. Hyphae with clamps. Probasidia broadly ellipsoid, 8–
12 × 6–8. Basidia cylindrical, bent, 4-celled, 18–33 × 2.5–4; sterigmata cylindrical, up to 15 long. Spores narrowly ellipsoid, somewhat bent,
7–9 × 2.5–3.5, forming secondary spores. On dead branches of deciduous
trees, e.g. *Fraxinus.*

Achroomyces peniophorae (Bourd. and Galz.) Wojewoda
Fruitbodies discrete, orbicular, 1–3 mm diam., or sometimes more widely
effused, pale cream to ochraceous with a narrow white fringe. Hyphae with
clamps. Basidia cylindrical, sometimes bent, 4-celled, 30–60 × 5–7; sterigmata cylindrical to subulate, 40–80 long. Spores broadly ellipsoid, 8–
10 × 5–6. Conidia 6–12 × 3.5–4.5, formed on conidiophores which bear
clamps. On the hymenium of various Corticiaceae, e.g. *Botryohypochnus,*
Gloeocystidiellum, *Grandinia* and *Hyphoderma*; also reported on *Dacrymyces*
and *Poria.*

Achroomyces sebaceus (Berk. and Br.) Wojewoda
Fruitbodies initially pustulate, softly gelatinous or almost slimy, whitish, coalescing to form masses up to 1 cm diam., when dry thin and greyish. Hyphae with clamps. Basidia cylindrical, stalked, 80–300 × 4–8; sterigmata cylindrical, 20–60 × 4. Spores ellipsoid, 9–15 × 5–9. Conidia 3–6 × 2.5–5.5. On *Botryosphaeria stevensii* on *Fraxinus* and on other pyrenomycetes and coelomycetes.

Achroomyces vestitus (Bourd. and Galz.) Wojewoda
Fruitbodies scarcely visible to the naked eye, slimy gelatinous, hyaline or greyish. Hyphae 3–10 thick, with erect, thick-walled, cystidium-like branches up to 110 × 10 in the hymenium. Basidia cylindrical or narrowly clavate, 40–70 × 7–10, 2–4-celled; sterigmata up to 20 long. Spores cylindrical to narrowly ellipsoid, somewhat bent, 15–30 × 5–7. On wood and bark of various deciduous trees and shrubs, e.g. *Alnus, Quercus, Rosa* and *Cytisus.*

Albatrellus *cristatus* (Schaeff. ex Fr.) Kotl. and Pouz. (Fig. 4)
Fruitbodies erect, pileate, soft and fleshy when young, centrally or excentrically stalked, solitary or several fused together; cap 5–10 cm diam., flat, round or flabelliform with undulating margin, upper surface olivaceous, yellowish green or brownish olive, velvety, cracking when old to form squamules, lower surface white, cream or slightly greenish, pores 1–3 per mm, angular, thin-walled, tubes shallow, decurrent; stalk concolorous, up to 3 × 1 cm. No cystidia. Basidia 15–30 × 5–8. Spores ellipsoid, hyaline, J–, 6–8 × 4.5–5.5. On ground in deciduous woods, especially under *Fagus* and *Quercus.*

Aleurodiscus
Fruitbodies resupinate and effused or discrete and discoid to saucer-shaped. Hymenium smooth, rather brightly coloured. Basidia large, hyaline, with long sterigmata. Cystidia and sometimes also dendrohyphidia present. Spores large, hyaline, J+, warted or spiny.

KEY

Fruitbodies discoid or saucer-shaped .... *amorphus*
Fruitbodies resupinate, effused .... *aurantius*

Aleurodiscus amorphus (Pers. ex Purt.) Schroet. (Fig. 5)
Fruitbodies discrete, rather tough, solitary or gregarious, discoid, pulvinate, or saucer-shaped, 1–6 mm diam., disc pinkish orange to ochraceous or pinkish grey, with inrolled, paler, fringed margin. Cystidia cylindrical to moniliform, hyaline, up to about 200 × 5–10. Basidia 90–180 × 20–30 with sterigmata up to 25 long. Spores subspherical or broadly

ellipsoid, with short, blunt spines, 20–30 × 20–25. Erumpent through bark of standing and fallen trunks and branches of *Abies* and *Picea*, seen most easily in wet weather.

Aleurodiscus aurantius (Pers. ex Fr.) Schroet. (Fig. 6)
Fruitbodies resupinate, effused, orange to ochraceous. Cystidia narrowly clavate, often moniliform towards the apex. Dendrohyphidia abundant, richly branched, covered by crystals. Basidia up to 70 × 15 with sterigmata 14 long. Spores 18–22 × 12–16, warted to finely echinulate. Mostly on dead stems of *Rosa* and *Rubus*, March–May.

Amphinema byssoides (Pers. ex Fr.) Erikss. (Fig. 7)
Fruitbodies resupinate, effused, cobwebby or membranous, with occasional distinct and characteristic crater-like holes, cream to pale ochraceous, rather loosely attached, margin with rhizomorphs. Hymenium smooth, finely velvety from projecting cystidia. Hyphae with clamps at all septa, yellowish, loosely intertwined. Cystidia plentiful, encrusted, septate, with clamps, about 100 × 4–6. Basidia 20–23 × 4–5. Spores ellipsoid, hyaline, smooth, J–, C+, 4–5 × 2–3. Essentially a fungus of the woodland floor, overgrowing leaves, twigs and other debris but developing its basidia mainly on rotten wood.

Amylostereum

Fruitbodies narrow and bracket-like, effuso-reflexed or fully resupinate; resupinate forms waxy and soft when fresh, brackets leathery to corky, hymenium smooth or tuberculate. Brown skeletal hyphae present. Cystidia plentiful, yellowish brown, thick-walled, apically encrusted. Basidia narrowly clavate. Spores cylindrical or rather narrowly ellipsoid, hyaline, thin-walled, smooth, strongly J+.

KEY

    Spores 7–12 long .... *laevigatum*
    Spores not more than 7–8 long 1
1. Cystidia 4–6 wide .... *chailletii*
    Cystidia 7–9 wide .... *areolatum*

Amylostereum areolatum (Fr.) Boidin (Fig. 8)
Fruitbodies narrow, bracket-like, sometimes imbricate, with hymenium decurrent and growing also onto the wood, 3–10 cm long, 0.5–3 cm wide, 1–1.5 mm thick; upper surface shortly tomentose, zoned dark brown and yellowish or rusty brown, margin paler, undulating; lower surface greyish, brownish or ochraceous, often tinted violaceous. Flesh two-layered with layers separated by a black line. Cystidia about 50 × 7–9. Basidia 20–25 × 3–4. Spores 5–6 × 2.5–3. On dead wood of *Picea*.

Amylostereum chailletii (Pers.) Boidin (Fig. 9)
Fruitbodies mostly resupinate and effused, but occasionally forming very narrow (not more than 1 cm wide) brackets, variable in colour, mostly ochraceous to brown or greyish brown but sometimes cinnamon, with dark brown margin, often cracking into squares when dry. Cystidia about 50 × 4–6. Basidia 20–25 × 4–5. Spores 5–8 × 2.5–3.5. On dead wood, mostly of *Abies*, rarely of *Picea*.

Amylostereum laevigatum (Fr.) Boidin (Fig. 10)
Fruitbodies resupinate, effused, firmly attached, pale brown, ochraceous or greyish brown, cracking when dry. Cystidia up to 140 × 7–9, encrusted part about 30 long. Basidia 20–30 × 4–6. Spores 7–12 × 3–4. On living and dead wood and bark of *Cupressus, Juniperus, Taxus* and *Thuja*.

Antrodia

Fruitbodies mostly resupinate but sometimes pileate or effuso-reflexed; brackets when present narrow, shelf-like, mostly white or pale brown, in one species becoming chrome yellow, pores usually round or angular. Hyphae with clamps. Skeletal hyphae hyaline, thick-walled or solid. Cystidioles sometimes present in hymenium, occasionally apically encrusted. Basidia 4-spored. Spores variable in shape, hyaline, smooth, thin-walled, J–. Only *A. albida* and *A. lenis* are found on wood of deciduous trees, all the other species and sometimes *A. lenis* also occur on conifer wood.

KEY

Spores mostly more than 6 long 1
Spores mostly less than 6 long 3
1. Spores ellipsoid-fusiform, 6–10 × 2–4 .... *serialis*
   Spores cylindrical to oblong-ellipsoid 2
2. Fruitbodies partly resupinate but also forming narrow brackets, spores 8–14 × 3.5–6 .... *albida*
   Fruitbodies wholly resupinate, discrete, spores 6–8.5 × 2.5–3.5 .... *ramentacea*
3. Spores broadly ellipsoid, straight, 2.5–3.5 wide 4
   Spores seldom more than 1.5 wide, often curved 5
4. Spores 5–7 long, found in cellars and mines .... *vaillantii*
   Spores 4–5.5 long, on dead conifer wood, occasionally on fence-posts .... *gossypium*
5. Spores all very strongly curved .... *lenis*
   Spores slightly curved or straight 6
6. Fruitbody surface becoming divided roughly into squares separated by deep cracks, often chrome yellow .... *xantha*

Not so 7

7. Fruitbodies resupinate, large, often over 30 cm long, pores 1–3 per mm
    .... *sinuosa*

    Fruitbodies resupinate but not more than 7–8 cm long, pores
    smaller, 3–4 per mm .... *albobrunnea*

Antrodia albida (Fr.) Donk (Fig. 11)

Fruitbodies resupinate or pileate and then forming imbricate brackets up to
2 cm wide, tough; upper surface white or cream, rarely ochraceous, velvety
at first, becoming partly smooth but with tufts remaining particularly along
the margin; pore surface white or cream, sometimes yellowing with age,
pores round or angular, 2–4 per mm in resupinate forms, rather larger in
pileate ones where they frequently become split and somewhat lamellate at
the margin, tubes 3–15 mm long. Spores oblong-ellipsoid, 8–14 × 3.5–6.
On dead wood of deciduous trees including *Acer*, *Betula*, *Corylus*, *Fagus*,
Fraxinus and *Quercus*, also often found on fence-posts.

Antrodia albobrunnea (Romell) Ryv.

Fruitbodies resupinate, up to 8 cm long, about 4 mm thick, rather soft,
especially when young, white or brownish with narrow margin, surface
sometimes patterned, pores round or angular, 3–4 per mm. Spores straight
or slightly curved, 4.5–6 × 1.5–2. On wood and bark of *Pinus* and sometimes
other conifers in open woods.

Antrodia gossypium (Speg.) Ryv. (Fig. 12)

Fruitbodies often discrete and rather small, resupinate, 2–4 mm thick, soft
and easily separable at first, later with a rather brittle surface, sometimes
with rhizomorphs; pore surface white at first, becoming unevenly pale
brown, pores angular, rather irregular, 3–6 per mm, resinous when old.
Spores broadly oblong-ellipsoid, 4–5.5 × 2.5–3.5. On dead conifer wood,
fence-posts etc.

Antrodia lenis (P. Karst.) Ryv. (Fig. 13)

Fruitbodies resupinate, rather soft, small and discrete at first but soon
becoming confluent, effused and up to 30 cm long, 1–3 mm thick; pore
surface rather shiny, white to creamy, pores round or angular, more irregular and sometimes lacerate on oblique or vertical surfaces, 4–8 per mm.
Cystidioles often present. Spores allantoid, often strongly curved, 3–5 × 1–
1.5. On deciduous trees and conifers, fairly common.

Antrodia ramentacea (Berk. and Br.) Donk

Fruitbodies resupinate, tough, discrete, almost round, up to about 6 cm
diam., 4 mm thick, with narrow white margin which may loosen and turn
up; pore surface white at first then pale brown, with pores round or
angular, thin-walled, about 0.5 mm diam., broadening to 1–3 mm and
becoming resinous and more irregular with age. Cystidioles sometimes

present. Spores cylindrical to ellipsoid, 6–8.5 × 2.5–3.5. Mostly on bark of *Pinus sylvestris.*

Antrodia serialis (Fr.) Donk (Fig. 14)
Fruitbodies partly resupinate but with narrow, projecting, shelf-like, sometimes imbricate brackets not as a rule more than 2 cm wide, tough, leathery, easily detached from wood when young; upper surface finely hairy, white at first, then yellowish brown to brown or even occasionally blackish brown; pore surface white at first, becoming pale ochraceous brown, pores round or angular, 2–4 per mm, elongated and split on vertical surfaces. Spores ellipsoid-fusiform, with distinct apiculus, 6–10 × 2–4. On conifer wood including worked timber, often on *Picea* but also on *Pinus, Larix* etc.

Antrodia sinuosa (Fr.) P. Karst.
Fruitbodies resupinate, often over 30 cm long, up to 5 mm thick, soft and easily separable when fresh, drying corky and brittle, with a narrow, thin, white margin; pore surface white at first but becoming brownish with age or on drying, pores mostly 1–3 per mm, thin-walled, angular becoming labyrinthine and somewhat dentate at the mouths of the tubes. Cystidioles clavate or subulate, seen occasionally in the hymenium. Spores narrowly cylindrical to allantoid, 4–5.5 × 1–1.5. Mostly on bare and sometimes burnt wood of conifers, e.g. *Picea* and *Pinus.*

Antrodia vaillantii (Fr.) Ryv.
Fruitbodies resupinate, effused, about 4 mm thick, soft, cottony, with wide margin and conspicuous white rhizomorphs; pore surface white or cream, pores 2–4 per mm, round or angular. Spores oblong-ellipsoid, 5–7 × 2.5–3.5. Mainly on worked conifer timber in mines and cellars.

Antrodia xantha (Fr.) Ryv. (Fig. 15)
Fruitbodies resupinate, 1–5 mm thick, widely effused and sometimes extending for several metres along the underside of fallen trunks and large branches, at first soft, when dry becoming chalky and forming deep cracks which split them up in a cubical manner; pore surface white to cream or chrome yellow, pores round or somewhat angular, 4–6 per mm. Spores allantoid but only slightly curved, 4–5.5 × 1–1.5. Mostly on conifers but recorded also on *Salix.* In the f. *pachymeres* Erikss. found mainly on *Larix,* the fruitbodies differ in having numerous projecting knobby or hoof-like, whitish, slightly velvety pilei.

Antrodiella *semisupina* (Berk. and Curt.) Ryv. and Johansen (Fig. 16)
Fruitbodies mostly pileate and imbricate, attached laterally, fan- or shell-shaped, up to 3 cm diam. and 5 mm thick, cream or yellowish with narrow, darker concentric zones, waxy or translucent, tough when fresh, hard when dry; pore surface whitish to ochraceous, pores 4–7 per mm. Hyphae with clamps. Skeletal hyphae thick-walled, hyaline. Cystidia none. Basidia

8–15 × 4–5, 4-spored. Spores ellipsoid, hyaline, thin-walled, smooth, 2.5–3.5 × 2–2.5. Mostly on deciduous trees, *Alnus, Betula, Carpinus, Corylus, Quercus* etc., rarely on conifers, uncommon.

Aphanobasidium *filicinum* (Bourd.) Jülich (Fig. 17)
Fruitbodies effuse, gelatinous to waxy, very thin, smooth, whitish grey or ochraceous. Hyphae hyaline, with clamps. Cystidia none. Basidia (pleuro-) cylindrical, hyaline, 10–25 long. Spores narrowly ellipsoid, hyaline, thin-walled, smooth, J−, 5–10 × 3–5. On ferns, herbaceous plants and wood.

Artomyces *pyxidatus* (Pers. ex Fr.) Jülich (Fig. 18)
Fruitbodies erect, gregarious or clustered, many times verticillately branched, stalked, 2–12 cm high, pale yellow, ochraceous or yellowish brown, main branches 1–3 mm thick. Hyphae up to 16 thick, with clamps, some containing oil drops and ending as 3–5 wide cystidia (g) in the hymenium. Basidia clavate, 20–30 × 4–5, with clamps. Spores ellipsoid, hyaline, finely verruculose, J+, 4–5.5 × 2–3. On rotten wood, mostly of deciduous trees, rarely on conifers.

Asterophora
Fruitbodies 'piggyback' toadstools, gregarious, growing on old, rotting specimens of *Russula* and occasionally *Lactarius*. Cap often almost spherical or hemispherical, not more than 2.5 cm diam., hymenium smooth or covering thick, ridge-like veins on lower surface; stalk central. Basidia containing carminophilic granules. Basidiospores often found only in young specimens, hyaline, smooth. Chlamydospores pale brown, formed abundantly either on the hymenium or on the upper surface of the cap.

KEY

Chlamydospores smooth-walled .... *parasitica*
Chlamydospores covered with large blunt protuberances .. *lycoperdoides*

Asterophora lycoperdoides (Bull.) Ditm. ex S.F. Gray (Fig. 19)
Cap almost spherical, 0.5–2 cm diam., upper surface white at first but soon with a mealy, fawn, clay or slightly reddish brown coating of chlamydospores; stalk mostly 5–10 × 2–5 mm, at first whitish but blackening from the base up. Basidiospores ellipsoid, 5–5.5 × 3.5–4. Chlamydospores pale brown, subspherical, 12–16 diam., covered with long blunt protuberances. Always on *Russula nigricans*, common only in wet seasons.

Asterophora parasitica (Bull. ex Fr.) Sing. (Fig. 20)
Cap hemispherical, convex or bell-shaped, 1–2.5 cm diam., upper surface whitish at first then pale grey or tinged lilac or pale brown, silky with thin flesh; stalk 1–3 cm × 1–3 mm, often curved or twisted, white or tinged lilac

or pale brown, frequently hairy at base. Basidiospores 5–6 × 3–4. Chlamydospores formed on the hymenium, pale brown, ovoid, thick-walled, smooth,
12–15 × 9–10. On old specimens of *Russula* and *Lactarius*.

Asterostroma

Fruitbodies resupinate, loosely attached to the substrate, effused or confluent forming large patches, softly membranous. Hyphae hyaline, without
clamps. Asterosetae embedded, brown, plentiful. Basidia hyaline, narrowly
clavate. Cystidia (g) hyaline, fusiform. Spores hyaline, smooth or tuberculate.

KEY

Spores smooth .... *laxum*
Spores coarsely tuberculate .... *cervicolor*

Asterostroma cervicolor (Berk. and Curt.) Mass. (Fig. 21)
Fruitbodies white to pale ochraceous. Asterosetae stellate, branches 40–
80 × 2–4. Cystidia up to 60 × 10–20. Spores coarsely tuberculate, 5–7.5
diam., tubercles up to 1.5 long. On dead wood of deciduous trees and
conifers.

Asterostroma laxum Bres. (Fig. 22)
Fruitbodies with smooth or tuberculate surface, yellowish, ochraceous or
cinnamon with paler margin. Asterosetae with arms dichotomously or
trichotomously branched, branches 20–40 × 2–3. Cystidia 40–70 × 5–9.
Spores spherical or subspherical, 6–9 diam., J+. On conifer wood and old
polypores.

Athelia

Fruitbody consisting of a thin white, whitish or cream, cobwebby skin
easily separated from the substrate. Hyphae hyaline, with or without
clamps, often encrusted with crystals, those of the subhymenial layer
frequently broader and with thicker walls. Cystidia none. Basidia often
clustered, clavate, tapered gradually towards the base, mostly short, with 2–
4 sterigmata. Spores of various shapes, hyaline, thin-walled, smooth, J–.

KEY

Spores often more than 10 long .... *teutoburgensis*
Spores never more than 10 long 1
1. Clamps at all or almost all septa 2
     Clamps none or few and confined to basal layer 3
2. Spores narrowly ellipsoid, 4.5–6 × 2.5–3 .... *bombacina*

Spores mostly subspherical, 6–9 × 5–7 .... *neuhoffii*
3. Spores not more than 6 long .... *decipiens*
    Spores mostly or often more than 6 long **4**
4. Spores 2–3 wide .... *acrospora*
    Spores 3–5.5 wide **5**
5. Basidia 2-spored, 20–30 long .... *arachnoidea*
    Basidia 4-spored, not more than 20 long **6**
6. Spores mostly pyriform with basal apiculus .... *pyriformis*
    Spores ellipsoid, rounded at base .... *epiphylla*

Athelia acrospora Jülich
Subhymenial hyphae without clamps, basal hyphae with few, scattered
Basidia 12–15 × 5–6. Spores cylindrical to narrowly ellipsoid tapered
towards the base, 5–8 × 2–3, often sticking together in clumps of 2–4.
Mostly on conifer wood and on leaves.

Athelia arachnoidea (Berk.) Jülich
Subhymenial hypahe without clamps, basal hyphae with few, scattered
clamps. Basidia 20–30 × 5–8, 2-spored, without clamps. Spores ellipsoid or
ovoid, 8–10 × 4–5. On leaves and branches of deciduous trees, often overgrowing and perhaps parasitic on algae and lichens.

Athelia bombacina Pers.
Hyphae with clamps at nearly every septum, frequently encrusted with
crystals. Basidia 10–15 × 5, with basal clamps, 4-spored. Spores narrowly
ellipsoid tapered towards base, 4.5–6 × 2.5–3, often sticking together in
clumps of 2–4. On dead conifer wood and on leaves and ferns.

Athelia decipiens (Höhn. and Litsch.) J. Erikss.
Hyphae all without clamps. Basidia 10–15 × 5–6, 4-spored or 2-spored.
Spores broadly ellipsoid, 4.5–6 × 3–3.5, often sticking together in clumps of
2–4. On rotten conifer wood.

Athelia epiphylla Pers. (Fig. 23)
Hyphae with or without encrustation, basal hyphae with a few scattered
clamps. Basidia 15–20 × 5–6, 4-spored. Spores cylindrical or ellipsoid, with
rounded, not tapered base, 6–10 × 3–5. On fallen branches, leaves, needles,
dead lichens etc.

Athelia neuhoffii (Bres.) Donk
Hyphae with clamps at nearly all septa. Basidia 20–30 × 6–8, mostly
4-spored. Spores subglobose or very broadly ellipsoid, 6–9 × 5–7. On
decayed wood and debris, common.

Athelia pyriformis (Christ.) Jülich
Hyphae all without clamps. Basidia 18–20 × 6–8, mostly 4-spored. Spores
usually pyriform and all with a distinct basal apiculus, 7–10 × 4–5.5. On
dead and living ferns, herbaceous plants, grasses and deciduous trees.

Athelia teutoburgensis(Brinkm.) Jülich
Clamps only on a few basal hyphae. Basidia 20–35 × 8–12. Spores ellipsoid, 9–14 × 4–6. On deciduous trees and conifers.

Athelopsis

Fruitbodies pellicular to thinly membranous, loosely attached to substrate, smooth, pale yellowish or ochraceous, often with a greenish tinge. Hyphae with clamps. Cystidia none. Basidia clavate, abruptly constricted below into a stalk. Spores hyaline, thin-walled, smooth, J–.

KEY

Spores ellipsoid, 6–8 × 4–5 .... *subinconspicua*
Spores cylindrical, 8–12 × 2–3 .... *glaucina*

Athelopsis glaucina (Bourd. and Galz.) Oberw. ex Parm. (Fig. 24)
Basidia 12–20 × 5–7. Spores cylindrical, 8–12 × 2–3. On wood of deciduous trees.

Athelopsis subinconspicua (Litsch.) Jülich
Basidia 15–25 × 4–7. Spores ellipsoid, 6–8 × 4–5. On wood of both deciduous trees and conifers.

Aurantioporus

Fruitbodies rather large brackets, most broadly attached, fleshy at first but drying hard. Pores 2–3 per mm. Hyphae with clamps, none skeletal. Cystidia none. Spores broadly ellipsoid, smooth, hyaline, J–.

KEY

Fruitbodies orange .... *croceus*
Fruitbodies mostly white to cream or pale ochraceous .... *fissilis*

Aurantioporus croceus (Pers. ex Fr.) Murrill
Fruitbodies broadly attached brackets 5–15 × 4–10 cm, about 6 cm thick at base, soft and watery fresh, hard and resinous when dry, red or carmine when touched with KOH; upper surface at first bright orange and velvety, turning brownish orange and smooth or with tufts of agglutinated hyphae; pore surface similarly coloured, tubes 5–14 long. Flesh bright orange. Spores broadly ellipsoid, 4–7 × 3–4.5. On living and dead deciduous trees, most commonly on *Quercus*.

Aurantioporus fissilis (Berk. and Curt.) Jahn (Fig. 25)
Fruitbodies brackets, either broadly attached or narrowed to a stem-like base, sometimes imbricate, flattened or somewhat hoof-like, up to 18 × 10 cm and 2–10 cm thick, fleshy but rather tough; upper surface tomentose, white then creamy or pale ochraceous, sometimes tinged with pink; pore surface similarly coloured, tubes 1–2.5 cm long. Spores 4–6 × 3–4. Mostly on living apple trees (*Malus*), occupying knot holes, recorded also on *Aesculus* and *Ulmus*; smell unpleasant, acidic.

Aureoboletus *cramesinus* (Secretan) Watling (Fig. 26)
Fruitbodies centrally stalked, pileate, cap 2.5–6 cm diam., convex, pinkish to ochraceous peach or like crushed strawberries, glutinous, flesh mostly white or whitish not changing colour when cut, pore surface lemon or golden yellow, not changing colour when bruised; stalk 5–7 × 0.5–1 cm, tapered towards base, mostly yellow but often somewhat reddish at base, glutinous to viscid. Cystidia on tube margin 30–70 × 10–15, often filled with bright yellow sap. Spores in mass ochraceous buff, 11–17 × 4.5–5.5. In deciduous woods, occasionally on bonfire sites, rather uncommon.

Auricularia
Fruitbodies either resupinate or pileate and then either ear- to shell-shaped or forming narrow, imbricate brackets, flabby elastic or tough gelatinous; hymenial surface smooth, wrinkled or veined, often purplish. Basidia cylindrical, with 1–3 transverse septa. Spores narrowly ellipsoid to allantoid, hyaline, smooth, J–.

KEY

Fruitbodies ear- to shell-shaped .... *auricula-judae*
Fruitbodies resupinate or forming narrow brackets .... *mesenterica*

Auricularia auricula-judae (Bull. ex StAmans) Wettst. (Fig. 27)
Jew's Ear
Fruitbodies 3–9 cm across, shaped like an ear or a shell, attached laterally and sometimes by a very short stalk, flabby gelatinous, becoming horny when dry but when wetted quickly returning to the original state; upper, sterile surface reddish brown or tan, covered with greyish down; lower, hymenial surface reddish or purplish brown, wrinkled or veined. Basidia with three transverse septa, 60–70 × 4–7, sterigmata 4, long. Spores 14–20 × 6–8. On dead or dying branches of deciduous trees and shrubs, especially common on *Sambucus nigra.*

Auricularia mesenterica (Dicks.) Fr. (Fig. 28)
Tripe Fungus
Fruitbodies resupinate or pileate with narrow and imbricate brackets

extending over an area of several centimetres to decimetres, loosely attached, elastic, gelatinous; sterile surface hairy, zoned, with grey and olivaceous bands; hymenium wrinkled or smooth, purplish brown with a whitish bloom. Hairs thick-walled, up to 0.5 mm long. Basidia 40–70 × 4–6, with 1–3 sterigmata. Spores allantoid, 15–18 × 6–7. On stumps and logs of deciduous trees, especially *Ulmus*.

Auriculariopsis ampla (Lév.) Maire (Fig. 29)
Fruitbodies pendent, cupulate or campanulate, about 1–1.5 cm diam.; outer (upper) surface white, felted, with closely interwoven, narrow, twisted, thick-walled non-septate hairs; hymenium lining cup gelatinous to waxy, smooth, radially folded or ridged, pale ochraceous to cinnamon brown. Tramal hyphae gelatinous, with clamps. Cystidia none. Basidia rather narrowly clavate, 25–30 × 4, 4-spored. Spores allantoid, hyaline, thin-walled, smooth, J–, 8–12 × 2–3. Mainly on dry branches of *Populus* and *Salix*.

Auriscalpium vulgare S.F.Gray (Fig. 30)
Ear-pick Fungus
Fruitbodies erect, pileate, brown, composed of a more-or-less kidney-shaped cap with pointed teeth 1–3 mm long on the lower surface and a mostly laterally attached stalk; cap 1–2 cm diam., thin, leathery, with hairy upper surface; stalk 2–12 cm long 1–2 mm thick, swollen at base to 6 mm. Cystidia (g) with granular contents. Basidia 4-spored. Spores broadly ellipsoid, hyaline, minutely spinulose, J+, 4.5–5.5 × 3.5–4.5. On fallen, often buried or half-buried cones of *Pinus*, common.

Bankera fuligineo-alba (Schmidt ex Fr.) Pouzar (Fig. 31)
Fruitbodies erect, pileate, centrally stalked, fleshy, with pointed, white or grey teeth 1–6 mm long covering the lower surface of the cap and running a little way down the stalk; cap soon flat or centrally depressed, velvety-scaly, whitish when young, becoming brown, 5–15 cm diam., flesh not zoned; stalk 3–6 × 1–3 cm, brown below, white near apex. Hyphae often distinctly swollen. Basidia 4-spored. Spores broadly ellipsoid or subglobose, hyaline, verrucose to echinulate, 4.5–5.5 × 3–3.5. Smells of fenugreek. On ground, often partly buried by litter, in pine forests, mainly in the highlands of Scotland.

Basidiodendron
Fruitbodies saprophytic on wood and ferns, resupinate, flat, effused, thin, waxy or gelatinous; hymenium smooth or reticulate-poroid. Hyphae hyaline, with clamps. Cystidia (g) yellowish brown, with resinous contents which later break up into large pieces. Basidia hyaline, borne on short side-branches of the erect main hyphae, longitudinally septate, remaining after

the spores have been dispersed as 2-4-celled empty cases. Spores hyaline, thin-walled, smooth or finely verruculose, J–.

KEY

Spores ellipsoid, 8–12 × 6–8 .... *cinereum*
Spores spherical or almost so 1
1. Spores 4–7 diam., apiculus small .... *eyrei*
Spores 7–9 diam., apiculus prominent .... *caesiocinereum*

Basidiodendron caesiocinereum (Höhn. and Litsch.) Luck-Allen (Fig. 32)
Fruitbodies firmly attached to substrate, whitish or pale grey, waxy, pruinose, appearing finely reticulate-poroid under a lens. Cystidia cylindrical to subulate, somewhat sinuous, hyaline to yellowish brown, mostly 20–60 × 4–12. Basidia broadly ellipsoid, 10–17 × 7–10, sterigmata mostly 4, 6–10 long. Spores spherical and very clearly apiculate, smooth or minutely warted, with guttules or oil drops, 7–9 diam. On rotten wood and ferns.

Basidiodendron eyrei (Wakef.) Luck-Allen
Fruitbodies waxy, pale grey to pale ochraceous, pruinose. Cystidia 30–100 × 5–9, hyaline to yellowish or brown. Basidia broadly ellipsoid, 12–18 × 10–13, sterigmata up to 25 long. Spores ellipsoid, 8–12 × 6–8. On wood.

Basidodendron eyrei (Wakef.) Luck-Allen
Fruitbodies waxy, pale grey, slightly pruinose. Cystidia up to 45 × 4–8, cylindrical, hyaline to yellowish brown. Basidia broadly ellipsoid, 10–15 × 6–8, sterigmata 4–8 long. Spores spherical or almost so, with small apiculus, smooth, 4–7 diam. On rotten wood, causing saffron-red staining.

Bjerkandera

Fruitbodies imbricate brackets or resupinate, pores 2–6 per mm, flesh pale, often paler than pore surface, separated from tube layer by a thin dark line which represents a gelatinized zone. Hyphae and basidia with clamps. Basidia 4-spored. Cystidia none. Spores oblong-ellipsoid, hyaline, smooth, J–. On dead wood of deciduous trees causing white soft rots.

KEY

Fruitbodies grey or greyish, spores 4.5–5.5 long .... *adusta*
Fruitbodies ochraceous or milky coffee, spores 5–8 long .... *fumosa*

Bjerkandera adusta (Fr.) P.Karst. (Fig. 33)
Fruitbodies often resupinate on underside of fallen trunks and branches but more commonly pileate and formed as imbricate, wavy, leathery brackets, 2–6 × 1–3 cm and up to 8 mm thick on vertical or sloping substrates; upper

surface finely velvety but not markedly zonate and often only faintly so, pale grey or greyish brown with white margin when young, blackish when old and hard; underside pale or mid grey, bruising blackish, pores 4–6 per mm. Spores 4.5–5.5 × 2–3. A very common species on dead wood of deciduous trees, especially *Fagus*.

Bjerkandera fumosa (Fr.) P. Karst.
Fruitbodies roughly semicircular, sessile, flat, wavy-edged, often imbricate brackets 10–15 × 4–8 cm and 2–3 mm thick; upper surface smooth or slightly velvety, ochraceous to milky coffee coloured, often darker along the edge; lower surface white or cream, bruising brownish, pores 2–4 per mm. Spores 5–8 × 2.5–3. On dead wood or damaged deciduous trees, mostly of *Populus* and *Salix*.

***Boidinia** furfuracea* (Bres.) Stalpers and Hjortst (Fig. 34)
Fruitbodies resupinate, effused, smooth, softly membranous, white or greyish white, furfuraceous or farinose, loosely attached. Hyphae with clamps. Basidia narrowly clavate, often with clamps, 25–35 × 4–6. Cystidia (g) flexuous, with yellowish oil drops, 40–90 × 6–10. Spores spherical or subspherical, hyaline, verrucose to echinulate, J+, 6–7 diam. On decayed wood of conifers.

***Boletinus** cavipes* (Opat.) Kalchbr. (Fig. 35)
Fruitbodies erect, pileate, stalked; cap 3–10 cm diam., convex or slightly umbonate, upper surface usually rusty tawny to cinnamon brown but sometimes yellow, suede to fibrillose-scaly; pore surface lemon, sulphur, or greenish yellow, tubes short, distinctly decurrent, pores large, angular and radially orientated; stalk 5–7 × 1–2 cm, with white floccose ring, lemon above ring, same colour as cap below; flesh yellow in cap, white or slightly pinkish in stalk. Cystidia cylindrical, fusiform or lageniform, 50–80 × 6–10. Spores in mass olivaceous, 7–10 × 3–4. Under *Larix*, apparently rare.

***Boletopsis** leucomelaena* (Pers.) Fayod (Fig. 36)
Fruitbodies erect, pileate, stalked, fleshy, drying soft or brittle; cap 4–15 cm diam., convex, undulating, upper surface greyish brown or olivaceous, smooth, or slightly scaly in centre, lower surface white drying pale grey, pores decurrent, polygonal, irregular, 1–3 per mm; stalk central or lateral, 4–7 × 1–3 cm, pale grey or olivaceous brown, smooth or with darkish scales; flesh white at first, turning pink then darkening to grey on exposure to air. Hyphae with clamps. Basidia mostly clavate. Spores irregularly tuberculate, pale brown when mature, J−, 4–6 diam. On ground amongst needle litter in pine forests.

Boletus
Fruitbodies fleshy-spongy, composed of a hemispherical to convex or flattened cap and a central stalk; cap with upper surface not viscid except

when young, dry, velvety or smooth; pore surface brightly coloured, mostly yellow or red, often bruising blue or green, tubes in most species rather long, occasionally decurrent, pores round or angular; stalk cylindrical, fusiform or obclavate, without a ring, frequently with a coloured network; flesh usually pale or brightly coloured, often blue on cutting. Cystidia always present in hymenium, variable in number and shape, ampulliform, clavate, cylindrical, fusiform or lageniform. Basidia usually 4-spored. Spores in mass usually olivaceous to snuff-brown, occasionally ochraceous or cinnamon buff, in one view mostly fusiform or subfusiform, in the other narrowly ellipsoid; in *B. rubinus* they are ellipsoid in one and broadly ellipsoid in the other and in *B. porosporus* one end is truncate with a pore.

KEY

Fruitbodies parasitic on *Scleroderma citrinum* .... *parasiticus*
On coniferous wood, stumps, trunks or sawdust 1
On the ground, mostly under trees 2
1. Cap sulphur to straw or lemon yellow .... *hemichrysus*
   Cap reddish ochraceous .... *lignicola*
2. Spores broadly ellipsoid, 5.5–7 × 4–6 .... *rubinus*
   Spores fusiform/narrowly ellipsoid, always more than 8 long 3
3. One end of spore truncate, with pore .... *porosporus*
   Not so 4
4. Most parts of the fruitbody readily turning dark blue, blue–black or dark violet when bruised or cut 5
   Most parts of the fruitbody either unchanged or turning slowly or rather faintly blue or bluish green when bruised or cut; if pore surface bruising blue as in *P. badius*, other parts little changed 11
5. Cap upper surface distinctly yellow 6
   Cap upper surface not yellow 7
6. Pore surface yellow or orange, stalk covered with orange or red floccules or granules .... *junquilleus*
   Pore surface lemon to chrome yellow, stalk covered with yellow floccules .... *pseudosulphureus*
7. Stalk 1–1.5 cm thick, pore surface yellow .... *pulverulentus*
   Stalk 2–6 cm thick, pore surface orange or red 8
8. Stalk with distinct red or dark red network 9
   Stalk not so 10
9. Flesh when cut with persistent red line at base of tubes .... *luridus*
   No persistent red line at base of tubes, fruity smell .... *purpureus*
10. Cap upper surface burnt-sienna to yellowish or orange brick, pore surface orange, peach or burnt sienna .... *queletii*
    Cap upper surface bay, chestnut or olive brown, pore surface vermilion

or reddish orange .... *erythropus*
11. Cap upper surface bright green or olivaceous .... *citrinovirens*
    Cap upper surface other colours 12
12. Stalk slender, mostly 0.5–1.5 cm, and not more than 2 cm thick 13
    Stalk 2–10 cm thick 19
13. Pore surface rust-coloured, taste peppery .... *piperatus*
    Pore surface yellow, taste not peppery 14
14. Cap upper surface cherry red or scarlet .... *versicolor*
    Cap upper surface other colours 15
15. Cap upper surface cracking to reveal red flesh below .... *chrysenteron*
    Cap upper surface not so 16
16. Stalks with coarse, elongated brick-red network .... *lanatus*
    Stalks with distinct dark ribs, especially in the upper part .... *spadiceus*
    Stalks with neither network nor ribs 17
17. Cap upper surface distinctly velvety, olivaceous brown or hazel
    .... *subtomentosus*
    Cap upper surface not so 18
18. Cap upper surface orange–brown or reddish brown, margin inrolled
    .... *moravicus*
    Cap upper surface ochraceous or brownish yellow, margin not inrolled
    .... *leonis*
19. Pore surface blood red, reddish orange or orange 20
    Pore surface yellow, whitish or greenish yellow 22
20. Stalk with only a slight network near the apex, 2–4 cm thick, often
    bruising blue .... *satanoides*
    Stalk with distinct network, not bruising blue 21
21. Stalk 3–10 cm thick, often swollen at base, network red .... *satanas*
    Stalk 2–5 cm thick, network purplish red .... *rhodoxanthus*
22. Smell of iodoform, iodine or phenol, especially in stem base, cap upper
    surface clay, greyish ochre or tawny .... *impolitus*
    Strong smell of ink, cap whitish grey or olivaceous grey .... *calopus*
    Smell pleasant, strong fruity, cap upper surface umber to chestnut
    .... *fragrans*
    Smell otherwise, pleasant but not fruity, earthy or none 23
23. Cap upper surface pinkish or reddish with wine-coloured streaks
    .... *regius*
    Cap upper surface pale, whitish, silver grey or clay buff 24
    Cap upper surface other colours, mostly darker 25
24. Cap upper surface whitish to silver grey, bruising brown, tomentose to
    silky .... *fechtneri*
    Cap upper surface dingy white to clay buff, often cracking, especially
    towards the centre .... *albidus*
25. Pore surface bruising wine or rust colour .... *aereus*
    Pore surface quickly bruising blue, pores large, angular .... *badius*

Pore surface bruising bluish green, pores small, round ... *appendiculatus*
Pore surface only slowly turning blue, cap upper surface dark maroon
to purplish bay with a distinct bloom .... *pruinatus*
Pore surface not changing colour when bruised 26
26. Cap upper surface rather pale hazel, straw- or snuff-coloured, cracking
in dry weather to form small squares .... *aestivalis*
Cap chestnut brown, greasy or viscid when wet 27
27. Stalk up to 7 cm thick or 10 cm at base, flesh all white, unchanging
when cut .... *edulis*
Stalk 3–4.5 cm thick, flesh white, unchanging except in cap and stem
cortex where it becomes dark wine-coloured .... *pinicola*

Boletus aereus Fr.
Cap 6–15 cm diam., dark or very dark cigar-brown to sepia, surface matt or
slightly downy with numerous minute cracks; pore surface white to cream
or yellowish, bruising wine or rust colour. Stalk 6–10 × 4–8 cm, obclavate
or swollen in the middle, off-white or buff colour, but appearing darker
because of a close network which varies in colour from brown at the apex to
clay-pink in the middle and rust below. Flesh white. Cystidia usually few,
30–40 × 8–12. Spores 12–15 × 4–5.5. Found under *Castanea*, *Fagus* and
Quercus, uncommon, summer and autumn.

Boletus aestivalis Fr.
Cap 6–18 cm diam., convex, pale hazel-brown, straw- or snuff-coloured,
rather like chamois leather, in dry weather cracking to form a mosaic of
small squares; pore surface white to greenish yellow, pores small. Stalk 6–
15 × 3–5 cm., buff or cap colour, covered completely or almost so by a
dense white network. Flesh white or slightly yellowish, with pleasant strong
smell. Spores 14–16 × 4–5. Under *Quercus* and *Fagus* in woods or sometimes
under solitary trees, early summer–autumn.

Boletus albidus Rocques
Cap 8–15 cm diam., convex, dingy white to clay-buff with pale smoke-grey
margin, velvety to smooth, often cracking, especially towards the centre;
pore surface lemon yellow or slightly greenish, blue when cut or bruised,
pores small. Stalk obclavate, 5–8 × 3–7 cm, sometimes tapered to a short,
root-like base, apex lemon yellow covered with a lemon or straw-yellow
network, the rest pale ochraceous occasionally flushed with red. Flesh
pulpy, white or pale yellow, faintly blue for a short time when cut. Cystidia
30–40 × 8–12, few. Spores 10–16 × 5–6. With *Fagus* and *Quercus*, summer
and autumn.

Boletus appendiculatus (Fr.) Secretan
Cap 8–18 cm diam., semi-globose to convex, pale to fairly dark bay or rusty
brown, finely tomentose, often cracking towards centre; pore surface
lemon yellow, bruising pale bluish green, pores very small. Stalk 10–12 ×

3–4 cm, upper part lemon yellow, lower part darker, ochraceous with reddish blotches and a pale creamy network. Flesh white or pale yellow, when cut turning bluish at apex of stem, rusty and spotted red towards the base. Cystidia 40–50 × 10–15. Spores 11–15 × 4–5.5. Under *Quercus*, late summer–early autumn, uncommon.

Boletus badius Fr. (Fig. 37)
Bay Bolete
Cap 5–14 cm diam., convex, chestnut brown, when quite young tomentose but soon smooth and shiny, viscid in wet weather; pore surface cream or dirty lemon yellow, readily bruising blue, pores large, angular. Stalk 4–12 × 2–4 cm, brownish, usually paler than cap, longitudinally striate, fibrillose. Flesh white or very pale yellow, when cut turning only slightly blue at the stem apex and above the tubes. Cystidia 40–50 × 8–12. Spores 13–16 × 4–6. Common in conifer woods, especially with *Pinus sylvestris*, autumn.

Boletus calopus Fr.
Cap 8–14 cm diam., semi-globose to convex, whitish, grey or olivaceous grey, soon smooth; pore surface sulphur yellow, bruising bluish green. Stalk 3–9 × 2–5 cm, yellow at the apex, otherwise mostly carmine, brownish at base, with a pale or sometimes wine-coloured network. Flesh bitter tasting and smelling like ink, white, cream or in places yellowish, quickly turning blue, darkest in the stem. Cystidia 30–40 × 5–8. Spores 12–15 × 4–5. Mostly with *Fagus* and *Quercus*, late summer–early autumn, not uncommon.

Boletus chrysenteron Bull. ex StAmans (Fig. 38)
Red-cracked Bolete
Cap 4–11 cm diam., convex at first, flattening, hazel, pale sepia or olivaceous buff, velvety to smooth, the surface cracking irregularly to reveal the carmine to coral flesh below; pore surface sulphur or lemon yellow, sometimes bruising bluish green, pores large, angular. Stalk 4–7 × 1–1.5 cm, lemon yellow near the apex, streaked below with red, fibrous or furrowed. Flesh cream or pale in cap except just below the skin where it is carmine or coral, somewhat reddish or pale brown in the stem, faintly blue when cut. Cystidia fusiform or lageniform, 40–50 × 8–10. Spores 12–15 × 4–5. Amongst grass in or near deciduous woods, early summer–autumn, very common.

Boletus citrinovirens Watling
Cap 3–4 cm diam., bright green especially at margin, becoming olivaceous with yellowish areas, velvety to smooth; pore surface lemon to chrome yellow. Stalk 5–15 × 1–1.5 cm, expanded to 2.5 cm at base, pale ochraceous at apex, the remainder with date brown or purplish date ribs and dots. Flesh mainly white but date brown or purplish date beneath cap cuticle and above tubes. Cystidia plentiful, 45–55 × 7–9. Spores 12–13 × 5–5.5. Under *Fagus*, few records.

Boletus edulis Fr.

Cep or Penny Bun

Cap mostly 6–20 cm diam., although occasionally larger, hemispherical to convex, chestnut brown, greasy, slightly viscid when damp, at first often with a whitish bloom, smooth and shiny, matt when quite dry; pore surface white at first, becoming creamy to greenish or greyish yellow, pores small, round. Stalk obclavate to cylindrical, 5–20 × 3–7 cm or at base up to 10 cm wide, white turning to tan or brown with a whitish network towards the apex. Flesh white, firm, sweet tasting, not changing colour when cut. Cystidia 30-50 × 5-10. Spores 14–16 × 4-5. In conifer and deciduous woods, summer–autumn, common.

Boletus erythropus (Fr.) Secretan

Cap 5–20 cm diam., convex, bay, chestnut or olivaceous brown, margin often ochraceous, bruising black or blackish blue, soon smooth, slightly viscid when wet; pore surface vermilion or reddish orange, at margin often yellowish, pores small, bruising dark blue to black. Stalk 5–15 × 2-5 cm, cylindrical or slightly swollen at base, yellowish brown, lemon towards apex, reddish towards base, with reddish or orange dots, smooth, base tomentose, blackening readily when handled. Cystidia 25–35 × 8-10, when young embedded in dark amorphous matter. Spores 12–17 × 4-6. In woods and pastures, mainly on sandy soils, late summer–autumn.

Boletus fechtneri Velen.

Cap 6–15 cm diam., convex, whitish to silver grey, darkening with age and bruising brown, tomentose or silky; pore surface lemon yellow, bruising bluish green. Stalk cylindrical to obclavate, 6–15 × 3-6 cm, lemon to chrome yellow, with reddish central zone and some red spotting, network yellow near apex reddish below. Flesh pale yellow, becoming bluish at stem apex and above tubes when cut. Cystidia few. Spores 9–15 × 5-6. Mostly under *Fagus*, uncommon.

Boletus fragrans Vitt.

Cap 6–12 cm diam., convex, umber to chestnut brown, velvety to smooth; pore surface lemon yellow, bruising faintly blue. Stalk 7–10 × 3-5 cm, broadly fusiform, apex pale yellow, middle part reddish or reddish brown, below this whitish, base black, no network. Flesh lemon yellow, when cut very slowly turning blue, greenish or slightly pink, smell pleasant, rather strong fruity. Cystidia few. Spores 10–16 × 4.5-5.5. In deciduous woods, rare, has been recorded under *Quercus*.

Boletus hemichrysus Berk. and Curt.

Cap 5–15 cm diam., convex, sulphur to straw or lemon yellow, bruising and ageing somewhat orange or rusty, tomentose; pore surface pale lemon yellow, sometimes with rusty spots, pores rather small, tubes slightly decurrent. Stalk 5–15 × 1-5 cm, sometimes with bulbous base, pale lemon yellow, smooth. Mycelium yellow. Flesh lemon yellow, turning bluish when cut. Cystidia few. Spores 6–10 × 3–4. Caespitose on stumps, trunks and sawdust of conifers, rare.

Boletus impolitus Fr.
Cap 6–15 cm diam., convex, clay, greyish ochraceous, tawny or somewhat olivaceous, slightly velvety to smooth; pore surface lemon to chrome yellow, pores small. Stalk 6–12 × 3–5 cm, slightly fusiform, apex and base yellowish, middle part brownish with reddish or rusty dots or streaks, no network. Flesh cream to pale lemon yellow, when cut turning slightly blue after a long time, smell of iodine, iodoform or phenol. Spores 9–15 × 4–6. Mostly in deciduous woods on boulder clay, summer and autumn, mostly associated with *Quercus.*

Boletus junquilleus (Quél.) Boud.
Cap 5–7 cm diam., convex, lemon to sulphur yellow, immediately bruising dark blue to black; pore surface yellow to orange, bruising blue, pores small. Stalk 8–12 × 1–1.5 cm, with slight swelling at base, yellow, covered with orange or red floccules or granules, no network. Flesh yellow, dark blue when cut, red in stem base. Cystidia few. Spores 10–15 × 5–6. Under *Quercus,* rare.

Boletus lanatus Rostk.
Cap 4–10 cm diam., convex or somewhat flattened, fulvous, cinnamon or buff, bruising darker, markedly velvety; pore surface lemon to chrome yellow, bruising blue very easily but colour soon fades, pores large, angular. Stalk 5–8 × 1–1.5 cm, sometimes 2 cm at base, buff, paler towards the apex, yellow in the middle, with a coarse, elongated brick-red network. Flesh white in cap, yellowish or buff in stem, blueing only slightly when cut. Cystidia plentiful. Spores 9–12 × 4–4.5. In deciduous and mixed woods, most commonly under *Betula,* autumn.

Boletus leonis Reid
Cap 3–5 cm diam., convex, ochraceous or warm brownish yellow, tomentose scaly near centre, otherwise smooth; pore surface lemon to chrome yellow, not bruising blue or green. Stalk cylindrical to fusiform, 4–7 × 1–1.5 cm, cream to ochraceous. Flesh pale cream, yellowish in stem base. Cystidia few. Spores 9–13 × 4.5–6, in mass ochraceous. Under *Quercus,* autumn, rare.

Boletus lignicola Kallenbach
Cap 4–10 cm diam., reddish ochraceous with rust-coloured scales, sometimes appearing areolate; pore surface yellow, bruising bluish, pores rather shallow, large and angular, tubes decurrent. Stalk sometimes excentric, about 5 × 1 cm, pointed at base, reddish ochraceous or rust coloured. Flesh creamy buff. Spores 7–11 × 3–4. On coniferous wood, rare.

Boletus luridus Schaeff. ex Fr.
Cap 9–13 cm diam., hemispherical to convex, greyish ochraceous to olivaceous brown, often pinkish at margin, bruising brownish or blackish blue; pore surface orange-red, bruising dark blue, pores small. Stalk 8–12 × 2–5 cm, cylindrical or obclavate, upper part orange or yellowish, wine-coloured and darkening towards the base, bruising blue, network coarse-meshed, distinct, darkening with age. Flesh pale yellowish, when cut turning blue-green or greyish violet, with persistent red line at base of tubes. Cystidia numerous. Spores 10–16 × 5–7. In deciduous woods on calcareous soils, often with *Fagus* and *Quercus*, summer and autumn.

Boletus moravicus Vacek
Cap 3–7 cm diam., hemispherical to convex, with an inrolled margin, orange-brown or reddish brown, tomentose, becoming cracked; pore surface creamy yellow to lemon or slightly olivaceous. Stalk 3–6 × 1–2 cm, cylindrical to fusiform, tapered to a point at base, upper part ochraceous, lower brownish, tomentose. Flesh whitish or pale ochraceous orange, not changing colour when cut. Cystidia few. Spores 8–12 × 5–6. Mostly under *Quercus*, rare.

Boletus parasiticus Fr. (Fig. 39)
Cap 2–6 cm diam., convex, ochraceous or olivaceous brown, slightly downy; pore surface yellow or sometimes rust-coloured when old, tubes slightly decurrent. Stalk 1–4 × 0.5–1 cm, tapered towards base, ochraceous or olivaceous brown. Flesh pale lemon yellow, not blue when cut. Spores 12–22 × 4–5. Parasitic on *Scleroderma citrinum*, gregarious, usually attached to base with the stalks curving upwards. Fairly common, at least in Essex and Suffolk and sometimes found in very large numbers.

Boletus pinicola (Vitt.) Venturi
Cap 8–18 cm diam., convex, bright reddish or purplish chestnut, sometimes darker in centre, greasy at first, then dry and slightly tomentose; pore surface whitish, then pale greenish yellow or olivaceous, not bruising blue, pores small. Stalk 8–10 × 3–4.5 cm, obclavate, pale reddish buff to chestnut with whitish or cinnamon network. Flesh mostly white or whitish, not changing colour when cut except in cap and stem cortex where it becomes dark wine coloured. Cystidia few. Spores 12–17 × 4–5. In conifer woods, late spring–autumn, rare.

Boletus piperatus Bull. ex Fr.
Peppery Bolete
Cap 2–8 cm diam., convex to flattened, cinnamon to burnt sienna, viscid or slightly greasy at first, soon becoming dry and glossy; pore surface rust-coloured, pores broad, angular, tubes somewhat decurrent. Stalk 4–8 × 0.5–2 cm, fusiform, silky, cinnamon or burnt sienna except at base where it is bright saffron yellow. Flesh bright yellow in stem base, otherwise pale

ochraceous, slight fruity smell, strong peppery taste. Spores 8–12 × 3–4. Mostly with *Betula* and *Pinus* on sandy soil, late summer–autumn, fairly common.

Boletus porosporus (Imler) Watling (Fig. 40)
Cap 4–8 cm diam., dark olivaceous brown to sepia, with yellowish tomentum when young, cracking when old to expose yellowish flesh; pore surface lemon to sulphur yellow, often with greyish tinge, bruising blue, pores angular, compound. Stalk 4–6 × 1–1.5 cm, cylindrical or slightly tapered towards base, upper part lemon yellow, lower part all brown or with a central reddish zone, somewhat fibrillose. Flesh pale lemon yellow or pale buff, brownish or brick-coloured in stem base, turning blue or bluish when cut. Cystidia few. Spores 13–15 × 4–5, truncate with a pore at the apex. In deciduous woods, mostly under *Quercus*, not common.

Boletus pruinatus Fr. and Hök
Cap 8–10 cm diam., dark maroon to purplish bay, with a distinct bloom, not cracking; pore surface lemon or chrome yellow, slowly bruising blue. Stalk 8–10 × 2–3 cm, lemon yellow at apex, the rest dotted or streaked with red. Flesh lemon to chrome yellow, slowly turning bluish green when cut. Spores 12–14 × 5–5.5. In deciduous woods, mostly under *Quercus*.

Boletus pseudosulphureus Kallenbach
Cap 9–15 cm diam., lemon to chrome yellow, sometimes tawny towards centre, smooth, slightly sticky in wet weather, bruising dark blue; pore surface lemon to chrome yellow, becoming slightly rusty or olivaceous with age, bruising dark blue or violet. Stalk 7–18 × 1.5–2.5 cm, obclavate-fusiform, lemon yellow to olivaceous with rusty spots when old, covered with yellow floccules, bruising blue–black. Flesh lemon to chrome yellow, brownish in stem base, immediately dark blue or bluish green when cut. Spores 12–17 × 4–5.5. Under *Fagus*, *Quercus* and *Tilia*.

Boletus pulverulentus Opat.
Cap 4–9 cm diam., convex, mostly dingy brown or snuff-coloured, sometimes ochraceous towards margin, tomentose to smooth, bruising blue–black; pore surface lemon to sulphur yellow, bruising blue or blackish violet. Stalk 5–7 × 1–1.5 cm, cylindrical or tapered towards base, lemon yellow at apex, other parts streaked or spotted red or wine colour, bruising blackish blue. Flesh lemon yellow, turning blue immediately when cut. Cystidia few. Spores 10–15 × 4–6.5. Amongst grass along the sides of paths in deciduous woods, mostly under *Quercus*.

Boletus purpureus Pers.
Cap 6–20 cm diam., convex, off-white to olivaceous buff or with coral tints, slightly velvety, occasionally cracking, bruising dark blue; pore surface reddish orange with yellow margin, pores small. Stalk 7–9 × 2–6 cm, cylin-

drical or obclavate, lemon yellow to reddish with red network, bruising purplish or blackish purple. Flesh yellow or greenish yellow, red in stem base, immediately turning blue or blue–black when cut, smells fruity. Spores 11–17 × 4–5.5. In mixed woods.

Boletus queletii Schulzer
Cap 6–20 cm diam., convex, mostly burnt sienna to yellowish or orange brick, variable, tomentose or smooth, bruising blackish blue; pore surface orange, peach or burnt sienna, bruising blackish blue. Stalk 5–10 × 2.5–4.5 cm, cylindrical, often tapered abruptly towards the base, yellow at apex then reddish or somewhat orange, becoming progressively darker downwards, bruising blue. Flesh lemon yellow in cap and upper part of stalk, purplish red in stem base, blue or greenish blue when cut. Spores 12–15 × 5–7. Fairly common in deciduous woods, mostly with *Fagus*, *Quercus* and *Tilia*.

Boletus regius Krombh.
Cap 10–15 cm diam., convex, pinkish or reddish with wine-coloured streaks, becoming pinkish brown, smooth or almost so; pore surface yellow. Stalk 7–12 × 2.5–3.5 cm, cylindrical to fusiform, lemon yellow, pinkish, brick or brownish at base, network sulphur yellow. Flesh pale lemon yellow, blueing only slightly or not at all when cut. Spores 10–15 × 4–5. In grass under *Fagus* and *Quercus*, there are only a few British records of this beautiful species.

Boletus rhodoxanthus (Krombh.) Kall.
Cap 7–20 cm diam., hemispherical to convex, off-white with pinkish areas more pronounced near margin; pore surface yellow at first but soon turning blood red. Stalk 6–15 × 2–5 cm, yellowish orange, base greyish olive, network purplish red. Flesh lemon yellow, blueing only slightly and temporarily in the cap when cut. Spores 10–15 × 4–5.5. In deciduous woods, with *Fagus* and *Quercus*, autumn, rare.

Boletus rubinus W.G.Smith (Fig. 41)
Cap 3–8 cm diam., convex or flattened, yellowish brown to cinnamon, sometimes with a reddish flush towards the margin, often cracking and then appearing adpressed scaly; pore surface coral or carmine, pores up to 1 mm diam., tubes slightly decurrent. Stalk 2–4 × 0.5–1 cm, tapered towards base, reddish at apex, lemon yellow below, often with red spots or splashes. Flesh yellow in stem base, otherwise mostly whitish with a little pink here and there. Spores broadly ellipsoid, 5.5–7 × 4–6, cinnamon buff in mass. Amongst grass under *Quercus*, fairly common.

Boletus satanas Lenz
Satan's Bolete, Devil's Bolete
Cap 10–20 cm diam. or occasionally even larger, hemispherical to convex, whitish to silvery grey or olivaceous grey, ochraceous when old, smooth or

minutely cracked; pore surface blood red or reddish orange, bruising green-ish, pores small. Stalk 8–12 × 3–10 cm, often swollen at base, yellow at apex, reddish below this and olivaceous at base, network red. Flesh whitish, pale straw or pale lemon yellow, turning pale blue when cut. Spores 10–14 × 4–6. On calcareous soils, usually with *Fagus* and *Quercus.*

Boletus satanoides Smotlacha
Cap 5–20 cm diam., convex, at first off-white to milky coffee colour, becoming greyish olivaceous with a distinct touch of red, especially around the edge; pore surface orange or reddish orange, more yellowish towards margin, pores small. Stalk 8–15 × 2–4 cm, often with slightly bulbous base, orange with slight network near apex, middle part rather dark red, base often yellowish, bruising blue. Flesh white to pale lemon, bluish when cut. Under *Quercus*, summer, uncommon.

Boletus spadiceus Fr.
Cap 3–9 cm diam., convex, cinnamon to brick- or snuff-coloured, with olivaceous tomentum when young; pore surface lemon to chrome yellow, bruising blue, pores angular, large. Stalk 3–9 × 1–1.5 cm, cylindrical or tapered towards base, buff to fawn or fulvous, rust-coloured towards base, with distinct darker ribs most pronounced in the upper part. Flesh white or creamy, often with clay-pink mottling in the upper part of the stem. Spores 11–14 × 4–5. In mixed woods.

Boletus subtomentosus Fr.
Yellow-cracked Bolete
Cap 4–12 cm diam., convex, olivaceous brown or hazel, velvety, skin very firmly attached, often cracking to show pale flesh; pore surface lemon yellow becoming olivaceous. Stalk 3–8 × 0.5–1.5 cm, cylindrical or tapered towards base, pale yellowish brown, occasionally very faintly ribbed. Flesh white to pale yellow. Spores 10–13 × 4–5. In deciduous and mixed woods, common.

Boletus versicolor Rostk.
Cap 3–7 cm diam., convex, flattening, sometimes with a wavy margin, cherry red or scarlet, often with an olivaceous flush towards the centre; pore surface lemon yellow, bruising blue, pores angular. Stalk 2–7 × 0.5–1 cm, lemon yellow at apex, the rest rather pale red with darker reddish spots. Flesh pale straw yellow except towards the base of the stem where it is red or brownish. Spores 12–14 × 4–5.5. Amongst grass in parks and decid-uous woods, mostly under *Quercus.*

Botryobasidium
Fruitbodies thin, cobwebby or cottony and reticulate, easily separated from the substrate, mostly whitish or grey, sometimes yellowish when old.

Hyphae broad, usually branching more or less at right angles, hyaline or yellowish, basal ones often wider and thicker-walled, clamps present in two species. No cystidia. Basidia often with 6 or more sterigmata, cylindrical to suburniform, broad, clustered, staining deeply with cotton blue. Spores hyaline, smooth, thin-walled, J–, C+. Conidial states often present with conidia borne on short pegs.

KEY

Hyphae all or almost all with clamps at septa 1
Hyphae without clamps 2
1. Spores 1.5–2 wide .... *intertextum*
   Spores 2.5–4 wide .... *subcoronatum*
2. Hyphae mostly with verrucose walls .... *pruinatum*
   Hyphae with smooth walls 3
3. Spores 8–12 long 4
   Spores 5–9 long 7
4. With conidial states 5
   Without conidial states 6
5. Basidia 16–18 long, conidia 11–13 wide .... *ellipsosporum*
   Basidia 10–12 long, conidia 17–22 wide .... *vagum*
6. Spores navicular .... *botryosum*
   Spores ellipsoid, obliquely attenuated at base .... *obtusisporum*
7. No conidial state, basidia up 25 long .... *laeve*
   Conidial states present, basidia not more than 18 long 8
8. Conidia not in chains, borne singly on pegs .... *conspersum*
   Conidia mostly in chains which are often branched 9
9. Conidial state golden yellow to orange .... *aureum*
   Conidial state greyish white .... *candicans*

Botryobasidium aureum Parm. (Fig. 42)
Hyphae without clamps, basal ones yellowish, 6–9 thick. Basidia 12–18 × 7–10, mostly 6-spored. Spores cylindrical to navicular, with oil drops, 6–9 × 3–4. Conidial state golden yellow to orange. Conidia yellowish, in branched chains, limoniform to broadly fusiform, 18–28 × 10–12. On rotten wood of deciduous trees, fairly common.

Botryobasidium botryosum (Bres.) J. Erikss. (Fig. 43)
Hyphae without clamps, basal ones yellowish, 9–10 thick. Basidia 20–25 × 10–12, mostly 6-spored. Spores navicular, 9–12 × 4.5–6. No conidial state. Mostly on coniferous wood, common.

Botryobasidium candicans J. Erikss. (Fig. 44)
Hyphae without clamps. Basidia 12–18 × 5–7, mostly 6-spored. Spores limoniform to navicular, 6–8 × 3–4. Conidial state has conidia in short chains, each conidium measuring 15–17 × 8–9. Colonies effused, greyish white. On wood of deciduous trees, very common.

Botryobasidium conspersum J. Erikss. (Fig. 45)
Hyphae without clamps. Basidia 12–15×6–8, 6-spored. Spores slender, navicular to cylindrical, 7–9×2.5–3. Conidial state has ellipsoid conidia borne singly on pegs along the sides of conidiophores, each conidium measures 15–20×9–14. Colonies fulvous or snuff-coloured. On decaying wood, very common.

Botryobasidium ellipsosporum Hol.-Jech.
Hyphae without clamps. Basidia 16–18×7–8, mostly 6–8-spored. Spores ellipsoid to fusiform, 8–10×3.5–4. Conidial state has conidia 20–25×11–13, with finely warted inner walls. On rotten wood, not common.

Botryobasidium intertextum (Schw.) Jülich and Stalpers
Fruitbodies ochraceous when old. Hyphae with clamps. Basidia 15–30×5–7. Spores narrowly navicular, 7–9×1.5–2.5. On coniferous wood.

Botryobasidium laeve (J. Erikss.) Parm.
Hyphae without clamps, basal ones 15–20 wide, all smooth. Basidia 17–25×7–9, 6-spored. Spores ellipsoid, obliquely attenuated at base, 5–8×2.5–3.5. Mostly on wood of deciduous trees but occasionally also on conifer wood.

Botryobasidium obtusisporum J. Erikss. (Fig. 46)
Hyphae without clamps. Basidia 18–25×8–10, mostly 6-spored. Spores ellipsoid, obliquely attenuated at base, 8–12×3.5–5. Mostly on conifer wood.

Botryobasidium pruinatum (Bres.) J. Erikss. (Fig. 47)
Hyphae without clamps, nearly all with verruculose walls, basal ones at least 15–20 wide. Basidia 18–25×7–10. Spores ellipsoid, obliquely attenuated at base, 5–8×2.5–3.5. Mostly on wood of deciduous trees.

Botryobasidium subcoronatum (Höhn and Litsch.) Donk (Fig. 48)
Hyphae all with clamps, basal ones 7–10 wide. Basidia with basal clamps, 20–30×7–10, 6-spored. Spores navicular, 6–10×2.5–4. On all sorts of decaying wood, very common.

Botryobasidium vagum (Berk. and Curt.) Rogers
Hyphae without clamps, basal ones 6–10 wide. Basidia 10–12×7–8.5, mostly 6-spored. Spores fusiform, 9–11×3–4. Conidia broadly ellipsoid to subspherical, yellowish brown, 20–30×17–22. On rotten wood.

Botryohypochnus *isabellinus* (Fr.) Erikss. (Fig. 49)
Fruitbodies resupinate, effused, thin, soft, cottony, hypochnoid or membranous, pale yellow to ochraceous, loosely attached to substrate. Hyphae broad, short-celled, more or less at right angles to one another, some yellowish and thicker walled, without clamps. No cystidia. Basidia clustered, 14–22×8–11, 4-spored, staining deeply with cotton blue. Spores

spherical, about 8–10 diam., echinulate, spines 1–3 long, J–, C+. On lower surface of fallen, rotten trunks and branches.

***Bourdotia** galzinii* (Bres.) Trott. (Fig. 50)
Fruitbodies resupinate, waxy or somewhat gelatinous, when young hyaline to pale grey or ochraceous, when old dark reddish brown. Hyphae difficult to see. Cystidia (g) 50–250 × 3–12, cylindrical or irregular, hyaline, yellowish or brownish. There are also slender gnarled hyphidia. Basidia solitary, ellipsoid, 12–22 × 8–12, with longitudinal septa, 4-spored, sterigmata up to 20 long. Spores ellipsoid to broadly cylindrical, 9–14 × 5–7.5, hyaline, J–. On wood of deciduous trees.

***Brevicellicium** olivascens* (Bres.) Larss. and Hjortst. (Fig. 51)
Fruitbodies resupinate, effused, firmly attached, membranous, smooth or minutely tuberculate or toothed, with protuberances not more than 0.2 mm tall, thin, somewhat waxy, pale greyish or yellowish to ochraceous with greenish tinge. Hyphae mostly 2–4 wide but with some cells inflated up to 10, with clamps. No cystidia. Basidia 12–20 × 6–7, with basal clamps, 4-spored. Spores subspherical, 4–5 diam., apiculate, thin-walled, smooth, J–. On dead wood, fairly common.

***Buglossoporus** pulvinus* (Pers.) Donk (Fig. 52)
Fruitbodies annual, 5–15 cm wide, 5 cm thick, flat or convex, sometimes tongue-shaped brackets, sessile or extended into a short, thick, stem-like base, fleshy when fresh, leathery to corky when dry; upper surface pale yellowish brown and finely hairy when young, becoming smooth and blackish, with a thin cuticle; pore surface white, bruising brown, 2–3 pores per mm. Flesh white or cream. Hyphae with clamps; hyaline skeletal ones frequent in flesh but absent from tube layer. Spores narrowly ellipsoid with a small, curved, oblique apiculus at base, hyaline to pale yellowish, smooth, J–, 6–9 × 2.5–4. On standing trees of *Quercus*, usually in depressions in bark, rare. This fungus has also been called *Piptoporus quercinus*. It has been found in Suffolk.

***Bulbillomyces** farinosus* (Bres.) Jülich (Fig. 53)
Fruitbodies resupinate, effused, firmly attached to substrate, thin, softly membranous, watery white or greyish and rather waxy when fresh, smooth or minutely papillate under a lens; accompanied or replaced by its *Aegerita* state consisting of spherical bulbils 0.1–0.4 mm diam., which look like little white eggs scattered over the surface. Hyphae hyaline, with clamps. Cystidia (m) projecting, hyaline, thick-walled, apically encrusted, 60–100 × 8–10. Basidia more or less urniform, 20–30 × 6–8, mostly 4-spored. Spores broadly ellipsoid, 6–9 × 5–7, hyaline, smooth, J–. *Aegerita* bulbils complex in structure, made up of dichotomously branched hyphae with clamps at some of the septa, the branches terminating in pyriform cells

10–20 × 10–12. On rotten wood and stumps in damp places, found most commonly in its *Aegerita* state.

Butlerelfia *eustacei* Weres. and Illman (Fig. 54)
Fruitbodies resupinate, effused, floccose or pellicular, smooth, white to ochraceous. Hyphae hyaline with clamps. Cystidia none. Basidia narrowly clavate, hyaline, 20–45 × 3.5–5.5, 4-spored. Spores ellipsoid, 4.5–7 × 2.5–4, hyaline, smooth, J–, C+. Causes an eye-spot disease of apples in storage.

Byssocorticium

Fruitbodies resupinate, effused, thinly membranous or pellicular, smooth or with pores, yellowish, ochraceous, bluish or bluish green. Hyphae mostly narrow, with or without clamps. No cystidia. Basidia clavate with numerous oil drops. Spores rather small, smooth, thick-walled, slightly coloured, J–, C+.

KEY

Fruitbodies poroid, yellowish	....	*mollicula*
Fruitbodies without pores, bluish	....	1
1. Basidia 20–28 × 4–5	....	*atrovirens*
Basidia 25–35 × 7–8	....	*pulchrum*

Byssocorticium atrovirens (Fr.) Bond. and Sing.
Fruitbodies resupinate, effused, bluish green or blue, smooth, thin, membranous or pellicular, loosely attached to substrate. Basidia 20–28 × 4–5, 4-spored, some with clamps. Spores spherical, with apiculus, 3–4 diam., bluish, thick-walled, each with a large oil drop. On decayed wood, leaves and other litter in deciduous and conifer woods.

Byssocorticium mollicula (Bourd.) Jülich
Fruitbodies resupinate, not more than 3 mm thick, poroid, soft, spongy, easily separated from substrate, yellowish to ochraceous, pores irregularly angular, 2–3 per mm. Hyphae narrow, no clamps. Basidia 4-spored, containing oil drops, 15–20 × 5. Spores spherical, 3.5–4.5 diam., thick-walled, smooth, yellowish, each with one oil drop. On leaf litter and decayed wood.

Byssocorticium pulchrum (Lundell) Christ. (Fig. 55)
Fruitbodies resupinate, effused, bluish, bluish grey or greenish blue, thin, cottony, smooth. Hyphae sometimes with clamps. Spores subspherical to pyriform, 4–6 × 4–5, bluish, each with one oil drop. On leaf litter and rotten wood.

Calocera

Fruitbodies solitary or gregarious, erect, tough gelatinous, surface often slimy when moist, clavarioid, with or without a sterile stalk, cylindrical, spatulate, clavate or lanceolate, in one species repeatedly dichotomously branched. Flesh 3-layered, with a compact central core surrounded by loosely interwoven hyphae and on the outside a closely compacted hymenial layer. Hyphae without clamps. Hyphidia cylindrical or narrowly clavate. Basidia deeply divided, resembling tuning forks. Spores oblong-ellipsoid to allantoid, smooth, hyaline or pale yellowish, with 0–3 septa, J−, germinating to form conidia or germ tubes.

KEY

Fruitbodies 3–10 cm high, repeatedly dichotomously branched . *viscosa*
Fruitbodies not more than about 2 cm high, unbranched or almost so
.... 1
1. Spores when mature with 0–1 septa, not more than 11 long ...... *cornea*
    Spores when mature with 1–3 septa, often more than 11 long 2
2. On wood of deciduous trees, spores 12–16 long .... *glossoides*
    On wood of conifers, spores 8–15 long 3
3. Fruitbodies with cylindrical stalk and broadly clavate to spatulate fertile
    upper part .... *pallido-spathulata*
    Fruitbodies without distinct stalk .... *furcata*

Calocera cornea (Batsch ex Fr.) Fr. (Fig. 56)
Fruitbodies rarely more than 1 cm high, 1–2 mm wide, gregarious and sometimes with 2 or 3 joined together at the base, subulate or cylindrical, rarely forked at tips, always rather pale yellow when fresh, drying orange, most of upper part fertile. Basidia 20–35 × 3–4.5. Spores hyaline, ellipsoid to allantoid, eventually becoming 1-septate, with two guttules, 7–11 × 3–4.5. On dead twigs and branches of deciduous trees all the year round, very common, especially on *Quercus*.

Calocera furcata (Fr.) Fr.
Fruitbodies up to about 2 cm high, in groups, often tufted, pale yellow or yellowish orange, mostly cylindrical to clavate, without a differentiated stalk, rarely dichotomously branched at apex. Basidia 30–60 × 3–4. Spores hyaline, narrowly ellipsoid to allantoid, 8–13 × 3–4, when ripe with 1–3 somewhat thickened septa. On conifers, especially *Abies* and *Pinus*.

Calocera glossoides (Pers. ex Fr.) Fr.
Fruitbodies up to 1 cm high, with a fairly distinct stalk which is blackish-brown when dry, upper, fertile part lanceolate, clearly broader than stalk, yellow when fresh, becoming golden brown. Basidia 30–50 × 3–5. Spores hyaline, 12–16 × 3–4.5, narrowly ellipsoid to somewhat allantoid, with 1–3

slightly thickened transverse septa. On wood of deciduous trees such as *Acer*, *Fagus* and *Quercus*, uncommon.

Calocera pallido-spathulata Reid
Fruitbodies up to 1 cm high, with a distinct cylindrical, whitish stalk, upper fertile part clavate or flatly spatulate and often clearly longitudinally wrinkled, whitish or pale yellowish. Basidia 25–40 × 3–4. Spores hyaline, 10–15 × 3.5–4, narrowly ellipsoid or somewhat allantoid, when mature with 1–3 rather thin transverse septa. On wood of conifers, especially *Larix* and *Picea*.

Calocera viscosa (Pers. ex Fr.) Fr. (Fig. 57)
Yellow Antler Fungus
Fruitbodies 3–10 cm high, stalked, repeatedly and mostly dichotomously branched, golden yellow to orange–yellow, paler or whitish at base, becoming distinctly orange when dry, branches rounded or flattened, sometimes longitudinally grooved, terminal ones short, tapered to a point and sometimes in threes as well as twos. Basidia 40–50 × 3–4.5. Spores oblong-ellipsoid, flattened on one side, often becoming 1-septate, with 2 guttules, hyaline or pale yellow, 8–12 × 3–4.5. On conifers, especially *Pinus*, mostly on stumps and roots, very common in autumn and winter.

Calyptella

Fruitbody composed of an erect or pendent, bowl- or bell-shaped cap prolonged at its base into a short stalk. Cap outer layer containing vertical, branched, coral-like hyphae which sometimes protrude; inner, hymenial surface smooth. Basidia clavate, sometimes with clamp at base, 4-spored. Cystidia none. Spores ellipsoid, sometimes flattened on one side or pip-shaped, hyaline, J–, smooth.

KEY

	Spores 5–6 wide	1
	Spores 3–4 wide	2
1.	Fruitbodies sulphur or greenish yellow	*campanula*
	Fruitbodies white to pale cream	*capula*
2.	Fruitbodies pale yellow	*laeta*
	Fruitbodies brownish	*cernua*

Calyptella campanula (Nees ex Pers.) Cooke (Fig. 58)
Cap 1–2 mm diam., with stalk up to 1.5 mm long, sulphur yellow or greenish yellow, outer surface almost smooth. Basidia amyloid, 20–25 × 6–7. Spores mostly about 9 × 5, with granular contents. On parts of dead herbaceous stems near the ground, recorded on *Beta*, *Solanum*, *Urtica* etc., uncommon.

54   Non-gilled hymenomycetes

Calyptella capula (Holmsk. ex Fr.) Quél. (Fig. 59)
Cap 2–7 mm diam., margin rather irregular, white to pale cream or pale brownish when quite old, slightly pruinose on the outside, stalk concolorous, up to 2 mm long. Basidia $20–30 \times 7–8$. Spores ellipsoid, flattened on one side, $7–11 \times 5–6$. On dead herbaceous plants and also occasionally on leaves of trees, especially common on *Anthriscus, Heracleum, Smyrnium* and other Umbelliferae, also on *Urtica*, April–December.

Calyptella cernua (Schum.) W.B. Cooke
Cap 1–2 mm diam., with very short stalk, brownish. Basidia with clamps, $25–28 \times 7–9$. Spores apiculate, flattened on one side, $8–9 \times 3–4$. On bark of *Sambucus nigra*.

Calyptella laeta (Fr.) W.B. Cooke
Cap 3–5 mm long and wide, pale yellow, stalk 3–10 mm long. Basidia $14–27 \times 4–6$, 4-spored, with basal clamps. Spores pip- or tear-shaped, $8–10 \times 3–4$. On dead herbaceous stems.

Cantharellus

Fruitbodies pileate, stipitate, erect, fleshy, cap at first convex, sometimes top-shaped but soon becoming depressed in the centre, or irregularly funnel-shaped with a wavy margin; hymenium covering decurrent, irregularly branched and sometimes interconnected, thick, blunt ribs or veins which are rarely more than 1 mm deep; stalk central or excentric, expanding above to pass gradually into the cap. Hyphae with clamps. No cystidia. Basidia 2–8-spored, narrowly clavate. Spores ellipsoid, hyaline but in mass creamy, yellowish or pinkish, smooth, thin-walled, J–.

KEY

Stalk solid, fruitbodies not deeply funnel-shaped 1
Stalk hollow, fruitbodies soon deeply funnel-shaped 3
1.   Cap 1–3 cm wide .... *friesii*
     Cap 3–15 cm wide 2
2.   Stalk concolorous with cap, egg-yolk colour to yellow–ochre .. *cibarius*
     Similar but cap margin violaceus .... *cibarius* var. *amethystea*
     Stalk ivory white, bruising and ageing rusty ochraceous .. *ferruginascens*
3.   Cap and stalk greyish brown, hymenial surface whitish grey .... *cinereus*
     Cap, stalk and hymenial surface other colours 4
4.   Hymenial surface pinkish yellow to orange, smooth or shallowly veined
     .... *lutescens*
     Hymenial surface yellowish becoming grey pruinose, veins thicker
     .... *tubaeformis*

Cantharellus cibarius Fr. (Fig. 60)

Chanterelle

Cap 3–15 cm wide, top-shaped, convex then flattened and finally depressed in the centre, margin wavy, at first inrolled, the colour of egg-yolk to yellow–ochre, hymenium yellow; stalk tapered towards base, concolorous with cap, stuffed solid, 2–8 × 1–4 cm. Basidia 60–100 × 6–9. Spores mostly 8–10 × 5–6, containing granules or oil drops, hyaline, pale orange–yellow in mass. Smell rather like that of apricots. Common in deciduous woods, especially under *Fagus* and *Quercus*, but found also with *Betula* and conifers where ground is damp and mossy. In the var. *amethystea* Quél. the margin of the cap is violaceous.

Cantharellus cinereus Pers. (Fig. 61)

Cap 2–5 cm wide, soon becoming funnel-shaped, greyish brown, hairy or slightly scaly in the centre, smooth and darker at the edge, hymenial surface whitish grey, ribs widely separated; stalk 4–8 × 0.5–0.9 cm, hollow in the middle, greyish brown, blackish when old. Basidia up to 100 × 10. Spores 7–12 × 5–8. Mostly in beech woods.

Cantharellus ferruginascens P.D. Orton

Cap 3–6 cm diam., similar in shape to that of *C. cibarius*, yellowish buff, darker at centre, with pale cream or whitish margin and silky matt surface, hymenium pale yellow or cream; stalk 2–4 × 0.5–1.5 cm, ivory white or creamy, bruising and ageing rusty ochraceous, solid. Basidia 60–80 × 8–10. Spores 8–10 × 5–6. According to Orton this grows on basic soils in mixed woods under *Bromus racemosus*, *Mercurialis perennis*, *Anenome nemorosa* and *Circaea lutetiana*.

Cantharellus friesii Quél.

Cap in shape similar to *C. cibarius* but smaller, 1–3 cm wide, orange to pinkish or brownish orange, velvety with small, darker scales, hymenium pinkish or yellowish orange; stalk 1–3 × 0.3–0.5 cm, concolorous with cap or more yellowish, solid. Basidia 60–70 × 6–10. Spores 7–11 × 4–6, yellow-ochre in mass. Mostly with *Fagus* on bare or mossy soil.

Cantharellus lutescens Fr. (Fig. 62)

Cap 2–6 cm wide, soon deeply funnel-shaped, yellow to brownish orange with paler margin, fibrillose to scaly, radially wrinkled, hymenium pinkish yellow to orange, smooth or very shallowly veined; stalk 2–6 × 0.5–1 cm, rather flattened, yellow or yellowish orange, hollow. Basidia 70–100 × 9–11. Spores 10–12 × 7–9. On the ground in mixed woods.

Cantharellus tubaeformis Fr. (Fig. 63)

Fruitbodies trumpet-shaped, with deep funnel and swollen, incurved margin; cap 2–6 cm wide, ochraceous brown or greyish brown; hymenial surface yellowish becoming grey-pruinose; stalk 4–8 × 0.5–1 cm, yellowish, hollow. Basidia 60–100 × 8–12. Spores 8–12 × 6–9. In deciduous woods and conifer woods.

Cejpomyces *terrigenus* Svrček and Pouzar (Fig. 64)
Fruitbodies resupinate, hypochnoid or thinly membranous, smooth, pale ochraceous. Hyphae hyaline, 8–10 thick, without clamps. No cystidia. Basidia cylindrical to clavate, 12–35 × 8–13, sterigmata 2–6, mostly 4, subulate, 10–30 long. Spores navicular to narrowly ovoid, hyaline, thin-walled, smooth, J–, 12–20 × 6–8. No secondary spores formed. On soil and rotten wood.

Cellypha *goldbachii* (Weinm.) Donk (Fig. 65)
Fruitbodies deep bowl-shaped, 1–2 mm diam., white or creamy with smooth or slightly wrinkled hymenium covering the inner surface; outer surface downy, with numerous capitate or clavate, non-septate, smooth, hyaline hairs 2 wide swollen to 4–6 at apex. Hyphae with clamps. No cystidia. Basidia clavate, 30–35 × 7–8. Spores pip-shaped or ellipsoid, obliquely attenuated at base, 10–15 × 3–5, hyaline, white in mass, smooth, J–. Common on dead grasses and sedges, especially in damp places. Recorded on *Agropyron, Arrhenatherum, Deschampsia, Festuca, Glyceria* and *Carex acutiformis*, June–August.

Ceraceomyces
Fruitbodies resupinate, usually easily separated from the substrate, effused, membranous to waxy, in some species with conspicuous rhizomorphs, surface smooth or merulioid, hymenium increasing in thickness with age and forming basidia at different levels. Hyphae with clamps at many septa. Cystidia absent or hypha-like. Basidia narrowly clavate, 4-spored. Spores subspherical or ellipsoid, hyaline, smooth, J–, often with oil drops.

KEY

Hymenium merulioid .... *serpens*
Hymenium smooth or submerulioid 1
1. Spores 2.5–3.5 diam. .... *sublaevis*
    Spores 6–10 × 4–5 .... *tessulatus*

Ceraceomyces serpens (Fr.) Ginns (Fig. 66)
Fruitbodies membranous to soft waxy, hymenium merulioid, ochraceous, pale brownish or tinged with lilac. Basidia 20–30 × 4–5. Spores 4–5 × 1.5–3. No cystidia. On wood and bark of deciduous trees and conifers.

Ceraceomyces sublaevis (Bres.) Jülich (Fig. 67)
Fruitbodies resupinate, smooth, membranous to soft waxy, whitish or cream. Cystidia few, hypha-like, projecting. Basidia 20–30 × 4–5. Spores

subspherical or very broadly ellipsoid, 2.5–3.5 diam. On rotten wood and sometimes on old polypores.

Ceraceomyces tessulatus (Cooke) Jülich (Fig. 68)
Fruitbodies resupinate, effused, large, with conspicuous rhizomorphs up to 14 thick, with clamps, white or cream, membranous to waxy. No cystidia. Basidia 25–35 × 5–7. Spores ellipsoid, obliquely attenuated at base, 6–10 × 4–5. On decaying wood of deciduous trees and conifers.

Ceratellopsis aculeata (Pat.) Corner (Fig. 69)
Fruitbodies erect, thread-like, unbranched, 0.5–2 × 0.05–0.1 mm, smooth, white, with sterile stalk and tip which is at first sterile but later sometimes becomes fertile. No sclerotia. Hyphae with clamps. No cystidia. Basidia 10–15 × 4–5, 4-spored. Spores narrowly ellipsoid, 4–6 × 2–3, hyaline, smooth, J–. On rotting leaves.

Ceratobasidium

Fruitbodies resupinate, membranous to waxy, smooth. Hyphae hyaline, without clamps, often branched at right angles to one another. Cystidia none. Basidia broadly cylindrical, clavate or ellipsoid, usually short, hyaline, mostly 2–4-spored, sterigmata stout, curved, not cut off from basidia by septa. Spores thin-walled, hyaline, smooth, J–, forming secondary spores.

KEY

No sclerotia formed, saprophytic on wood .... *cornigerum*
Brownish sclerotia formed, parasitic on phanerogams .... *anceps*

Ceratobasidium anceps (Bres. and Syd.) Jacks.
Fruitbodies very thin, mealy or membranous, white or pale grey, smooth. Sclerotia pale at first, becoming brownish. Basidia 10–18 × 10–12, sterigmata up to 15 long. Spores ellipsoid, hyaline, 9–13 × 5–7.5 An important parasite of phanerogams.

Ceratobasidium cornigerum (Bourd.) Rogers (Fig. 70)
Fruitbodies very thin, mealy or waxy, smooth, yellowish to pale grey, almost invisible when dry. Basidia 12–16 × 7–11, 4-spored, sterigmata up to 14 long. Spores ellipsoid, 7–12 × 4–6. Saprophytic on deciduous trees and conifers.

Ceriporia

Fruitbodies resupinate, poroid, firmly attached, soft and waxy when fresh, brittle when dry, various colours, often white or cream at first but becoming pink, cinnamon, purple etc., tubes often rather short. Hyphae hyaline, without clamps. No cystidia. Basidia 4-spored. Spores mostly cylindrical to allantoid, hyaline, smooth, thin-walled, J−.

KEY

Spores 3.5–5 long 1
Spores mostly more than 5 long 2
1. Pore surface pink or pinkish violet .... *excelsa*
    Pore surface white or cream, when old cinnamon or greenish brown .... *viridans*
2. Pore surface white or cream, spores 6–10 × 2.5–4 .... *reticulata*
    Pore surface becoming purple or wine-coloured, spores 5–8 × 1.5–2 .... *purpurea*

Ceriporia excelsa (Lund.) Parm. (Fig. 71)
Fruitbodies mostly rather small, soft and cottony when fresh, becoming brittle when old, at first whitish or pale ochraceous bruising pink and later becoming all pink or pinkish violet, about 2 mm thick; pores round or angular, 1–3 per mm, in patches at first but later covering most of the surface. Basidia 12–16 × 4–5. Spores ellipsoid-cylindrical or allantoid, 3.5–5 × 2–2.5. On the lower surface of rotting trunks and branches of deciduous trees and conifers.

Ceriporia purpurea (Fr.) Donk (Fig. 72)
Fruitbodies usually up to about 8 cm across but sometimes more widely effused, about 2 mm thick, soft when fresh, pinkish turning purple or wine-coloured, pores 3–5 per mm. Basidia 12–22 × 4–5. Spores allantiod, 5–5 × 1.5–2. On fallen dead wood of deciduous trees.

Ceriporia reticulata (Fr.) Dom. (Fig. 73)
Fruitbodies about 4–7 cm across and up to 1 mm thick, soft when fresh, white or cream becoming cinnamon but only when old, pores angular, 2–4 per mm, shallow, walls thin and forming a reticulum. Basidia 15–20 × 5–6. Spores cylindrical to allantoid, 6–10 × 2.5–4. On rotting wood, mostly of deciduous trees.

Ceriporia viridans (Berk. and Br.) Donk
Fruitbodies effused, soft, waxy, up to 3 mm thick, white or cream when young, turning cinnamon, greenish brown or occasionally pink, pores round or angular, 3–5 per mm. Basidia 10–15 × 4–5. Spores 3.5–5 × 1.5–2.5, cylindrical or allantoid. On rotting wood, mostly of deciduous trees.

Ceriporiopsis

Fruitbodies resupinate, effused, soft waxy when fresh, brittle when dry, with round, angular or sometimes labyrinthine pores. Hyphae hyaline, with clamps, stain deeply with cresyl blue. No cystidia. Spores ellipsoid or cylindrical, sometimes curved, hyaline, smooth, thick-walled, J–.

KEY

Spores cylindrical, 4–5 × 1.5–2 .... *gilvescens*
Spores ellipsoid, 5.5–7 × 3.5–4.5 .... *aneirinus*

Ceriporiopsis aneirinus (Sommerf. ex Fr.) Dom. (Fig. 74)
Fruitbodies up to 4 mm thick, cream or ochraceous, drying brownish, margin white, fimbriate, pores 1–3 per mm, angular, thin-walled, sometimes becoming labyrinthine. Basidia 20–25 × 5–7. Spores ellipsoid, 5.5–7 × 3.5–4.5. On wood, mostly of deciduous trees and especially on that of *Populus tremula.*

Ceriporiopsis gilvescens (Bres.) Dom. (Fig. 75)
Fruitbodies up to 5 mm thick, white often flushed pink when fresh, bruising and ageing reddish brown or greyish brown, pores round or angular, 3–5 per mm, sometimes labyrinthine; narrow, floccose or fimbriate, white or pinkish margin present only in young colonies. Hyphae frequently encrusted. Basidia 10–12 × 4. Spores cylindrical to allantoid, 4–5 × 1.5–2. On bare wood, mostly of deciduous trees.

Cerocorticium

Fruitbodies resupinate, initially often small and round but soon becoming confluent and then covering large areas, firmly attached except when old, waxy when fresh, with smooth, tuberculate or distinctly toothed hymenium. Hyphae with many clamps. No true cystidia, hyphidia present in some species. Basidia large, clavate, containing oil drops, with basal clamps, 4-spored. Spores ellipsoid or spherical, hyaline, smooth, containing small oil drops or granules, J–.

KEY

Hymenium with teeth .... *molare*
Hymenium without teeth 1
1. Hymenium smooth, spores spherical or almost so, 10–14 diam. *hiemale*
    Hymenium often tuberculate, spores mostly ellipsoid, 7–12 × 6–9
    .... *confluens*

Cerocorticium confluens (Fr.) Jülich and Stalpers (Fig. 76)
Fruitbodies resupinate, smooth or tuberculate, initially small and round but soon becoming confluent and then covering large areas, waxy soft, hygrophanous, pale greyish, cream or ochraceous, often somewhat opalescent, with slight violaceous or bluish tints, up to 1 mm thick. Hyphae all with

clamps. Basidia clavate, stalked, flexuous, with basal clamp, containing numerous oil drops, 30–60×7–10. Spores broadly ellipsoid or sub-spherical, 7–12×6–9. On wood and bark, mostly of deciduous trees, especially in damp places. A very common species.

Cerocorticium hiemale (Laurila) Jülich and Stalpers (Fig. 77)
Fruitbodies resupinate, often small but sometimes confluent and covering several square centimetres, smooth, membranous to waxy, white or cream, thin. Hyphae all with clamps at septa. Hyphidia cylindrical, flexuous, up to 3 wide, projecting slightly, rarely branched, sometimes coated with acicular crystals in old specimens. Basidia 60–70×10–15, containing oil drops. Spores spherical or subspherical, 10–14 diam., containing oil drops. Mostly on conifer wood.

Cerocorticium molare (Chaill. ex Fr.) Jülich and Stalpers (Fig. 78)
Fruitbodies resupinate, initially small and round but soon confluent and then covering large areas, 30 cm or more across, cream to ochraceous, waxy, hymenium toothed, teeth 1–5 mm long, fimbriate at tips when young. Basidia clavate, stalked, 35–50×6–9, with conspicuous oil drops and basal clamps. Spores ellipsoid, 9–12×6–8, containing small oil drops or granules. On wood and bark of deciduous trees, especially *Quercus*, some-times on dead but still attached branches.

Cerrena unicolor (Fr.) Murr. (Fig. 79)
Fruitbodies leathery, mostly pileate and forming sessile, imbricate brackets up to 10 cm long, 2–5 cm wide and 0.5–1 cm thick, occasionally effuso-reflexed or resupinate; upper surface zoned, hairy, whitish, grey or greyish brown, often made greenish by algae; lower surface pale greyish brown with elongated, sinuous, labyrinthine pores. Hyphae with clamps, thick-walled skeletal ones present. No cystidia. Flesh separated from tomentum by distinct black line. Basidia 13–15×4–5. Spores broadly ellipsoid, 5–7×3–3.5, hyaline, smooth, J–. On stumps and dead trunks of deciduous trees, most frequently on *Betula*.

Chaetoporellus latitans (Bourd. and Galz.) Singer (Fig. 80)
Fruitbodies resupinate, up to 10 cm long, 3 mm thick, soft when fresh, brittle when dry, white or cream becoming ochraceous to ochraceous brown; on horizontal surfaces pores round and 1–2 per mm, on vertical surfaces irregular or labyrinthine. Hyphae with clamps. Cystidia cylindrical or subulate, hyaline, up to 35×5. Spores allantoid, 3–4×0.5–1, hyaline, smooth, thin-walled, J–. On conifer wood.

Chondrostereum purpureum (Pers. ex Fr.) Pouz. (Fig. 81)
Silver-leaf Fungus
Fruitbodies resupinate but discrete and often with upturned margins when growing on horizontal surfaces, on vertical surfaces forming small brackets,

thin, softly leathery when fresh, horny when dry; sterile surface white or greyish, hairy; hymenium smooth or slightly wrinkled, violet or violaceous brown. In section a black line can be seen between the tomentum and the flesh; bladder-like vesicles formed at the ends of hyphae are abundant in the subhymenium. Hyphae with clamps. Cystidia few, cylindrical or fusiform, up to 40×4–6, hyaline. Basidia up to 50×5. Spores oblong-ellipsoid, usually flattened on one side, 5–10×2.5–4, hyaline, smooth, thin-walled, J–. On dead wood of deciduous trees, also parasitic especially on rosaceous plants e.g. *Prunus* where it causes silver-leaf disease. Very common.

Chromocyphella *muscicola* (Fr.) Donk. (Fig. 82)
Fruitbodies cup- or shell-shaped, up to 3 mm diam., short-stalked or sessile; outer surface white or greyish, silky or downy, with short, irregular, cystidium-like hairs, with clamps; hymenium lining inside of cup smooth or wrinkled, at first white but becoming reddish brown from the spores as these mature. Basidia 4-spored, 20–25 long. Spores spherical or sub-spherical, 7–10 diam., verruculose, reddish brown in mass. On mosses, especially those which grow on bark.

Cinereomyces *lindbladii* (Berk.) Jülich (Fig. 83)
Fruitbodies resupinate, effused, poroid, 1–8 mm thick, loosely attached, tough and leathery to corky, pore surface turning grey quite soon but edge remains white, pores round or angular, 3–4 per mm. Hyphae with clamps; skeletal ones abundant, strongly gelatinized, with walls dissolving readily in KOH. Cystidia hypha-like or thin subulate. Basidia 10–20×4–6. Spores cylindrical to allantoid, 4–6.5×1.5–2, hyaline, thin-walled, smooth, J–. Most records on bark of *Picea* but not confined to that host.

Clavaria

Fruitbodies terrestrial, erect, often rather slender clubs, solitary, gregarious or in clumps, mostly unbranched and only in *C. zollingeri* regularly branched, often fragile, white or various colours. Hyphae hyaline, without clamps, frequently forming secondary septa. Cystidia only in one species *C. purpurea*. Basidia with in some species a large, widely bowed clamp at the base. Spores mostly ellipsoid to spherical, hyaline, thin-walled, usually smooth, spiny only in *C. asterospora*, J–, sometimes full of small oil drops or granules but normally without large guttules.

KEY

Fruitbodies up to 5–12 cm tall 2
2. Spores with large spines .... *asterospora*
   Spores smooth 3
3. Basidia with large basal clamps, spores 7–10 × 6–9 .... *acuta*
   Basidia without clamps, spores 5–8 × 3–4.5 .... *vermicularis*
4. Fruitbodies mostly branched, with repeated forking towards the tips
   .... *zollingeri*
   Fruitbodies not or very rarely branched 5
5. Basidia with large basal clamps, fruitbodies rose or flesh-coloured,
   discolouring yellowish etc. .... *incarnata*
   Basidia without clamps, fruitbodies not discolouring 6
6. Cystidia present, fruitbodies purple or purplish .... *purpurea*
   No cystidia, fruitbodies bright rose pink .... *rosea*
7. Fruitbodies very small, 6–15 mm tall .... *lithocras*
   Fruitbodies much taller, up to 3.5–12 cm tall 8
8. Fruitbodies lemon yellow to yellow–ochre, spores spherical or
   subspherical .... *straminea*
   Fruitbodies duller colours, spores other shapes 9
9. Basidia without basal clamps 10
   Basidia with basal clamps 11
10. Basidia 30–45 × 7–9, spores 5.5–8 × 3.5–4 .... *fumosa*
    Basidia 20–25 × 4–5, spores 4–6 × 2.5–3.5 .... *crosslandii*
11. Fruitbodies dirty yellowish .... *argillacea*
    Fruitbodies mostly grey or greyish brown 12
12. Spores 6–12 × 4–5.5 .... *tenuipes*
    Spores 8–9 × 6–7 .... *greletii*

Clavaria acuta Sow. ex Fr. (Fig. 84)
Fruitbodies pure white, solitary or in clumps, unbranched, cylindrical to narrowly clavate, sometimes bent over, often slightly flattened, 1–8 cm × 1–3 mm. Hyphae up to 30 wide. Basidia clavate, 40–55 × 8–10, with 2 or 4 sterigmata, with basal clamps. Spores broadly ellipsoid or subspherical, 7–10 × 6–9. Amongst grass or litter in woods and fields.

Clavaria argillacea Pers. ex Fr. (Fig. 85)
Field or Moor Club
Fruitbodies rather pale yellow, greyish yellow or greenish yellow, darker and more yellow towards the base, mostly in clumps, unbranched or occasionally forked, cylindrical to clavate, sometimes flattened, stalked, 3–7 cm × 2–8 mm. Hyphae up to 25 wide. Basidia with clamps. Spores ellipsoid, often flattened on one side, 9–12 × 4.5–6. Mainly on heaths and moorland.

Clavaria asterospora Pat. (Fig. 86)
Fruitbodies pure white, solitary or a few together, unbranched, stalked,

often bent over, slenderly clavate with the upper part cylindrical and usually rounded at the tip, hollow and brittle, 1.5–5 cm × 1–5 mm. Hyphae up to 20 wide. Basidia 30–50 × 7–12, with 4 sterigmata. Spores spherical or subspherical, 7–10 × 6–9, with spaced-out spines up to 4 long. On bare soil, uncommon.

Clavaria corbierii Bourd. and Galz.
Fruitbodies white, solitary or a few together, unbranched, clavate, stalked, 7–12 × 0.5–1 mm, at apex flattened or somewhat indented. Hyphae up to 6 wide. Basidia without clamps, 14–18 × 3–5. Spores broadly ellipsoid, smooth, with oil drops, 3–5 × 3–4. On soil.

Clavaria crosslandii Cotton
Fruitbodies pale grey or pale brownish, in small groups, unbranched, laterally compressed, more or less grooved, stalked, flattened apically, brittle, 2–3.5 cm × 3–7 mm. Hyphae up to 15 wide. Basidia without clamps, 20–25 × 4–5. Spores ellipsoid, shortly apiculate, without oil drops, 4–6 × 2.5–3.5. Amongst short grass or mosses in woods, uncommon.

Clavaria fumosa Fr. (Fig. 87)
Fruitbodies cream, pale greyish brown or pale ochraceous brown, usually in rather dense clumps, unbranched, cylindrical or fusiform, often slightly compressed laterally and grooved, 2–12 cm × 2–7 mm. Hyphae up to 30 wide. Basidia without clamps, 30–45 × 7–9. Spores ellipsoid, 5.5–8 × 3.5–4, packed with granules or small oil drops. Amongst grass in open situations.

Clavaria greletii Boud.
Fruitbodies grey, greyish brown or brown, solitary or in small groups, stalked, more or less cylindrical, tapered towards the apex, laterally flattened, with one or two longitudinal grooves, 6–8 cm × 3–6 mm. Hyphae up to 14 wide. Basidia 30–60 × 6–10, 4-spored, with basal clamps. Spores broadly ellipsoid, apiculate, 8–9 × 6–7, packed with small granules or oil drops. Amongst grass and mosses along edges of pine woods.

Clavaria incarnata Weinm.
Fruitbodies at first rose or flesh-coloured, becoming yellowish then changing to white or pale grey from the tip downwards, solitary or in clumps, unbranched, cylindrical or compressed, 2–8 cm × 1–1.5 mm. Hyphae up to 15 wide. Basidia with basal clamps. Spores ellipsoid or ovoid, 7–10 × 4–6.5. On soil or amongst grass in woods.

Clavaria lithocras Reid
Fruitbodies creamy ochraceous, the upper part with a hard, resinous, orange-coloured secretion, clavate, stalked, 6–15 × 1–2.5 mm. Hyphae up to 15 wide. Basidia about 16 × 5, 4-spored, without clamps. Spores ellipsoid, 4–6 × 2.5–4. On ground, under moss.

Clavaria purpurea Fr.
Fruitbodies purple, purplish brown or purplish grey, usually in clumps, unbranched, cylindrical or compressed, 3–12 cm × 2–5 mm. Hyphae up to 15 wide. Cystidia present, 50–130 × 5–15. Basidia without clamps. Spores 6–15 × 3–5, ellipsoid, somewhat flattened on one side. Amongst grass or on bare soil, mostly under conifers.

Clavaria rosea Fr.
Fruitbodies bright rose pink, solitary or in clumps, unbranched, cylindrical or somewhat fusiform, sometimes compressed, 2–5 cm × 1–5 mm. Hyphae up to 11 wide. Basidia without clamps. Spores ellipsoid, somewhat flattened on one side, 5–8 × 2.5–3.5. Amongst grass and in woods.

Clavaria straminea Cotton
Fruitbodies at first pale lemon yellow, then yellow–ochre and when quite old brownish, solitary or in clumps, unbranched or rarely forked at apex, distinctly stalked, cylindrical, sometimes tapered towards the tip, compressed laterally, 2–4.5 cm × 1–4 mm, very brittle. Hyphae up to 15 wide. Basidia with clamps, 30–50 × 5–10. Spores spherical or subspherical, apiculate, 6–9 diam., full of granules or small oil drops. Amongst grass and moss, under bushes.

Clavaria tenuipes Berk. and Br.
Fruitbodies pale clay-coloured or pale grey, solitary or in clumps, cylindrical or clavate, often compressed laterally, stalked, 1.5–6 cm × 2–10 mm. Hyphae up to 30 wide. Basidia 30–40 × 7–9, with clamps. Spores ellipsoid or somewhat broader towards the base, 6–12 × 4–5.5. Amongst short grass or on bare soil or on bonfire sites, heaths, pastures and woods.

Clavaria vermicularis Fr. (Fig. 88)
Fruitbodies white, occasionally with pale yellow tips, mostly in clumps, cylindrical to fusiform, sometimes flattened and with a longitudinal groove, rarely forked at apex, 6–12 cm × 3–7 mm. Hyphae up to 15 wide. Basidia 40–50 × 6–8, without clamps. Spores broadly ellipsoid, to pyriform, with granular contents, 5–8 × 3–4.5. Amongst grass in fields and woods.

Clavaria zollingeri Lév. (Fig. 89)
Fruitbodies amethyst to deep pinkish violet, in clumps, mostly branched and dichotomously or trichotomously forked towards the branch ends, tips acute, 4–8 cm tall, branches 1–5 mm thick, often compressed laterally. Hyphae up to 23 wide. Basidia without clamps, 50–60 × 7–9. Spores broadly ellipsoid, subspherical or pyriform, 4.5–7 × 3–5. In woods and parks.

Clavariadelphus

Fruitbodies erect, found singly or several together but not in clumps, mostly remaining unbranched, 3–30 cm tall, cylindrical to clavate or turbinate, rounded or flattened at apex, hollow when old, cream, ochraceous or brownish yellow, turning green with ferrous sulphate. Hyphae with clamps, often swollen at septa. Basidia 4-spored, usually with a basal clamp. Spores 9–15 long, mostly ellipsoid, hyaline, smooth, thin-walled, J–. On soil or humus.

KEY

Fruitbodies with broad, flat tops .... *truncatus*
Fruitbodies with rounded tops 1
1. Fruitbodies slender, 0.5–1.5 cm wide .... *ligula*
   Fruitbodies broadly clavate, 2–7 cm wide .... *pistillaris*

Clavariadelphus ligula (Schaeff. ex Fr.) Donk (Fig. 90)
Fruitbodies cylindrical to narrowly clavate, often slightly compressed laterally, 3–10 × 0.5–1.5 cm, cream, ochraceous or yellowish brown. Basidia 40–60 × 6–8. Spores cylindrical to narrowly ellipsoid, 9–15 × 3–5. In conifer woods.

Clavariadelphus pistillaris (Fr.) Donk (Fig. 91)
Giant Club
Fruitbodies clavate, rounded at apex, sometimes longitudinally wrinkled, 10–30 × 2–7 cm, pale yellow at first, turning dark ochraceous or orange-brown, bruising brown, with white mycelium at base. Basidia 50–90 × 9–12. Spores ellipsoid, with large guttules, 11–15 × 6–9. On chalky soil in beech woods.

Clavariadelphus truncatus (Quél.) Donk (Fig. 92)
Fruitbodies clavate to turbinate with flattened tops which are sometimes indented in the middle, margin inflated, often somewhat tuberculate when old, ochraceous or golden brown, 5–15 × 2–8 cm. Basidia 70–150 × 9–12. Spores ellipsoid, 10–13 × 6–7. On ground in conifer and mixed woods.

Clavicorona taxophila (Thom) Doty (Fig. 93)

Fruitbodies erect, clustered, unbranched, when young clavate, becoming trumpet-shaped and somewhat sunken at the apex, white at first, then yellowish, drying ochraceous orange, 0.5–3 cm tall, 1–2 mm wide at base, 3–9 mm wide at apex. Hyphae without clamps, up to 12 wide, some containing large oil drops. Basidia 20–30 × 3.5–4.5, with basal clamps. Spores broadly ellipsoid or subspherical, 3–4 × 2–3, smooth or slightly verruculose, not or weakly amyloid. On dead branches and other debris of *Taxus* and *Juniperus*; and occasionally other trees.

Clavulina

Fruitbodies erect, unbranched or branched, not fragile; branches cylindrical or flattened, blunt, pointed or crested at apex. Hyphae with or without clamps. Basidia cylindrical to narrowly clavate, mostly with two sterigmata which are large and strongly incurved. Spores subspherical or broadly ellipsoid, smooth, thin-walled, J–, each with one large oil drop or guttule. Terrestrial.

KEY

Fruitbodies violet or amethyst .... *amethystina*
Fruitbodies white or other colours 1
1. Fruitbodies with few thick, rugose branches, themselves only a few times branched, creamy, pale ochraceous or grey .... *rugosa*
Fruitbodies richly branched, with primary branches subdividing many times 2
2. Tips of branches crested with short, pointed teeth, mostly white or cream but sometimes grey or other colours .... *cristata*
Tips of branches rounded or bluntly toothed, without crests, mostly grey or greyish .... *cinerea*

Clavulina amethystina (Fr.) Donk (Fig. 94)
Fruitbodies solitary or in clumps, violet or amethyst, branched from the pale or whitish base, with or without a short stalk, 2–6 cm tall; branches numerous and themselves repeatedly branched, with tips blunt or pointed. Basidia 40–60 × 5–8. Spores broadly ellipsoid, 7–11 × 6–8. On the ground under deciduous trees.

Clavulina cinerea (Fr.) Schroeter (Fig. 95)
Grey Coral Fungus
Fruitbodies solitary or in clumps, pale to dark grey sometimes tinged with violet, brownish when old, 3–11 cm tall, stalk when present up to 3 × 1 cm, densely and repeatedly branched; branches cylindrical or flattened, rugose, tips rounded or bluntly toothed. Basidia 40–60 × 5–9. Spores subspherical, 7–10 diam. On ground, mostly in deciduous woods.

Clavulina cristata (Fr.) Schroeter (Fig. 96)
White or Crested Coral Fungus
Fruitbodies solitary or in clumps, mostly white, whitish or cream, but occasionally grey or other colours, 2–8 cm tall, with a variable number of cylindrical or flattened, sometimes rugose primary branches arising from a short stalk and these become much subdivided and have crests of short, pointed teeth at their tips. Basidia 30–60 × 6–8. Spores broadly ellipsoid, 7–10 × 6–8. On soil in woods or occasionally growing on rotten wood.

Clavulina rugosa (Fr.) Schroeter (Fig. 97)

Fruitbodies solitary or in clumps, creamy, pale ochraceous or greyish, 5–12 cm tall, occasionally without but mostly with relatively few thick branches which are themselves only a few times branched; branches often longitudinally grooved or rugose, narrowly clavate or flattened, blunt at the tips or with a few teeth. Basidia 40–80 × 6–9. Spores 10–16 × 8–12. On the ground, often amongst mosses, in conifer and mixed woods.

Clavulinopsis

Fruitbodies erect, solitary, gregarious or in clumps, branched or unbranched, sometimes stalked, white or various colours. Hyphae mostly swollen, with clamps but without secondary septa. Basidia mostly 4-spored, with normal basal clamps. Spores ellipsoid to almost spherical, or pyriform, hyaline with smooth or rough nodulose walls, J–, often with one large oil drop. Terrestrial.

KEY

Fruitbodies unbranched or very rarely with one or two short side-branches 1

Fruitbodies regularly branched 7

1. Spore walls, bearing large tubercles .... *helvola*

   Spore walls smooth 2

2. Spores 8–12 × 2–3.5 .... *vernalis*

   Spores never more than 8 long 3

3. Spores 5–8 long 4

   Spores not more than 5 long 6

4. Fruitbodies yellow with white or whitish tips .... *luteo-alba*

   Fruitbodies yellow or ochraceous without white tips 5

5. Spores rather irregular and often with a lateral apiculus .... *laeticolor*

   Spores regularly ellipsoid to subspherical .... *fusiformis*

6. Growing amongst *Sphagnum*, stinking .... *luteo-ochracea*

   Neither growing amongst *Sphagnum* nor stinking .... *microspora*

7. Fruitbodies remaining white .... *dichotoma*

   Fruitbodies grey or predominantly so .... *cinereoides*

   Fruitbodies other colours, sometimes very pale but not remaining white 8

8. Spores up to 7 long 9

   Spores not more than 5 long 10

9. Fruitbodies yellow to ochraceous .... *corniculata*

   Fruitbodies brown or greyish brown .... *umbrinella*

10. Fruitbodies growing amongst *Sphagnum*, stinking .... *luteo-ochracea*

    Fruitbodies not growing amongst *Sphagnum*, not stinking .... *subtilis*

Clavulinopsis cinereoides (Atk.) Corner (Fig. 98)
Fruitbodies pale grey or slightly pinkish grey, without stalks, richly branched from the base, tough, 5–7 cm tall; branches erect, 3–4 times dichotomously branched, terminating in two long, rather blunt teeth. Basidia 50–80 × 6–7. Spores spherical or subspherical, apiculate, 4–7 diam., each with one large oil drop. On the ground in woods, uncommon.

Clavulinopsis corniculata (Fr.) Corner (Fig. 99)
Fruitbodies bright yellow to ochraceous, 2–7 cm tall, with white, downy stem base, often several times dichotomously branched, tough, terminal branches antler-like, their tips incurved; stalk 0.5–4 cm × 1–4 mm. Basidia 40–60 × 7–9. Spores spherical or almost so, 4.5–7 diam. In grassy places in woods, fairly common.

Clavulinopsis dichotoma (Godey) Corner
Fruitbodies always white, up to 6 cm tall, brittle to tough, dichotomously branched, branches sometimes flattened. Basidia 20–35 × 4–6. Spores 4–7 × 4–5. Usually growing amongst moss by sides of paths in woods, uncommon.

Clavulinopsis fusiformis (Fr.) Corner (Fig. 100)
Fruitbodies yellow to ochraceous, usually in dense clumps, unbranched, narrowly fusiform with stalk-like base, sometimes twisted, laterally compressed and longitudinally grooved, 5–15 cm × 2–10 mm. Basidia 40–60 × 6–9. Spores broadly ellipsoid or subspherical, 5–8 × 4.5–7.5. Amongst grass in fields and woods, sometimes on poor soil with bracken, fairly common.

Clavulinopsis helvola (Fr.) Corner (Fig. 101).
Fruitbodies pale yellow to yellowish orange, solitary or gregarious, unbranched or just occasionally with a short branch near the apex, cylindrical to narrowly clavate, with rounded tips, 3–6 cm × 1–4 mm. Basidia 30–50 × 6–7. Spores subspherical, 4–7 × 4–6, walls rough, with blunt tubercles 1–2 long. In woods and meadows and on heaths, fairly common.

Clavulinopsis laeticolor (Berk. and Curt.) Petersen (Fig. 102)
Fruitbodies yellow to yellowish orange, solitary or gregarious, unbranched, stalked, cylindrical or narrowly clavate, sometimes compressed laterally, 2–10 cm × 1–6 mm. Basidia 50–60 × 8–10. Spores ellipsoid to subspherical, with distinct apiculus which is often lateral, 5–8 × 4–6. In woods and meadows.

Clavulinopsis luteo-alba (Rea) Corner (Fig. 103)
Fruitbodies yellow or creamy yellow with white or very pale tips, solitary, gregarious or in clumps, stalked, usually unbranched, occasionally with 1 or 2 short branches near the apex, cylindrical to narrowly clavate, sometimes laterally compressed, 3–7 cm × 1–3 mm. Basidia 50–60 × 6–7. Spores ovoid

or ellipsoid, shortly apiculate, 5–8 × 3–4.5. In fields and woods and on heaths, fairly common.

Clavulinopsis luteo-ochracea (Cavara) Corner
Fruitbodies pale ochraceous, with a greyish brown or ochraceous brown stalk, solitary or in small groups, unbranched or once or twice branched at apex, cylindrical or somewhat flattened and spathulate, 1–5 cm tall. Basidia 25–30 × 5–6. Spores broadly ellipsoid, 4–5 × 3–3.5. With strong, unpleasant smell. On ground amongst *Sphagnum.*

Clavulinopsis microspora (Josserand) Corner
Fruitbodies at first pale cream or ochraceous than becoming fawn-brown from the base upwards, solitary or gregarious, stalked, unbranched, 0.5–2 cm × 1–1.5 mm. Basidia 25–35 × 4–6. Spores ellipsoid, 3.5–5 × 2–3. On leaf-mould and other litter.

Clavulinopsis subtilis (Fr.) Corner (Fig. 104)
Fruitbodies white to very pale ochraceous, solitary or in clumps, stalked, dichotomously branched above stalk to become bushy or antler-like, 1–4 cm tall. Basidia 30–35 × 4–6. Spores mostly subspherical, 3–4.5 diam. Growing amongst grass, usually in woods, not common.

Clavulinopsis umbrinella (Sacc.) Corner
Fruitbodies at first white but soon becoming pale brown or greyish brown with darker brown tips, shortly stalked, dichotomously branched, 2–4 cm tall. Basidia 70–90 × 8–9. Spores subspherical or pyriform, apiculate, 4–7 × 3–6, with oil drops. Amongst grass.

Clavulinopsis vernalis (Schw.) Corner
Fruitbodies pale ochraceous to orange, with very short, whitish stalk, in clumps, mostly unbranched, rarely with one short side-branch, clavate, not more than 1.5 cm tall. Basidia 20–35 × 6–8. Spores 8–12 × 2–3.5. On the ground.

Coltricia

Fruitbodies erect, consisting of a cap and a stalk which is usually central, all parts blackening with KOH; cap flat or depressed in centre, upper surface velvety or hairy, often conspicuously concentrically zoned, pores on lower surface mostly angular. No setae. No clamps on hyphae. Spores ellipsoid or oblong-ellipsoid, mostly smooth, yellow or pale yellowish brown. Terrestrial.

KEY

Spores 3.5–4.5 wide .... *perennis*
Spores 5–8 wide 1

1. Spores 6–10 long. Found in deciduous woods .... *cinnamomea*
    Spores 9–14 long. Found in conifer woods .... *montagnei*

Coltricia cinnamomea (Pers.) Murrill
Fruitbodies solitary or in clusters and then often fused together; cap 2–4 cm diam., 2–3 mm thick near stalk, orbicular, with lobed margin, flat or depressed in centre and funnel-shaped, upper surface glossy, velvety, concentrically zoned, cinnamon to dark reddish brown, lower surface reddish brown, pores 2–4 per mm; stalk central, cylindrical or flattened, often swollen at base, 2–4 cm × 2–6 mm, velvety, rusty or dark reddish brown. Spores broadly ellipsoid, yellow, 6–10 × 5–7. In deciduous woods, mainly beech, on soil and sometimes very rotten wood, uncommon.

Coltricia montagnei (Fr.) Murrill
Fruitbodies stalked, fan-shaped or with orbicular head up to 12 cm diam., 6–10 mm thick near stalk, flat or depressed in the centre, upper surface velvety or, when old, with isolated tufts of hairs, pale yellowish or reddish brown with little or no zonation, lower surface ochraceous or rusty, pores up to 4 mm diam.; stalk 2–5 cm long, tapered towards base, reddish brown, finely tomentose. Spores oblong-ellipsoid, thick-walled, yellow, 9–14 × 5.5–8. On soil in conifer woods.

Coltricia perennis (Fr.) Murrill (Fig. 105)
Fruitbodies solitary or in clusters and then sometimes fused together, softly leathery or corky; cap 2–8 cm diam., roughly orbicular but with a wavy margin, usually depressed in the centre or somewhat funnel-shaped, 3–5 mm thick near stalk, upper surface velvety at first, becoming smooth, zonate, ochraceous, cinnamon brown or greyish, lower surface pale ochraceous or rusty brown, with narrow sterile margin, pores angular, 2–4 per mm, decurrent, becoming uneven; stalk usually central, 2–5 × 0.2–1 cm, velvety, rusty or dark brown. Basidia 10–25 × 6–8. Spores oblong-ellipsoid, dark golden brown, smooth, 6–9 × 3.5–4.5. Mostly on sandy soil, by the sides of paths in woods and on heaths.

Columnocystis *abietina* (Pers. ex Fr.) Pouzar (Fig. 106)
Fruitbodies effuso-reflexed, forming irregular, leathery patches several cm diam., 0.5–2 mm thick, closely attached but turned up at the edges all round, on vertical surfaces most distinctly pileate and *Stereum*-like, sterile surface, nearest the substrate, finely hairy, dark reddish brown or blackish brown, often zonate, hymenium smooth or undulating, pruinose, pale bluish grey or lilac grey when fresh, turning brownish with age. Hyphae with clamps at most septa; thick-walled, dark brown skeletal hyphae present. Cystidia cylindrical, up to 250 × 8–15, with very thick walls, pale to dark brown, some projecting well above the surface and finely encrusted. Basidia 50–80 × 6–8, clavate, 4-spored. Spores ellipsoid, hyaline, smooth,

thin-walled, J–, 9–13 × 4–5. On conifer logs and branches, mainly those of *Picea abies*.

Confertobasidium *olivaceo-album* (Bourd. and Galz.) Jülich (Fig. 107)
Close to *Athelia* but differs in having distinctly brown basal hyphae and rhizomorphs. Fruitbodies resupinate, effused, pellicular or thinly membranous, loosely attached to substrate, white or pale cream, sometimes tinted olive. Clamps at all septa. Basidia 10–15 × 3–4.5, clavate, 4-spored. Spores ellipsoid, 3.5–4 × 2, hyaline, thin-walled, smooth, J–. On wood and bark, mainly of conifers.

Coniophora

Fruitbodies resupinate, effused, membranous or fibrous-cottony, sometimes with brown rhizomorphs, initially pale, white, cream or yellow, finally becoming brown or olivaceous brown as spores mature. Clamps few but sometimes as many as 6 formed in a verticil at a single septum. Basidia 4-spored. Spores ellipsoid, ovoid, fusiform or navicular, brown, yellowish brown or olivaceous brown, with rather thick, smooth walls, J–, in some species dextrinoid or cyanophilous. On wood and bark.

KEY

	Hymenium with projecting, brown, thick-walled cystidia .... *olivacea*	
	Hymenium without cystidia 1	
1.	Spores broadly fusiform to navicular, 14–24 long .... *fusispora*	
	Spores ovoid to ellipsoid, not more than 15 long 2	
2.	Spores 10–15 long 3	
	Spores 7–11 long 4	
3.	Fruitbodies not more than 0.3 mm thick, hymenium smooth, spores strongly dextrinoid .... *arida*	
	Fruitbodies 0.5–2 mm thick, hymenium usually tuberculate, spores not or very weakly dextrinoid .... *puteana*	
4.	Thick rhizomorphs usually present .... *marmorata*	
	No rhizomorphs .... *hanoiensis*	

Coniophora arida (Fr.) Karst.
Fruitbodies thin, only up to 0.3 mm, hymenium smooth, ochraceous brown or pale brown, margin white to pale yellow. Hyphae without clamps or with several in verticils at a few septa, smooth or, in var. *suffocata* (Peck) Ginns, encrusted with crystals, hyaline or brownish, 6–12 wide. No cystidia. Basidia 40–60 × 7–10. Spores ovoid or broadly ellipsoid, yellowish or yellowish brown, strongly dextrinoid, 10–15 × 6–8. On dead wood, mainly of conifers, common.

Coniophora fusispora (Cooke and Ellis) Sacc. (Fig. 108)
Fruitbodies up to 0.5 mm thick, hymenium smooth, orange–yellowish to dark brown, margin white or pale ochraceous. Hyphae sometimes with clamps, singly or in verticils. Basidia 60–100 × 8–12. No cystidia. Spores broadly fusiform to navicular, yellowish brown or brown, 14–24 × 6–9. Mostly on conifer wood, occasionally on wood of deciduous trees and on litter.

Coniophora hanoiensis Pat.
Fruitbodies up to 0.8 mm thick, hymenium smooth or granular, yellowish brown to chocolate brown, without rhizomorphs. Hyphae hyaline or brown, no clamps. Spores ellipsoid, yellowish or brownish, 8–11 × 7–8. On wood, rare.

Coniophora marmorata Desm.
Fruitbodies membranous, up to 0.4 mm thick, greyish brown or olivaceous brown, with pale brown rhizomorphs up to 1 mm wide. Spores broadly ellipsoid, yellowish, 7–10 × 5–7. Growing on stones, cement and wood of buildings.

Coniophora olivacea (Fr.) Karst. (Fig. 109)
Fruitbodies thin, hymenium velvety, greyish brown to olivaceous brown, margin paler, sometimes with brown rhizomorphs. Cystidia projecting, cylindrical, closely septate, thick-walled, brown, encrusted, up to 250 × 8–16. Spores ovoid to ellipsoid, with one side sometimes straight, yellowish brown to brown, 8–12 × 4.5–7. Mostly on coniferous wood.

Coniophora puteana (Schum. ex Fr.) P. Karst (Fig. 110)
Fruitbodies up to 20 cm across, 0.5–2 mm thick, at first white to cream, then pale yellow to chrome yellow, finally dark brown or olivaceous brown, with margin remaining pale, hymenium smooth to tuberculate. No cystidia. Spores ellipsoid to ovoid, one side sometimes flattened, yellowish brown to brown, scarcely or not at all dextrinoid, 10–15 × 6–8. On wood and bark of deciduous trees and conifers, common on fallen trunks and branches.

Corticium *roseum* Pers. (Fig. 111)
Fruitbodies resupinate, at first small and orbicular, often becoming confluent and widely effused, when fresh pink, bruising darker pink or reddish, ochraceous when old and dry, membranous to waxy, up to 1 mm thick, firmly attached to substrate, smooth or tuberculate. Hyphae hyaline with clamps. Dendrohyphidia numerous, richly branched, 20–25 × 1. Basidia narrowly clavate, 50–70 × 5–7, 4-spored, when young bladder-like, with basal clamps. Spores ellipsoid, 10–16 × 6–10, hyaline, smooth, J–, rose pink in mass. On attached branches and fallen trunks, found most commonly on *Populus tremula* and *Salix*.

Cotylidia
Fruitbodies tough, soft or leathery, erect, fan-shaped and clustered to form rosettes or funnel-shaped and solitary to gregarious, tapered below to a stalk, yellowish, hymenium on lower surface smooth or wrinkled, pale. Hyphae without clamps. Cystidia protruding well above the level of the 4-spored basidia. Spores ellipsoid, hyaline, thin-walled, smooth, J–. Terrestrial.

KEY

Fruitbodies fan-shaped, forming rosettes, spores 6–9 long ..... *pannosa*
Fruitbodies funnel-shaped, spores 3.5–5.5 long .... *undulata*

Cotylidia pannosa (Sow. ex Fr.) Reid (Fig. 112)
Fruitbodies 1–5 cm tall, leathery, mostly fan-shaped, lobed or wavy at the edge, tapered down into a stalk, white or cream at first, becoming yellowish or orange–brown, often clustered in the form of rosettes, hymenium on lower surface pale. Cystidia cylindrical to narrowly clavate, 100–170 × 8–12, hyaline. Basidia about 50 × 7. Spores 6–9 × 3.5–4.5. On the ground, amongst litter in woods, recorded several times under *Fagus*.

Cotylidia undulata (Fr.) P. Karst. (Fig. 113)
Fruitbodies 0.5–1.5 cm tall and 0.4–1 cm wide, soft but tough, solitary or gregarious, funnel-shaped, smooth, fringed, undulating or deeply incised at the edge, upper surface yellowish or honey-coloured, somewhat fibrillose, hymenium on lower surface smooth, whitish to pale ochraceous; stalk whitish to pale yellowish grey. Cystidia cylindrical, 40–60 × 5–7. Basidia 15–20 × 4–5. Spores 3.5–5.5 × 2–2.5. Amongst mosses, in damp places or on old bonfire sites.

Craterellus cornucopioides(L.) Pers. (Fig. 114)
Horn of Plenty
Fruitbodies erect, gregarious, often in clumps, thin and leathery, rather limp, funnel-shaped or trumpet-shaped, turned over and wavy at the margin, 3–15 cm tall, 2–8 cm wide, upper or inner surface dark greyish brown to almost black, slightly grooved or scaly, hymenium on lower or outer surface paler, ash-grey, smooth when young, later becoming wrinkled with shallow grooves; stalk central, hollow to the base. Basidia 70–100 × 8–10, 2-spored. Hyphae without clamps. Spores ellipsoid, 10–17 × 7–10, hyaline, smooth, thin-walled, J–. On the ground, sometimes in large quantities amongst dead leaves and easily overlooked, in woods, especially under *Fagus*, autumn.

Creolophus cirrhatus (Pers. ex Fr.) P. Karst. (Fig. 115)
Fruitbodies pileate, horizontal, laterally attached, bracket-like and irregularly imbricate, roughly semicircular or shell-shaped, soft, thick-fleshed, 2–

8 cm across, whitish, cream or ochraceous, upper surface unevenly warted, scaly or with sterile spines towards the margin, hymenium on lower surface with pendent teeth 5–15 mm long, Hyphae with clamps. Cystidia flexuous, hyaline, 100–150 × 5–8. Basidia narrowly clavate, about 20 × 6. Spores ellipsoid, 3.5–4 × 3, hyaline, smooth, J+. On dead wood of *Betula*, *Fagus*, *Salix* etc., found sometimes on standing trees as a wound parasite.

Cristinia

Fruitbodies resupinate, effused, loosely attached to substrate, mostly with rhizomorphs, hymenium smooth, warted or with teeth. Hyphae thin-walled, with clamps. No cystidia. Basidia cylindrical, rather short, 4-spored, with cyanophilous drops when young. Spores subglobose, hyaline or yellowish, smooth, rather thick-walled, J–, C+.

KEY

Fruitbodies sulphur yellow or ochraceous, with blunt teeth up to 3 mm long .... *gallica*
Fruitbodies white to pale ochraceous, without teeth .... *helvetica*

Cristinia gallica (Pilát) Jülich (Fig. 116)

Fruitbodies sulphur yellow or ochraceous, with blunt teeth up to 3 mm long. Basidia 25–35 × 6–8. Spores 5–7 diam. On the lower surface of fallen trunks and branches of deciduous trees.

Cristinia helvetica (Pers.) Parm. (Fig. 117)

Fruitbodies white or pale ochraceous, membranous, smooth to tuberculate. Hyphal cords form rhizomorphs peripherally. Cells short, clamps at all septa. Basidia 20–25 × 5–7. Spores 3.5–5 diam. On rotten wood.

Crustoderma dryinum (Berk. and Curt.) Parm. (Fig. 118)

Fruitbodies resupinate, effused, waxy or crustose, up to 0.2 mm thick, firmly attached to substrate, smooth, pale ochraceous or yellowish. Hyphae yellowish, with clamps, much branched and densely packed together. Cystidia protruding a long way above the basidia, cylindrical, hyaline, smooth, 80–130 × 5–9, wall rather thinner at the apex. Basidia narrowly clavate, 25–35 × 5–6. Spores cylindrical to narrowly ellipsoid, 7–9 × 2.5–3.5, hyaline or yellowish, J–, C+. On conifer wood.

Cylindrobasidium evolvens (Fr.) Jülich (Fig. 119)

Fruitbodies mostly resupinate, rarely and only on vertical surfaces effuso-reflexed or pileate, membranous or waxy, smooth to tuberculate, whitish to cream or ochraceous, loosely attached to substrate, cracking when dry, at first thin but later thickening sometimes up to 1 mm or more. Hyphae hyaline, full of oil drops, with many clamps. Cystidia fusiform, thin-walled,

60–70 × 5–8, hyaline. Basidia slender, cylindro-clavate, 40–70 × 5–6, 4-spored. Spores ellipsoid, pyriform or lachrymoid, 8–12 × 4–6, hyaline, smooth, J−, often sticking together in clumps of 2 or 4. When pileate, brackets up to 1 cm across, often imbricate, thin, upper surface whitish, hairy, somewhat zonate. Very common on wood and bark.

Cylindrobasidium parasiticum Reid
Parasitic on sclerotia of *Typhula incarnata*. As in *C. evolvens* the spores are clumped in groups of 2–4, the hyphae have many clamps, the thin-walled cystidia are fusiform but narrower, up to 80 × 2–6, the basidia cylindro-clavate, but smaller, 30–45 × 4–7, and the spores often lachrymoid, 8–11 × 3–5. We have not seen this species but according to Reid it has a thin, granular hymenium and hyphae which turn brown.

Cyphellostereum laeve (Fr.) Reid (Fig. 120)
Fruitbodies laterally stalked, shallow cup-shaped or fan-shaped with inturned margin, about 1 cm diam. or less, white, cream or slightly brownish, wrinkled and downy on the outside; hymenium lining cup smooth or very slightly furrowed; stalk short broadening out into and concolorous with the cup. Cystidia up to 40 × 5, somewhat capitate, projecting. Basidia narrowly clavate, 12–22 × 3–5, 4-spored. Spores in mass white, ellipsoid, obliquely attenuated at base, 3.5–4.5 × 2–2.5, each with a guttule, J−. On mosses, especially *Polytrichum* species, on sandy soil.

Cytidia salicina (Fr.) Burt (Fig. 121)
Fruitbodies resupinate, soon becoming shallow bowl-shaped, 1 cm diam., 1–2 mm thick, sometimes joined together to form large patches, tough elastic, gelatinous or waxy, when fresh hymenium bright red or wine-coloured, sterile border whitish, somewhat hairy, outer surface white, farinose. Hyphae hyaline, with clamps. Dendrohyphidia abundant, yellowish brown. Basidia 80–150 × 8–10, with 4 curved sterigmata. Spores cylindrical to ellipsoid, slightly curved, 12–18 × 4–5, hyaline, smooth, J−, filled with small guttules. On wood and bark of *Salix* species, often on attached branches and standing trunks.

Dacrymyces
Fruitbodies solitary or gregarious, often confluent, softly or firmly gelatinous, mostly pulvinate or discoid with smooth or undulating surface, usually sessile or almost so with amphigenous hymenium and without superficial hairs, rarely turbinate and weakly hairy. In very wet weather fruitbodies often become soft and appear to dissolve. Hyphae and pro-basidia with or without clamps. Basidia deeply divided, forming 2 very long sterigmata and resembling tuning forks. Spores usually hyaline, with 0–10

transverse septa and sometimes also with 1 or several longitudinal septa; germinating to form germ tubes or conidia.

KEY

Ripe basidiospores muriform .... *ovisporus*
Ripe basidiospores often with more than 4 transverse septa 1
Ripe basidiospores never with more than 4 transverse septa 3
1. Hyphae and probasidia with clamps .... *variisporus*
Hyphae and probasidia without clamps 2
2. Spores 8–12 wide .... *estonicus*
Spores 4–7 wide .... *chrysospermus*
3. Ripe spores with rather thick walls and 3–4 thick transverse septa, very common .... *stillatus*
Ripe spores thin-walled, with 0–4 rather thin transverse septa, less common 4
4. Fruitbodies stalked .... *capitatus*
Fruitbodies sessile 5
5. Spores without septa, hyphae and probasidia with frequent large, arched clamps .... *macnabbii*
Spores with up to three transverse septa, clamps present or not 6
6. Hyphae and probasidia with clamps 7
Hyphae and probasidia without clamps 8
7. Fruitbodies chestnut brown or yellowish brown, on wood of deciduous trees, especially *Quercus* .... *enatus*
Fruitbodies hyaline to pale yellowish brown, on wood of conifers *tortus*
8. Fruitbodies yellowish orange when fresh .... *lacrymalis*
Fruitbodies hyaline to pale yellow-ochre .... *minor*

Dacrymyces capitatus Schw.
Fruitbodies pulvinate with a short stalk when young, later becoming clavate with a flat or concave head 1–3 mm diam. and a stalk 1–3 mm long but the head is not abruptly separated from the stalk as it is in *Ditiola*. Hyphae without clamps. Spores ellipsoid and somewhat bent, thin-walled, 11–17 × 3.5–7, when ripe with 3 or occasionally 4 transverse septa, germinating to form germ tubes or small spherical conidia. Mostly on wood of deciduous trees.

Dacrymyces chrysospermus Berk. and Curt.
Fruitbodies solitary or gregarious, seldom confluent, 1–4 mm diam., variable in shape, sometimes discoid or cyathiform, with a more or less distinct stalk, yellowish orange fresh, orange-brown when dry. Hyphae and probasidia without clamps. Spores narrowly ellipsoid, hyaline, 16–26 × 4–7, when ripe with 3–7 fairly thick septa. On wood of conifers and deciduous trees.

Dacrymyces enatus (Berk. and Curt.) Massee
Fruitbodies gregarious, when young pustulate, 1–3 mm diam., chestnut brown or yellowish brown, firmly gelatinous, when mature flattened and confluent forming large masses, when dry dark brown or blackish. Hyphae and probasidia with a few small clamps. Spores narrowly ellipsoid to allantoid, 9–16 × 3–4.5, when ripe with 1 or occasionally 3 transverse septa. On wood of deciduous trees, especially *Quercus.*

Dacrymyces estonicus Raitviir
Fruitbodies solitary or gregarious, seldom confluent, up to 4 mm diam., at first pustulate, then turbinate or bowl-shaped, firmly gelatinous, when fresh yellow or pale orange, when dry yellowish or reddish brown. Hyphae without clamps. Spores hyaline to pale yellowish, broadly ellipsoid, 18–26 × 8–12, when ripe with 7–10 transverse septa, germinating to form ellipsoid conidia 4 × 2. On conifer wood, uncommon.

Dacrymyces lacrymalis (Pers. ex S.F. Gray) Sommerf.
Fruitbodies gregarious, at first pustulate, then lobed or somewhat cyathiform and wavy edged, with a central point of attachment, seldom confluent, gelatinous, yellowish orange when fresh, yellowish brown when dry; the base bears solitary, thin-walled hairs 20–25 × 4–5. Hyphae without clamps. Spores narrowly ellipsoid to allantoid, 10–15 × 4.5–6, thin-walled, when ripe with 1–3 thin septa. On wood of deciduous trees, mainly *Quercus.*

Dacrymyces macnabbii Reid
Fruitbodies gregarious but not confluent, discoid, gelatinous, fairly thick, pale yellow fresh, yellowish brown and somewhat sunken when dry. Hyphae and probasidia with frequent large, arched clamps. Spores hyaline, narrowly ellipsoid, often slightly bent, 10–13 × 3.5–4, without septa. On wood of *Pinus.*

Dacrymyces minor Peck
Fruitbodies pulvinate, 0.5–2 mm diam., gregarious, gelatinous, hyaline to pale yellow–ochre, when young often with a greenish sheen, when dry yellowish brown. Hyphae without clamps. Spores narrowly ellipsoid to allantoid, thin-walled, 8–16 × 4–6, when ripe with 1–3 thin septa. On wood of conifers and deciduous trees.

Dacrymyces ovisporus Bref.
Fruitbodies up to 4 mm tall, firmly gelatinous, with a wrinkled orange-coloured head 2–5 mm wide and a short, paler, stem-like base, when dry dark brown. Hyphae and probasidia with clamps. Spores broadly ellipsoid or subglobose, 13–17 × 8–12, hyaline, muriform when ripe, germinating to form small spherical conidia or germ tubes. On conifer wood.

Dacrymyces stillatus Nees ex Fr. (Fig. 122)
Fruitbodies gregarious, 1–4 mm diam., pulvinate or discoid, often con-

fluent, softly to firmly gelatinous, reddish orange when young and forming conidia but yellow and somewhat wrinkled when forming basidiospores. Hyphae without clamps. Spores narrowly ellipsoid to allantoid, fairly thick-walled, 10–16 × 4–6, when ripe with 3 or occasionally 4 rather thick transverse septa. Conidia mostly 8–16 × 3–5, hyaline. On wood of deciduous trees and conifers and frequently also on rails and fence posts, very common everywhere but conspicuous only in wet weather.

Dacrymyces tortus (Willd.) Fr.
Fruitbodies gregarious but seldom confluent, pustulate, 0.5–2 mm diam., at first convex then sunken in the middle, hyaline or pale yellowish brown, often with a greenish sheen, when dry forming a thin blackish film. Hyphae and probasidia with clamps. Spores hyaline, narrowly ellipsoid to allantoid, 9–15 × 4–5, thin-walled, with 0–3 transverse septa. On conifer wood.

Dacrymyces variisporus McNabb (Fig. 123)
Fruitbodies pulvinate, up to 3–5 mm diam., somewhat sunken in the middle, often gregarious and sometimes confluent, gelatinous, when fresh pale orange to orange–brown, when dry yellowish brown to dark brown. Hyphae and probasidia with clamps. Spores hyaline, at first broadly ellipsoid and 14–17 × 8–9, when ripe narrowly ellipsoid or allantoid, 18–30 × 6–9, with 3–9 transverse septa and occasionally also 1–2 longitudinal septa. On wood of deciduous trees and conifers.

Dacryobasidium *coprophilum* (Wakef.) Jülich (Fig. 124)
Fruitbodies resupinate, *Athelia*-like, effused, pellicular, thin, smooth, with thin rhizomorphs, white or cream. Hyphae hyaline with few clamps. No cystidia. Basidia narrowly clavate, hyaline, containing cyanophilous granules or drops, 20–25 × 6–8. Spores spherical, 4–6 diam., yellowish, smooth, with rather thick walls, J–, C+. On dung and litter.

Dacryobolus
Fruitbodies resupinate, firmly attached, smooth, tuberculate or toothed. Hyphae with clamps. Cystidia present. Basidia narrow, 4-spored. Spores allantoid, small and slender, hyaline, smooth, thin-walled, J–.

KEY

Fruitbodies smooth or tuberculate .... *karstenii*
Fruitbodies distinctly toothed .... *sudans*

Dacryobolus karstenii (Bres.) Oberw. ex Parm. (Fig. 125)
Fruitbodies effused, membranous to leathery, cream or ochraceous, smooth to tuberculate, becoming cracked when old. Thick-walled skeletal hyphae in subiculum. Cystidia protruding, hyaline, cylindrical, of two

kinds: (1) fairly thin-walled, 50–80 × 3–4, (2) thick-walled, up to 250 × 6–8. Basidia 40–45 × 2–3.5, often clearly constricted a short distance below the sterigmata. Spores 4–6 × 1–1.5. On conifer wood and bark, mostly on fallen trunks and branches of *Pinus sylvestris.*

Dacryobolus sudans (Fr.) Fr. (Fig. 126)
Fruitbodies effused, forming waxy, creamy or pale ochraceous crusts with conical teeth up to 1 mm long which are often capped by drops of liquid giving the hymenium a pearly look. Cystidia with basal clamps, short ones mixed with basidia and tufts of long, septate ones, coated with droplets, projecting from ends of teeth. Basidia 25–30 × 3–4. Spores 5–7 × 1.5, each containing a few small guttules. On conifer wood and occasionally bark, especially of *Pinus sylvestris.*

Daedalea *quercina* L. ex Fr. (Fig. 127)
Maze-gill
Fruitbodies perennial, pileate, tough, corky to woody, broadly attached, solitary or imbricate and fused brackets, roughly semicircular, commonly 5–20 × 6–10 cm, triquetrous, up to 8 cm thick where attached, thinning out to a sharp edge; upper surface slightly velvety or tomentose, pale brown or greyish brown, concentrically sulcate zoned, smooth or tuberculate, with smooth ochraceous margin; lower surface pale ochraceous or beige, pores mostly labyrinthine with some of the walls resembling forked gills. Thick-walled, yellowish brown skeletal hyphae often project into the hymenium. Basidia about 20 × 4–5. Spores ellipsoid, 5.5–7.5 × 3–3.5, hyaline, smooth, J–. Mainly on dead wood of *Quercus* but recorded also on *Castanea.*

Daedaleopsis *confragosa* (Bolt. ex Fr.) Schroet. (Fig. 128)
Blushing Bracket
Fruitbodies annual, pileate, corky, broadly attached, solitary or imbricate brackets, semicircular or fan-shaped, 5–20 × 4–10 cm, 2–5 cm thick, with thin, sharp edge; upper surface flat, concentrically zoned and ridged and often also radially striate, ochraceous to reddish brown or dark brown; lower surface whitish to pale ochraceous or greyish brown, bruising pink or red; pores radially elongated, slot-like, 1–2 per mm or, in var. *tricolor* (Bull.) Bond., labyrinthine with branched, anastomosing lamellae. Hyphae with clamps. No cystidia. Yellowish, thick-walled skeletal hyphae present. Dendrohyphidia in hymenium. Basidia 15–25 × 4–5, 4-spored. Spores cylindrical, slightly curved, 7–11 × 2–3, hyaline, smooth, J–. On dead standing and fallen trunks and branches of deciduous trees, especially *Salix,* but also *Alnus, Betula* etc.

Datronia
Fruitbodies annual, pileate, resupinate or effuso-reflexed; when pileate, brackets imbricate, narrow, shelf-like, upper surface dark brown, pores

angular and slot-like, pore surface pale. Hyphae with clamps. Skeletal hyphae brownish. No cystidia. Spores cylindrical to ellipsoid, hyaline, thin-walled, smooth, J–. This genus is closely related to *Antrodia*.

KEY

Frequently forming shelf-like brackets, pores 1–2 per mm .... *mollis*
Mostly resupinate, pores 4–6 per mm .... *stereoides*

Datronia mollis (Sommerf.) Donk (Fig. 129)
Fruitbodies sometimes resupinate and widely effused or effuso-reflexed but commonly forming imbricate, wavy, shelf-like brackets 2–7 cm long, 0.5–2 cm wide and up to 0.5 cm thick, sterile surface velvety wearing smooth, narrowly concentrically zoned, brown to blackish brown, pore surface pale ochraceous to grey, bruising brown, pores angular or slot-like, 1–2 per mm, longitudinally slit on vertical surfaces. In section a thin black line is seen above the flesh. Basidia 20–25 × 5–7. Spores 7.5–10 × 2.5–4. On fallen dead trunks and branches of deciduous trees, common.

Datronia stereoides (Fr.) Ryv. (Fig. 130)
Fruitbodies mostly resupinate, rarely effuso-reflexed, usually rather small, pore surface pale ochraceous to greyish brown, pores 4–6 per mm, margin whitish, sterile. Spores 9.5–12 × 3.5–4.5. On wood of deciduous trees, rare.

Delicatula *integrella* (Pers. ex Fr.) Fay. (Fig. 131)
Fruitbodies white, consisting of cap and central stalk; cap orbicular with irregular margin, 0.5–1 cm diam., shallow veins on lower surface forked and sometimes anastomosing, disappearing towards edge; stalk 1.5–2 cm × 0.5–1 mm, with velvety basal bulb. No cystidia. Spores broadly ellipsoid, hyaline, smooth, J+, 7.5–9 × 4–5. On rotting wood, leaves and humus.

Dendropolyporus *umbellatus* (Pers. ex Fr.) Jülich (Fig. 132)
Fruitbodies annual, compound, hemispherical, up to 50 cm across, arising from black sclerotia in the soil and made up of a number of umbrella-like caps formed at the ends of branches of a much-branched stalk; upper surface of caps ochraceous or pale brown, minutely scaly or fibrillose, stalks and decurrent pore surface white or cream, pores 1–3 per mm. Spores cylindrical, 8–10 × 2–3.5, hyaline, smooth, J–. Associated with roots of deciduous trees, especially *Quercus*.

Dendrothele *acerina* (Pers. ex Fr.) Lemke (Fig. 133)
Fruitbodies resupinate, orbicular, about 1 cm diam., sometimes confluent, membranous, mostly thin and firmly attached, smooth or tuberculate, white, greyish or pale ochraceous. Hyphae without clamps, surrounded by crystals. Dendrohyphidia abundant, also some swollen cystidia, about 30 × 10, with slender, sometimes branched tips, often obscured by crystals. Basidia 35–60 × 6–10, 4-spored. Spores broadly ellipsoid, 10–14 × 7–10,

hyaline, smooth, fairly thick-walled, J−, C+. On living bark, mostly of *Acer campestris* and *A. platanoides*.

Dicellomyces *scirpi* Raitv. (Fig. 134)
Fruitbodies pustulate, up to 1.5 cm long, buff, pruinose, waxy to gelatinous, swelling and becoming hyaline when wet. Hyphae without clamps. Basidia up to 140 × 6–8, slender-stalked, irregularly clavate, forked to form 2 curved sterigmata. Spores oblong-ellipsoid, often slightly curved especially towards the protuberant truncate base, hyaline, smooth, often becoming 1-septate, 12–24 × 5–7, germinating to form slender, allantoid or coiled conidia. Between the veins on the lower surface of leaves of *Scirpus sylvaticus*, especially near the tips; corresponding areas on the upper surface have yellow streaks, June–July.

Dichomitus *campestris* (Quél.) Domanski and Orliez (Fig. 135)
Fruitbodies resupinate, pulvinate, soft corky, oval to oblong, 10–20 × 2–5 cm, 0.5–1.5 cm thick in the middle, ochraceous to cream-coloured, margin smooth, blackish, pores angular, 1–2 per mm, sometimes split or irregularly elongated, tubes in one or more layers. Hyphae with clamps, skeletal ones dendroid with tapering branches. No setae and no cystidia. Basidia 20–30 × 6–9. Spores cylindrical to oblong-ellipsoid, 9–14 × 3.5–5, hyaline, smooth, thin-walled, J−. Most collections from Scotland on standing trunks and attached branches of *Corylus* but recorded also on other trees, including *Alnus*, *Fagus*, *Populus* and *Quercus*.

Digitatispora *marina* Doguet
Fruitbodies resupinate, flatly pulvinate, smooth, pale grey, up to 4 mm diam. and 0.3 mm thick. Hyphae with clamps. Basidia up to 50 × 5, 4-spored. Spores hyaline, smooth, tetraradiate, with cylindrical basal part 30–40 × 3–4, and 3 narrow, cylindrical arms 30–45 × 2–3. On driftwood in the sea.

Ditiola *radicata* (Alb. and Schw.) Fr. (Fig. 136)
Fruitbodies gregarious, firm, gelatinous, when fully developed turbinate, with distinct head and stalk, up to 7 mm tall, head an arched disc, smooth or wrinkled, 2–5 mm diam., orange; stalk rooting, 4 × 1–2 mm, covered by a whitish or cream-coloured tomentum composed of hyaline, thick-walled, rough, septate hairs. Hyphae without clamps. Basidia of tuning-fork type, narrowly clavate, without clamps, up to 55 × 3–5, forked to form 2 sterigmata. Spores hyaline, ellipsoid to somewhat allantoid, thin-walled, 8–14 × 3.5–5, when ripe often 1–3-septate. On fallen branches of conifers.

Donkioporia *expansa* (Desm.) Kotl. and Pouz. (Fig. 137)
Phellinus megaloporus
Fruitbodies perennial, resupinate, flat pulvinate, often covered with exuded drops, tough elastic when fresh, drying hard, up to 2 m long, 0.5–2.5 cm

thick, whitish then yellowish brown or darker brown with silvery sheen, margin pale, velvety, disappearing, pores round, 4–5 per mm, tubes in several layers, 2–6 mm long, tobacco brown. Hyphae with clamps, skeletal ones thick-walled, olivaceous brown. No setae. Cystidioles subulate, hyaline. Basidia 8–25 × 4–6. Spores ellipsoid, 4.5–7 × 3–4, hyaline, thin-walled, smooth, J–. Mostly on wood of *Castanea* and *Quercus*, uncommon.

Echinotrema clanculare Parker-Rhodes (Fig. 138)
Fruitbodies resupinate, soft, membranous, whitish, poroid or irpicoid, with more or less parallel, undulating and anastomosing plates up to 5 mm high and 1 mm thick. Hyphae without clamps. No cystidia. Basidia urniform, about 15 × 5, with 3–6 sterigmata. Spores spherical, 4–5 diam., with rather thick walls, echinulate. In rabbit burrows occupied by sea-birds.

Efibulobasidium albescens (Sacc. and Malbr.) Wells (Fig. 139)
Fruitbodies pustulate or drop-like, up to 3 mm diam., sometimes becoming confluent, softly or firmly gelatinous, hyaline to pale yellow, surface of hymenium granular or undulating. Hyphae without clamps. Hyphidia cylindrical, branched or unbranched. Basidia subspherical to broadly ellipsoid, 15–25 × 11–16, with longitudinal septa not easily seen, sterigmata cylindrical to narrowly clavate, 50–120 × 2–3. Spores fusiform to falcate, 15–25 × 6–7.5. On herbaceous plants and shrubs, e.g. *Epilobium angustifolium*, *Geranium pratense, Juncus, Rumex* and *Ligustrum*.

Eichleriella deglubens (Berk. and Br.) Reid (Fig. 140)
Fruitbodies resupinate, widely effused, firmly attached to the substrate, when fresh softly waxy or gelatinous, pale pink or flesh-coloured, when dry greyish brown, leathery, with a whitish margin which sometimes becomes free, most of surface smooth but with sparsely distributed lumps and spines up to 1 mm tall which have sterile tips. Hyphae with clamps. Hyphidia few, branched. Basidia subclavate, 30–50 × 10–14, longitudinally septate, with 2–4 thick sterigmata. Spores hyaline, smooth, cylindrical to narrowly ellipsoid, often slightly curved, with a few large guttules, 12–22 × 6–10, J–. On dead branches of deciduous trees, especially *Populus*, but also *Fagus* and *Fraxinus*, autumn–spring.

Eocronartium muscicola (Pers. ex Fr.) Fitzp. (Fig. 141)
Fruitbodies erect, 1–5 cm tall, gelatinous, thread-like or narrowly clavate, resembling a slender *Typhula*, fertile part 1–3 mm thick, whitish or pale yellowish, stalk narrower, smooth. Hyphae without clamps. Cystidia none. Basidia abundant, hyaline, cylindrical, bent, 30–40 × 5–8, with 3 transverse septa, sterigmata cylindrical, up to 20 × 3–4. Spores narrowly fusiform to sickle-shaped, 18–25 × 4–5, hyaline, smooth, thin-walled, J–. On living mosses, a perennial parasite in the gametophyte.

Episphaeria *fraxinicola* (Berk. and Br.) Donk (Fig. 142)
Fruitbodies sessile, 0.2–0.7 mm diam., shallow cup-shaped, with inrolled margin, white, hairy, hairs often finely encrusted but not swollen at their tips, hymenium lining cup cream. No cystidia. Basidia 20–25 × 7–8, 4-spored. Spores broadly ellipsoid, often obliquely attenuated at base, 6.5–8 × 4.5–6, in mass coloured ochre to brown. On stromatic pyrenomycetes such as *Diatrype stigma* on bark, March–April.

Epithele *typhae* (Pers.) Pat. (Fig. 143)
Fruitbodies resupinate, orbicular, becoming confluent, membranous to crustose, cream to ochraceous, with numerous sterile teeth, 0.1–0.2 mm long formed of bundles of hyphae. Hyphae with clamps. Cystidia none but slender hyphae with a few branches at the apex lie between the basidia. Basidia stalked, 60–70 × 8–12, with 4 curved sterigmata 10–15 long. Spores fusiform, 18–30 × 6–8, yellowish, smooth, with rather thick walls and several guttules or oil drops, J–. On dead leaves and stems of the larger sedges, e.g. *Carex acutiformis* and *C.riparia*, *Scirpus* and *Typha*.

Erythricium *laetum* (P. Karst.) Erikss. and Hjortst. (Fig. 144)
Fruitbodies resupinate, effused, loosely attached, softly membranous, smooth or slightly wrinkled, rose or salmon pink when fresh, drying creamy or white, margin white. Hyphae hyaline, without clamps, closely septate, up to 10–15 thick, with very short fat cells often anastomosing. No cystidia. Basidia 40–50 × 8–12, with 4 curved sterigmata. Spores broadly ellipsoid, apiculate, 11–15 × 6–7.5, hyaline, smooth, with rather thick walls, J–, C+. On the lower side of fallen trunks and branches of deciduous trees with bark still attached.

Exidia
Fruitbodies gelatinous or cartilaginous, variously shaped, e.g. turbinate, discoid or pulvinate, separate or confluent, with the hymenium formed on the side away from the substrate and the sterile side next to it; hymenium smooth, undulating, ridged or honeycombed, with or without glandular warts. Hyphae with clamps. Basidia spherical to ellipsoid, hyaline, with longitudinal septa and 2–4 sterigmata, mostly with basal clamps. Spores cylindrical and usually curved, hyaline, smooth, J–, often forming secondary spores and conidia.

KEY

Hymenium 2-coloured, or 2-toned, the centre darker than the periphery .... *cartilaginea*
Hymenium evenly coloured 1
1. Hymenium white, dirty white or pale grey .... *thuretiana*

Hymenium black or almost so 2
Hymenium other colours such as ochraceous or brown 3
2. Fruitbodies confluent, spores 10–15 × 4–5 .... *glandulosa*
    Fruitbodies remaining separate, spores 14–22 × 5–7 .... *truncata*
3. Fruitbodies confluent, forming large masses .... *saccharina*
    Fruitbodies remaining separate 4
4. Fruitbodies amber to dark brown .... *recisa*
    Fruitbodies cinnamon brown with pinkish or flesh-coloured sheen
    .... *repanda*

Exidia cartilaginea Lund and Neuh.
Fruitbodies closely gregarious and soon confluent, forming masses up to
20 cm long, 1–2 mm thick, very firmly gelatinous to cartilaginous,
hymenium 2-coloured, ochraceous to reddish or olivaceous brown in the
middle, paler towards the outside and with a whitish fringed edge, flat,
undulating or ribbed. Basidia 11–16 × 10–13, sterigmata up to 60 × 2–3.
Spores 10–13 × 4–5, forming secondary spores 7–8 × 3–4 and rod-like,
slightly curved conidia 5–6 × 2. On branches of deciduous trees including
Betula.

Exidia glandulosa Fr. (Fig. 145)
Witches' Butter
Fruitbodies mostly gregarious and confluent, forming masses up to 30 cm
long and 2 cm thick, flat, undulating and lobed or cerebriform, black or,
rarely, dark blackish brown, closely adhering to the substrate, gelatinous,
hymenium with numerous large glandular warts. Basidia 10–17 × 8–13.
Spores 10–15 × 4–5. On dead branches of deciduous trees, common,
especially on *Quercus*.

Exidia recisa (Ditm. ex S.F. Gray) Fr. (Fig. 146)
Fruitbodies solitary or a few together but not confluent, sessile or shortly
stalked, up to 2.5 cm high and 3.5 cm diam., flat, turbinate or shell-shaped,
amber to dark brown, gelatinous, fairly firm when fresh but becoming
flabby and drooping when old, hymenium undulating or honeycombed,
occasionally with a few scattered glandular warts. Basidia 10–16 × 5–11,
sterigmata up to 30 × 2.5. Spores 12–15 × 3–4, forming secondary spores 8–
10 × 3–3.5, conidia 5–6 × 2, rod-shaped. On branches of deciduous trees.

Exidia repanda Fr.
Fruitbodies remaining separate, roundish, flat, 1–2.5 cm diam. and up to
4 mm thick, with edge lobed or corrugated, hymenium smooth or slightly
undulating with a few glandular warts, hyaline at first but soon becoming
cinnamon brown with a pinkish or flesh-coloured sheen, sterile surface
darker, roughly punctate. Basidia 9–14 × 11–12, sterigmata 40–50 × 2.
Spores 10–15 × 3–4, forming rod-like conidia 4–6 × 1.5. On branches of
deciduous trees, especially *Alnus* and *Betula*.

Exidia saccharina (Alb. and Schw.) Fr.
Fruitbodies frequently confluent and forming masses up to 20×2.5 cm, lying flat on substrate with hardly any stalk, hymenium caramel to dark brown, irregularly undulating and wrinkled. Basidia $10–20 \times 8–16$, sterigmata up to $40 \times 2–3$. Spores $9–15 \times 4–6$, forming curved secondary spores $8–9 \times 4–4.5$ and rod-shaped conidia $5–7 \times 1–2$. On wood of conifers, e.g. *Larix*, *Picea* and *Pinus*.

Exidia thuretiana (Lév.) Fr. (Fig. 147)
Fruitbodies gregarious, 0.5–3 cm diam., confluent and then extending up to 12 cm, firmly attached except at margin, hymenium tuberculate or undulating, often indented at edge or with transverse furrows, white, dirty white or very pale grey. Basidia $16–24 \times 12–17$, sterigmata up to $120 \times 2–3$. Spores $15–23 \times 6–7$, forming secondary spores $9–12 \times 5–7$ and curved conidia $6–6.5 \times 2$. On fallen branches of deciduous trees, especially *Fagus*.

Exidia truncata Fr. (Fig. 148)
Fruitbodies solitary or gregarious but not confluent, flat turbinate or shell-shaped, largely free from the substrate, sometimes stalked, 2–8 cm diam., 1–3 cm thick, black, firmly gelatinous when fresh, flabby and drooping when old, hymenium clearly separate from sterile edge, smooth then ribbed, pitted or furrowed and bearing numerous glandular warts. Basidia $14–18 \times 5–7$, forming secondary spores $10–11 \times 4–4.5$ and rod-like conidia $4–5 \times 2$. On fallen branches of deciduous trees, commonly on *Quercus* and *Tilia*.

Exidiopsis
Similar to *Sebacina* but hyphae and basidia have clamps. Fruitbodies resupinate, effused, firmly attached to substrate. Hyphae with clamps. Basidia with longitudinal septa, hyaline, subspherical to ellipsoid, with basal clamps, sterigmata 2–4, cylindrical. Spores narrowly ellipsoid, cylindrical or allantoid, smooth, thin-walled.

KEY

    Fruitbodies with distinct rose pink tint .... *effusa*
    Fruitbodies without rose pink tint 1
1. Spores 5–7 long .... *fugacissima*
    Spores 15–20 long .... *calcea*
    Spores more than 7 but not more than 15 long 2
2. Spores cylindrical, curved, 4–6 wide .... *grisea*
    Spores ellipsoid, only slightly curved, 5–8 wide .... *opulea*

Exidiopsis calcea (Pers. ex StAm.) Wells
Fruitbodies resupinate, thin, whitish to pale greyish blue, cracking when

dry. Often with crystals in subhymenium. Basidia broadly ellipsoid, 15–28 × 10–15, sterigmata 12–13 × 2–3. Spores cylindrical, curved, 15–20 × 5–8. On wood, mainly of conifers, e.g. *Picea*, widespread. Has been found on *Salix*.

Exidiopsis effusa (Bref. ex Sacc.) Möller (Fig. 149)
Fruitbodies resupinate, gelatinous or waxy, pale greyish but tinted distinctly rose pink. Basidia 10–18 × 9–11, sterigmata up to 40 × 1–2. Spores broadly cylindrical, curved, 12–17 × 4–6. On dead branches of deciduous trees, especially *Fagus*.

Exidiopsis fugacissima (Bourd. and Galz.) Sacc. and Trott.
Fruitbodies resupinate, gelatinous to waxy, hyaline, thin, almost invisible when dry. Basidia subspherical, 5–8 × 4–7, sterigmata up to 16 × 1–2. Spores cylindrical, curved, 5–7 × 2.5–4. On dead branches of deciduous trees.

Exidiopsis grisea (Pers.) Bourd. and L. Maire
Fruitbodies resupinate, gelatinous to waxy, pale grey. Basidia subspherical to broadly ellipsoid, 11–15 × 8–10, sterigmata 30–50 × 2. Spores cylindrical, curved, 10–15 × 4–6, often with secondary spores. Mostly on dead branches of conifers, rarely on those of deciduous trees.

Exidiopsis opulea (Bourd. and Galz.) Reid
Fruitbodies resupinate, gelatinous, hyaline to ochraceous, almost invisible when dry. Basidia 10–13 × 9–12, sterigmata up to 35 long. Spores ellipsoid, sometimes curved, 9–13 × 5–8. On branches of deciduous trees and occasionally on old polypores.

Exobasidium

Parasitic fungi which cause leaf spots, distortion and galls. There is no obvious fruitbody; the mycelium is intercellular, it sends haustoria into the host cells and induces proliferation of the parenchyma. Basidia, formed singly or in tufts between the epidermal cells, eventually pierce the cuticle and form white, powdery masses of spores on the lower surface. Basidia hyaline, clavate, 2–6-spored, not septate, sterigmata subulate. Spores hyaline, smooth, ellipsoid to narrowly cylindrical, straight or curved, thin-walled, sometimes septate when mature, germinating to form germ tubes or rod-like conidia.

KEY

On *Camellia* .... *camelliae*
On *Rhododendron* 1
On *Vaccinium* 2

1. Galls fleshy, with narrow base, red, on *R. ferrugineum* and *R. hirsutum* .... *rhododendri*
   Galls involving part or whole of leaf, pale, on cultivated *Rhododendron* spp. incl. azaleas .... *japonicum*
2. On *V. vitis-idaea*, spores with up to 6 septa .... *vaccinii*
   On *V. myrtillus*, spores with not more than 3 septa 3
3. Leaf spots not or scarcely thickened, usually yellow–ochre, rarely reddish or purple .... *arescens*
   Leaf spots thickened, blistered or wrinkled, yellowish green or pale red, infection systemic .... *myrtilli*

Exobasidium arescens Nannf.
Leaf spots up to 5 mm diam., not or scarcely thickened, flat or bowl-shaped, upper surface usually yellow–ochre, rarely reddish or purple. Spores cylindrical, slightly curved, $10–14 \times 3–4.5$, when mature with 1–3 septa. Conidia rod-like, $5–10 \times 1–1.5$. On leaves of *Vaccinium myrtillus*.

Exobasidium camelliae Shirai
Induces the formation of large fleshy galls on *Camellia japonica*, involving the terminal flower bud and sometimes adjacent leaves. Spores ellipsoid to clavate, slightly curved, $14–22 \times 3–8$, when mature with 1–7 septa.

Exobasidium japonicum Shirai
Galls up to 3 cm diam., fleshy, irregular, involving either the whole or part of the leaf, pale. Basidia with up to 6 spores. Spores cylindrical, sometimes curved, $12–20 \times 3–4.5$, forming either germ tubes or rod-like or narrow fusiform conidia $5–25 \times 1–3$. On cultivated *Rhododendron* species including azaleas.

Exobasidium myrtilli Siegm.
Infection systemic. Twigs scarcely elongating, more upright than is usual. Leaves somewhat larger and thicker, blistered or wrinkled, the upper surface yellowish green or pale red, white spore mass on lower surface between the veins. Basidia 4-spored. Spores narrowly cylindrical, $10–16 \times 1.5–4$, when mature with 1–3 septa. Conidia rod-like, $5–11 \times 1–1.5$. On *Vaccinium myrtillus*.

Exobasidium rhododendri (Fuckel) Cram.
Galls fleshy, up to 3 cm diam., with narrower base, mostly on the underside of leaves at the end of shoots, pale reddish to red, eventually covered with white mass of spores. Basidia 4-spored. Spores narrowly ellipsoid to cylindrical, somewhat curved, $12–15 \times 2.5–4$, 1-septate when mature. Conidia $5–10 \times 1–2$. On *Rhododendron ferrugineum* and *R. hirsutum*.

Exobasidium vaccinii (Fuckel) Woronin (Fig. 150)
Thickened spots or galls up to 1 cm wide on leaves, the upper side rather bright red with a yellowish border, lower side rosy at first, whitening as

spores are formed. Basidia 40–50 × 4–5, 4-spored. Spores cylindrical, somewhat curved, 11–20 × 2–4, with 1–6 septa when mature. Conidia 8–11 × 1. On *Vaccinium vitis-idaea.*

***Femsjonia** pezizaeformis* (Lév.) P. Karst. (Fig. 151)
Fruitbodies gregarious, erumpent through bark, turbinate or discoid with a thick stalk; part of which is embedded in the host, 0.5–1.5 cm high, gelatinous, yellow, drying orange or reddish brown, head flat or slightly concave or convex, 0.5–1.5 cm diam., smooth or wrinkled, stalk and sterile part of head covered with slender white hairs. Hyphae, thick-walled hairs and basidia all with clamps. Basidia of tuning-fork type. Spores ellipsoid, slightly curved, hyaline, smooth, J–, 15–30 × 8–11, when mature with 3–15 transverse septa. On wood, mostly of deciduous trees, especially *Quercus*, rarely on conifers.

***Fibrodontia** gossypina* Parm. (Fig. 152)
Fruitbodies resupinate, effused, membranous to soft waxy, softly fibrous when dry, firmly attached to the substrate, up to 0.35 mm thick, white to ochraceous, with teeth about 1 mm long. Hyphae thin-walled with clamps and thick-walled skeletal ones. Cystidia few, clavate or with swollen tips, often with an oily mass at the apex. Basidia 12–20 × 4–5.5. Spores ellipsoid, 3.5–4.5 × 2.5–3.5, hyaline, smooth, thin-walled, J–. On wood and bark of deciduous trees.

Fibulomyces
Fruitbodies *Athelia*-like, resupinate, effused, thin, pellicular, loosely attached to the substrate, cottony, soft, often with narrow rhizomorphs running out from the margin. Hyphae hyaline, with clamps at most septa, branched and anastomosing freely. No cystidia. Basidia clavate, short, with basal clamps, 4-spored. Spores small, thin-walled, smooth, hyaline, J–.

KEY

Spores 3–5 long .... *mutabilis*
Spores 5–6.5 long .... *septentrionalis*

Fibulomyces mutabilis (Bres.) Jülich (Fig. 153)
Fruitbodies with white, smooth surface. Hyphae often encrusted with crystals. Basidia 12–18 × 4–5. Spores cylindric-ellipsoid, 3–5 × 1.5–2.5. On dead wood of conifers and on litter lying on the ground.

Fibulomyces septentrionalis (J. Erikss.) Jülich
Fruitbodies with white, cream or slightly pinkish, smooth to merulioid surface. Basidia 16–20 × 4–5. Spores cylindric-ellipsoid, 5–6 × 1.5–2. On decaying wood, mostly of conifers.

Fistulina hepatica Schaeff. ex Fr. (Fig. 154)
Beefsteak Fungus or Ox-tongue
Fruitbodies pileate, more or less tongue-shaped brackets, 10–20 cm wide, 2–6 cm thick, sessile or with short lateral stalk, fleshy or spongy with blood-red juice when fresh, upper surface purplish red or reddish brown, rough, furrowed, often sticky, lower surface off-white or pale cream, bruising reddish brown, pores round, 2–3 per mm, tubes separable. No cystidia. Spores ellipsoid, 3–6 × 3–4, hyaline, pinkish yellow in mass, smooth, thin-walled, J–. On *Quercus*, mostly rather low down on the trunks.

Flagelloscypha
Fruitbodies usually gregarious, bowl-, bell- or saucer-shaped, sessile or shortly stalked, white or pale yellowish, with hymenium forming a lining on the inside, outer surface hairy, hairs slender, thick-walled, encrusted with white crystals of calcium oxalate except towards the apex where there is a smooth section which is often whip-like. Basidia frequently 2-spored. Spores hyaline, white in mass, J–, often asymmetrical or lemon-shaped, smooth.

KEY

On ferns, e.g. *Blechnum spicant* .... *morlichensis*
On dead leaves, wood and bark or stems 1
1. Spores 9–15 long .... *faginea*
    Spores 7–10 long .... *minutissima*

Flagelloscypha faginea (Lib.) W.B. Cooke
Fruitbodies bowl-shaped, sessile or shortly stalked, white or slightly yellowish, less than 1 mm diam., hairs on outer surface with bald, whip-like terminal part up to 50 long. Basidia 2-spored. Spores 9–15 × 3–5, asymmetrical, widest above middle. Mostly on fallen dead leaves of *Alnus, Betula, Fagus* etc., rarely on wood.

Flagelloscypha minutissima (Burt.) Donk (Fig. 155)
Fruitbodies 0.3–1 mm diam., bowl- or saucer-shaped, sessile or shortly stalked, with margin inrolled when dry, white on the outside, hairs with whip-like portion 30–50 long, septate, hymenium greyish or ochraceous. Basidia mostly 2-spored, 20–30 × 5–6. Spores lemon-shaped to navicular, 7–10 × 3.5–5. On fallen wood and bark, also on dead woody stems of herbaceous plants, e.g. *Epilobium.*

Flagelloscypha morlichensis W.B. Cooke
Fruitbodies 0.5 mm wide and tall, white, covered with hairs 80–110 × 2.5–3.5, hyaline, coated with acicular granules, tapered towards the tip, whip-like. Hyphae with clamps. Basidia 20–25 × 6–8, 4-spored, with basal clamps.

Spores 9–13 × 3–4.5, smooth, hyaline, tear-shaped. On dead fronds of ferns, e.g. *Blechnum spicant.*

Flaviporus *brownei* (Humb. ex Steudel) Donk (Fig. 156)
Fruitbodies perennial, fan- or shell-shaped, often imbricate brackets up to 4 × 3 cm and 1–4 mm thick, fleshy when fresh, hard when dry, upper surface smooth or somewhat rough, blackish purple or chestnut with thin blackish crust, lower surface yellow to cinnamon, bruising darker, pores very small 8–11 per mm, tubes in several layers. Hyphae with clamps, skeletal ones thick-walled, hyaline to yellowish. Cystidia (m) 40–70 × 6–8, conical, thick-walled, gelatinous, encrusted. Basidia 7–10 × 3–5, 4-spored. Spores ellipsoid, 2–3 × 1.5–2, hyaline, smooth, thin-walled, J–. On timber in mines, greenhouses etc., sometimes effuso-reflexed or resupinate.

Fomes *fomentarius* (L. ex Fr.) Kickx (Fig. 157)
Hoof Fungus or Tinder Fungus
Fruitbodies mostly solitary, perennial, pileate, hoof-shaped brackets 6–50 × 4–25 cm and up to 25 cm thick, hard and woody, with a crust, upper surface greyish brown or grey, concentrically grooved, lower surface cream to ochraceous or pale greyish brown, bruising brown, pores 2–4 per mm, tubes stratified, flesh hard, fibrous, pale brown or cinnamon brown, turns black with KOH. Hyphae with clamps, skeletal ones hyaline to yellow. No cystidia. Basidia 20–30 × 8–11. Spores oblong-ellipsoid, 15–22 × 5–7, hyaline to lemon yellow, smooth, J–, apparently formed only in spring. On standing trunks of deciduous trees, mostly on *Betula* in Scotland and the north of England, on *Fagus* but uncommon in the south of England.

Fomitopsis
Fruitbodies solitary, a few together or imbricate, perennial, pileate, sessile, often dimidiate or hoof-like, hard, woody, encrusted brackets, pores on lower surface 3–6 per mm, tubes stratified, flesh white, cream, pink or wood-colured. Hyphae with clamps. No cystidia. Spores cylindrical to ellipsoid, hyaline, smooth, thin-walled, J–. Often on standing trunks of trees.

KEY

Fruitbodies small, 4–10 × 2–6 cm, pore surface pink .... *rosea*
Fruitbodies large, up to 40–50 × 25–40 cm, pore surface other colours 1
1. Spores 3.5–4 wide, mostly on conifers .... *pinicola*
    Spores 5–6.5 wide, on deciduous trees .... *cytisina*

Fomitopsis cytisina (Berk.) Bond. and Sing.
Fruitbodies flat, 10–50 × 5–40 cm, 2–6 cm thick, sessile, often growing together side by side, and extending up to 150 cm, or imbricate, upper

surface at first somewhat hairy but soon bald, very unevenly lumpy, greyish brown to blackish grey, with distinct crust, lower surface pale grey to ochraceous, sometimes tinged very slightly rose, bruising brown, pores round or slightly angular, 4–6 per mm, tubes stratified, flesh wood colour with slight rosy tint. Basidia $10-20 \times 8-10$. Spores broadly ellipsoid, 6–$9 \times 5-6.5$, hyaline or yellowish, C+. On living and dying deciduous trees.

Fomitopsis pinicola (Fr.) P. Karst. (Fig. 158)
Fruitbodies solitary or imbricate, dimidiate or hoof-shaped, sessile, 5–$40 \times 5-25$ cm, 5–15 cm thick near point of attachment, woody and hard or corky, upper surface concentrically grooved, smooth or slightly wrinkled, exuding clear droplets when growing actively, young pale yellowish, turning reddish orange and glossy with resinous crust and finally dull greyish to blackish, the changes taking place gradually so that there may be broad zones of different colours, margin pale, lower surface pale ochraceous or pale yellow, purplish grey when scratched or bruised, when old pale brownish, pores round, 3–4 per mm, tubes stratified. Basidia $15-20 \times 6-8$. Spores cylindrical to narrowly ellipsoid, apiculate, $6-8 \times 3.5-4$. Mostly on standing or fallen trunks or stumps of conifers, rarely on deciduous trees. Sour smell when fresh.

Fomitopsis rosea (Alb. and Schw. ex Fr.) P. Karst.
Fruitbodies solitary or a few together, dimidiate or hoof-shaped, woody, hard, $4-10 \times 2-6$ cm, 1–4 cm thick near point of attachment, upper surface concentrically grooved, pinkish grey at first but soon black or blackish, with crust, lower surface rose or greyish pink, brownish when old, pores round, 3–5 per mm, tubes stratified. Basidia $12-17 \times 4-5.5$. Spores cylindrical or narrowly ellipsoid, $6-7 \times 2-3$. On dead wood of conifers.

Funalia *gallica* (Fr.) Bond. and Sing. (Fig. 159)
Fruitbodies annual, pileate, tough, sessile, semicircular brackets, $2-15 \times 1-$7 cm, 0.5–1 cm thick, solitary, imbricate or in rows, upper surface dark brown, reddish brown or greyish brown, somewhat zonate and radially furrowed, covered with stiff hairs which are sometimes agglutinated to form tufts, lower surface brown or grey, pores angular, 1–3 mm diam., flesh reddish brown, turning black at first with KOH but colour soon fading. Hyphae with clamps. No cystidia. Basidia $15-25 \times 5-8$. Spores cylindric-ellipsoid, $7-15 \times 3-5$, hyaline, thin-walled, J−. On dead wood, most commonly of *Fraxinus*.

Ganoderma
Fruitbodies mostly rather thick brackets, solitary or in small groups, soft corky-tough when young, usually hardening and becoming woody when old, with small pores on the lower surface which is usually pale but bruises

brown, and a crust on the upper surface which may be smooth and shining as if lacquered or dull and somewhat lumpy; it is frequently coated with shed spores. Hyphae often with clamps, skeletal ones hyaline or brown. No cystidia. Spores of a kind found only in this genus, truncate at apex, brown, with two walls, the inner one thick and ornamented with transverse striations, the outer one thin and colourless, appearing reticulate in face view.

KEY

Upper surface smooth, resinous or appearing lacquered 1
Upper surface dull, often lumpy 5
1. Fruitbodies laterally stalked 2
    Fruitbodies sessile or almost so 3
2. Spores mostly 7–11 × 6–8, on roots and stumps of deciduous trees
    .... *lucidum*
    Spores 11–13 × 7.5–8.5, on stumps of *Abies* and other conifers
    .... *carnosum*
3. Flesh chestnut brown, on *Fagus* .... *pfeifferi*
    Flesh white, pale yellowish or tawny 4
4. Flesh white or pale yellowish, on *Larix* and *Taxus* .... *valesiacum*
    Flesh pale tawny or wood-coloured on *Quercus* .... *resinaceum*
5. Spores 6.5–8.5 × 4.5–6, fruitbody not more than 8 cm thick
    .... *applanatum*
    Spores 8–13 × 5.5–9, fruitbodies up to 30 cm thick .... *adspersum*

Ganoderma adspersum (Schulzer) Donk (Fig. 160)
Fruitbodies sessile, up to 60 × 25 cm and 30 cm thick, upper surface dull reddish brown to dark brown or blackish brown, knobbly and sometimes concentrically zoned, crust containing branching hyphae, pores on lower surface 3–4 per mm, flesh dark brown, thicker than tube layer. Spores 8–13 × 5.5–9, reddish brown in mass. Common on deciduous trees.

Ganoderma applanatum (Pers.) Pat.
The Artist's Fungus
Fruitbodies sessile, up to 60 × 30 cm and 8 cm thick, plate-like, upper surface dull greyish-brown or cocoa-brown, knobbly and grooved concentrically, crust hyphae unbranched, pores on lower surface 4–6 per mm, flesh cinnamon brown, thinner than tube layer. Numerous cone-shaped insect galls are seen often on the pore surface of this species but not *G. adspersum*. Spores 6.5–8.5 × 4.5–6. Mostly on *Fagus*.

Ganoderma carnosum Pat.
Fruitbodies laterally stalked, up to 30 cm diam, stalk up to 25 × 4 cm, both upper surface and stalk shining as if lacquered, mahogany or blackish red, lower surface pale, greyish-white or cream, 3–4 pores per mm, flesh pale

brown. Spores 11–13.5 × 7.5–8.5. Mostly on stumps of *Abies* and other conifers.

Ganoderma lucidum (Fr.) P. Karst. (Fig. 161)
Fruitbodies laterally stalked, up to 25 cm diam., stalk up to 25 × 3 cm., both upper surface and stalk shining as if lacquered, orange–brown to dark purplish brown, lower surface whitish or cream, 4–5 pores per mm, flesh pale brown, crust containing spheropedunculate or clavate hyphae. Spores mostly 7–11 × 6–8. On roots and stumps of deciduous trees.

Ganoderma pfeifferi Bres.
Fruitbodies sessile, up to 30 × 12 cm, upper surface smooth, resinous, copper-coloured, with yellowish zone towards the edge, especially in young specimens, darkening with age, pore surface whitish to cream or ochraceous, bruising brown, pores round, 3–5 per mm in fresh specimens, flesh chestnut brown, thick-walled clavate or spheropedunculate hyphae in crust. Spores 9–13 × 7–9. On lower part of trunks, mainly of *Fagus*.

Ganoderma resinaceum Boud.
Lacquered Bracket
Fruitbodies sessile or almost so, up to 45 × 20 cm, 10 cm thick, upper surface concentrically zoned, smooth, resinous, with a varnished look, reddish brown to maroon or almost black, with a yellow or creamy edge, lower surface whitish to cream, pores 3 per mm, flesh pale tawny or wood-coloured, thick-walled, clavate or spheropedunculate hyphae in crust. Spores 9–11 × 4.5–7. Parasitic on *Quercus*, mostly near base of trunk.

Ganoderma valesiacum Boud.
Fruitbodies sessile or almost so, upper surface smooth, shining, resinous, dark chestnut brown to black, the crust sometimes splits exposing the white or pale yellowish flesh, lower surface pale yellowish, pores small; clavate or spheropedunculate hyphae present in crust. Spores 9–12 × 6–8. On *Larix* and *Taxus*, uncommon.

Geopetalum *carbonarium* (Alb. and Schw. ex Fr.) Pat. (Fig. 162)
Fruitbodies solitary or clustered, funnel-shaped, leathery; cap 3–6 cm diam., with wavy inflexed margin, upper surface grey, greyish brown or fuliginous, fibrillose, lower surface white to deep cream, or slightly greyish, with hymenium covering deeply decurrent, several times forked veins or shallow ridges; stalk typically central, up to 5 × 1 cm, rooting, whitish to rather pale grey. Cystidia numerous, lanceolate or fusiform, sharply pointed, thick-walled, hyaline, 90–120 × 13–14. Spores cylindrical, hyaline, 9–11 × 4–5. On bonfire sites.

Gloeocystidiellum

Fruitbodies resupinate, effused, membranous to somewhat waxy, smooth to tuberculate, often fissured when dry. Cystidia (g) thin-walled, tubular, sinuous, filled with granules or oil drops. Basidia 4-spored. Spores hyaline, thin-walled, smooth or warted, J+.

KEY

Spores smooth, fruitbodies up to 1 mm thick .... *ochraceum*
Spores verrucose, fruitbodies much thinner .... *porosum*

Gloeocystidiellum ochraceum (Fr.) Donk (Fig. 163)
Fruitbodies often large, up to about 1 mm thick, waxy to hard, much fissured when old, creamy to dark ochraceous. Hyphae without clamps, mostly vertical and forming compact tissue which is stratified. Cystidia 40–60 × 4–6. Basidia 20–30 × 4–5. Spores broadly ellipsoid to ovoid, 4.5–6 × 3–4. On conifer wood.

Gloeocystidiellum porosum (Berk. and Curt.) Donk (Fig. 164)
Fruitbodies thin, membranous or waxy, white to cream, cracking. Hyphae with clamps at all septa, subhymenial layer composed of closely packed vertical hyphae, hyphae of basal layer parallel to substrate. Cystidia up to 150 × 9–15, subulate, sometimes swollen at base. Basidia about 20 × 4. Spores ellipsoid, 4.5–6 × 3–3.5, verrucose. On wood of deciduous trees.

Gloeophyllum

Fruitbodies sessile, pileate, tough, leathery, semicircular or fan-shaped, rather thin, flat, often imbricate brackets with basidia borne at least in part on lamellae which anastomose freely and sometimes form labyrinths and angular pores, flesh brown. Hyphae with clamps, skeletal ones yellowish or reddish brown. Cystidia abundant but not protuberant, hyaline to yellowish or rusty brown. Spores cylindrical or oblong-ellipsoid, hyaline, smooth, J–.

KEY

Lamellae 8–12 per cm at margin .... *abietinum*
Lamellae 15–23 per cm at margin .... *sepiarium*
Lamellae 20–40 per cm at margin or with pores 2–4 per mm .. *trabeum*

Gloeophyllum abietinum (Bull. ex Fr.) P. Karst. (Fig. 165)
Fruitbodies often fan-shaped, imbricate brackets, frequently fused laterally, up to 8 × 5 cm and 1 cm thick, upper surface flat, concentrically zoned and furrowed, at first finely hairy but becoming smooth, reddish or greyish brown, darkening with age, with pale, sometimes yellowish margin when young, lower surface lamellate, ochraceous to greyish brown, lamellae up

to 1 cm deep, sinuous, anastomosing, with toothed edges, 8–12 per cm at margin, flesh dark brown, fibrous, 2=layered, turned black by KOH. Skeletal hyphae thick-walled, yellowish or reddish brown. Cystidia abundant, yellowish or rusty brown, thick-walled, subulate with clamp at base, smooth, 30–50 × 4–8. Basidia 20–35 × 5–7. Spores cylindrical, sometimes slightly curved, 8–12 × 3–4. On fallen trunks etc of conifers, especially *Picea*.

Gloeophyllum sepiarium (Wulf. ex Fr.) P. Karst. (Fig. 166)
Fruitbodies fan-shaped, imbricate and often laterally fused brackets, sometimes clustered from a common base, 4–12 × 2–7 cm, 0.5–1 cm thick, upper surface flat, concentrically furrowed and radially wrinkled, weakly zonate, softly hairy at first then somewhat bristly with agglutinated hairs, yellowish brown then rusty brown, darkening with age, margin often remaining pale yellowish, lower surface pale rusty ochraceous to tobacco brown, lamellate or sometimes in part daedaleoid, sinuous, anastomosing, close together, 15–23 per cm at margin, flesh dark brown, turning black with KOH. Cystidia abundant, subulate, hyaline, 35–40 × 2–7. Basidia 40–45 × 4–6. Spores cylindrical, 9–13 × 3.5–5. On dead wood of conifers, especially *Picea*.

Gloeophyllum trabeum (Pers. ex Fr.) Murrill
Fruitbodies fan-shaped or semicircular brackets, 3–8 × 1–3 cm, mostly less than 1 cm thick, imbricate, sometimes with several arising from a common base or fused laterally, upper surface velvety becoming smooth, scarcely zoned, uneven, somewhat tuberculate, brown or greyish brown, with a pale, sharp, undulating margin, lower surface ochraceous to brown, with pores 2–4 per mm or partly or wholly lamellate with lamellae anastomosing and 20–40 per cm at the margin, flesh brown. Skeletal hyphae yellowish brown. Cystidia cylindrical to clavate, often with conical, pointed tips, hyaline or pale yellowish, 20–35 × 5. Basidia 15–25 × 5–6. Spores oblong-ellipsoid, 7–9 × 3–4.5. On dead wood of deciduous trees and conifers.

Gloeoporus *dichrous* (Fr.) Bres. (Fig. 167)
Fruitbodies annual, pileate or effuso-reflexed, shelf-like brackets 1–4 × 0.5–3 cm, 3–5 mm thick, imbricate or joined laterally, soft when fresh, becoming horny, sterile upper surface white, cream or ochraceous, tomentose, sometimes slightly zonate, lower surface rubbery when fresh, at first greyish pink, becoming dark purplish or reddish brown, often with white margin, pores round to angular, 4–6 per mm, tubes shallow, gelatinous, easily separated from white, cottony or felt-like flesh. Hyphae with clamps. No cystidia. Basidia 10–15 × 4–8. Spores allantoid, 4–5 × 1–1.5, hyaline, thin-walled, smooth, J–. On wood of deciduous trees, e.g. *Betula*, *Corylus* and *Quercus* and sometimes on old polypores.

Gloiocephala *menieri* (Boud.) Sing. (Fig. 168)
Fruitbodies pileate, stipitate, cap up to 1 cm diam., pale reddish to

ochraceous brown, margin somewhat involute, hymenium on lower surface smooth or with veins or ridges, stalk excentric, somewhat curved, about 1 × 0.2 mm, black or blackish. Cap skin with fusiform, capitate, thick-walled cystidia 20–30 × 10–15 and conspicuous protruding, reddish brown cap cells. Spores fusiform, 15–25 × 3.5–7.5. On dead *Carex*, *Typha* etc.

Gomphus *clavatus* (Pers. ex Fr.) S.F. Gray (Fig. 169)
Fruitbodies solitary or in clumps, clavate when young, becoming turbinate with a flat top overlapping the edge, when old depressed in the centre and funnel-shaped, up to 10 cm tall and 7 cm wide, lilac, rose or flesh-coloured when fresh, eventually olivaceous, with hymenium covering deeply decurrent, forked and anastomosing vein-like ridges, flesh white, soft, fragile, stalk short, solid. Hyphae hyaline, with clamps. Basidia with basal clamps, 4-spored, 50–80 × 8–11. Spores ellipsoid, 10–16 × 4–6, in mass ochraceous, verrucose, J–, C+. On soil in conifer forests.

Grandinia
Fruitbodies resupinate, effused, membranous, firmly attached to substrate, mostly with fertile teeth but in some species smooth or tuberculate, whitish, cream or pale ochraceous. Hyphae with clamps, cyanophilous, rather loosely intertwined, often branching from or opposite clamps. Cystidia of various kinds, often conspicuous. Basidia in British species 4-spored, rather short and somewhat constricted in the middle. Spores small, mostly about 4–6 long and never more than 8, ellipsoid, cylindrical or allantoid, hyaline, smooth, thin-walled, J–. On rotten wood and bark.

KEY

Hymenial surface without teeth 1
Hymenial surface with teeth 4
1. Spores 6–8 long, cylindrical to allantoid .... *subalutacea*
   Spores shorter, ellipsoid 2
2. Cystidia of one kind only, septate with clamps .... *pallidula*
   Cystidia of two kinds 3
3. Numerous protruding subulate, sharply pointed cystidia present
   .... *hastata*
   No sharply pointed, subulate cystidia formed .... *alutaria*
4. Spores allantoid, 6–8 × 1.5 .... *stenospora*
   Spores not allantoid, wider 5
5. Cystidia over 100 and sometimes up to 300 long 6
   Cystidia shorter 7
6. Spores 3–3.5 wide, on conifer wood .... *abieticola*
   Spores 3.5–4.5 wide on wood of deciduous trees .... *barba-jovis*
7. Cystidia variable but often differing little from hyphae 8

Cystidia all differing distinctly from hyphae 9
8. Spores 4–6 × 2–2.5, cylindrical .... *nespori*
   Spores 6–7.5 × 2.5–3, flattened or slightly incurved on one side
   .... *quercina*
9. Cystidia of one kind only 10
   Cystidia of two kinds 11
10. Cystidia subulate .... *crustosa*
    Cystidia swollen at apex, capitate .... *granulosa*
11. One cystidium type tapered abruptly to a slender encrusted thread
    .... *arguta*
    Some cystidia capitate, others moniliform .... *breviseta*

Grandinia abieticola (Bourd. and Galz.) Jülich (Fig. 170)
Fruitbodies cream to ochraceous, densely covered with small, conical teeth which terminate in protruding cystidia. Cystidia up to 100 × 5–7, cylindrical, flexuous, thick-walled except near rounded apex, smooth, secondary septa without clamps. Spores almost ellipsoid, slightly broader at base, 5–6 × 3–3.5. On decaying wood of conifers.

Grandinia alutaria (Burt) Jülich (Fig. 171)
Fruitbodies pale ochraceous, surface tuberculate. Cystidia of two kinds, both plentiful: (1) septate, 50–100 × 5–7, with clamps at septa and apical and intercalary swellings which are often encrusted, (2) non-septate, 20–40 × 2–3, tapered rather abruptly to a long, slender, encrusted thread. Spores broadly ellipsoid, 4.5–5 × 3–3.5. On decaying wood, mostly of conifers.

Grandinia arguta (Fr.) Jülich (fig. 172)
Fruitbodies white or cream to clay-coloured or ochraceous, with conical to almost cylindrical teeth 0.5–2 mm long. Cystidia of two types: (1) 40–80 × 5–7, with swollen apex usually encrusted, occasionally septate with clamps, (2) 30–40 × 2–3, tapered abruptly to a slender encrusted thread. Spores broadly ellipsoid, 4.5–6 × 3.5–4. On very rotten wood, mostly of deciduous trees.

Grandinia barba-jovis (Fr.) Jülich (Fig. 173)
Fruitbodies white to ochraceous, with projecting teeth 1–3 mm long. Cystidia abundant in hymenium, including teeth, 100–300 × 6–8, cylindrical, flexuous, rather thick-walled except towards apex, protruding. Spores ellipsoid, 4.5–6 × 3.5–4.5. On wood and bark mainly of deciduous trees, including standing trunks and attached branches.

Grandinia breviseta (P. Karst.) Jülich (Fig. 174)
Fruitbodies whitish to pale yellow, clay-coloured or ochraceous, with conical teeth. Cystidia of two types: (1) 30–40 × 3–4, with swollen, encrusted tips, (2) 40–60 × 4–5, constricted at intervals to become moniliform. Spores ellipsoid, 4–5 × 3–3.5. On wood and bark, mostly of conifers.

Grandinia crustosa (Pers. ex Fr.) Fr. (Fig. 175)
Fruitbodies white to cream, with teeth about 1 mm long, surface polygonally cracked when old. Cystidia subulate, 20–40 × 2–3. Spores cylindric-
ellipsoid, 5–6 × 2.5–3. On wood and bark, mostly of deciduous trees.

Grandinia granulosa (Pers. ex Fr.) Fr. (Fig. 176)
Fruitbodies pale cream to pale ochraceous, with well spaced out teeth.
Cystidia 30–40 × 3–5, with distinctly swollen tips. Spores broadly ellipsoid,
5–6 × 3.5–5. On decayed wood of both conifers and deciduous trees.

Grandinia hastata (Litsch.) Jülich (Fig. 177)
Fruitbodies cream, smooth or somewhat tuberculate. Cystidia of two kinds:
(1) numerous, subulate, sharply pointed, protruding, 30–40 × 3–5, often
with encrusted tips, (2) few, not protruding, moniliform. Spores cylindric-
ellipsoid, 5–6 × 2–2.5. On wood and bark of deciduous trees and conifers.

Grandinia nespori (Bres.) Cejp (Fig. 178)
Fruitbodies whitish, cream or pale ochraceous, polygonally cracked when
old, densely covered with small, conical teeth. Cystidia often differing little
from hyphae, those at the tips of the teeth cylindrical, flexuous, 60–80 × 5–
5.5, those mixed with basidia swollen at tips, 40–50 × 3–4, smooth or
encrusted. Spores cylindrical, 4–6 × 2–2.5. On wood and bark, mostly of
conifers.

Grandinia pallidula (Bres.) Jülich (Fig. 179)
Fruitbodies creamy white to pale ochraceous, smooth or minutely tuberculate. Cystidia septate, with clamps, often constricted and somewhat
moniliform, protruding, up to 120 × 4–6, tips sometimes encrusted. Spores
3.5–5 × 2–3. On rotten wood, mostly of conifers.

Grandinia quercina (Fr.) Jülich (Fig. 180)
Fruitbodies whitish, cream or ochraceous, with more or less conical teeth
2–3 mm long, with fimbriate tips. Cystidia hypha-like, 20–40 × 3–5,
variable, some swollen at tips which may be encrusted, others moniliform
or subulate, those at tips of teeth mostly tapered. Spores cylindrical, more
flattened or slightly curved on one side, 6–7.5 × 2.5–3. On rotten wood of
deciduous trees.

Grandinia stenospora (P. Karst.) Jülich (Fig. 181)
Fruitbodies white to cream, with teeth only up to 1 mm long. Cystidia 50–
70 × 4–7, cylindrical or occasionally moniliform, smooth, thin-walled, some
protruding. Spores allantoid, 6–8 × 1.5. On wood and bark of conifers.

Grandinia subalutacea (P. Karst.) Jülich (Fig. 182)
Fruitbodies cream, yellowish or ochraceous, smooth or minutely tuberculate. Cystidia plentiful, cylindrical or almost so, thick-walled, encrusted
towards the slightly wider apex, 100–200 × 5–7, protruding. Spores cylindrical to allantoid, 6–8 × 1.5–2. On rotten wood, mostly of conifers.

Granulobasidium *vellereum* (Ell. and Crag.) Jülich (Fig. 183)
Fruitbodies resupinate, effused, thin, membranous, somewhat waxy, firmly attached to substrate, smooth or minutely tuberculate, whitish to rose, yellowish when dry, no rhizomorphs. Hyphae with clamps. Chlamydospores in subhymenium pyriform or broadly fusiform, thick-walled, C+, with central oil drop or guttule, 8–10×6–8. No cystidia. Basidia narrowly clavate, flexuous, 50–60×5–6, 4-spored, hyaline, with basal clamps, full of granules or tiny oil drops. Spores spherical, hyaline, wall thick with an uneven surface, 7–8 diam., each usually with one large guttule, J–, C+. Mostly on dead trunks of *Ulmus*, best collected January–February.

Grifola *frondosa* (Dicks. ex Fr.) S.F. Gray (Fig. 184)
Fruitbodies compound, 20–50 cm diam., each consisting of a large number of overlapping, fan-shaped or spathulate, lobed, pale grey or greyish brown, softly leathery, 0.5–1 cm thick caps attached laterally to a common stalk, lower surfaces of pilei white or pale cream, with round or angular pores 2–3 per mm. Hyphae sometimes with clamps. Basidia 20–25×6–8. Spores broadly ellipsoid, 5–7×3.5–5, hyaline, smooth, thin-walled. On stumps, mostly of *Quercus*, more rarely of *Carpinus*, *Castanea* etc.

Guepiniopsis
Fruitbodies at first pustulate, becoming discoid or cyathiform, sessile or stalked, firmly gelatinous. Hymenium confined to surface of disc or inside of cup; sterile surface with a palisade of thick-walled elements. Hyphae with or without clamps. Basidia of tuning-fork type. Spores hyaline, when mature with up to 8 transverse septa and in some species with a few longitudinal septa also, J–.

KEY

Spores 16–28×8–10 .... *chrysocoma*
Spores not more than 18×6 1
1. Palisade elements with terminal cell very large and often lanceolate
   .... *alpina*
   Palisade elements not so .... *buccina*

Guepiniopsis alpina (Tracy and Earle) Brasf. (Fig. 185)
Fruitbodies up to 1 cm tall, at first narrowly clavate, becoming turbinate, flat apical disc 3–4 mm diam., yellow to pale orange. Hyphae with clamps. Palisade elements at first clavate, then capitate, finally with terminal part up to 40 long, lanceolate, with greatly thickened wall often laminated. Spores 10–18×4–5.5, with 3 transverse septa when mature. On wood and bark of conifers.

Guepiniopsis buccina (Pers. ex Fr.) Kennedy
Fruitbodies stalked, up to 1 cm tall, apical part 5–9 mm diam., plate-shaped or cyathiform, with the lower surface ribbed, yellow or orange, stalk rooting. Hyphae all without clamps. Palisade elements of stalk and lower side of head clavate or capitate, 8–12 thick, sometimes constricted at septa. Spores 11–16 × 4–6, when mature with 1–3 somewhat thickened septa. On dead branches of deciduous trees.

Guepiniopsis chrysocoma (Bull. ex St Amans) Brasf. (Fig. 186)
Fruitbodies 2–5 mm tall, 1–4 mm wide, at first pulvinate, becoming cylindrical and then cyathiform, sessile or narrowed to a stem-like base, yellow or yellowish orange. Hyphae with clamps. Palisade elements narrowly clavate, 4–6 thick. Spores 16–28 × 8–10, when mature with 3–8 transverse septa and a few longitudinal septa. On wood and bark of conifers.

Gyroporus
Fruitbodies fleshy, firm, each composed of a convex or somewhat flattened cap and a central stalk; cap with upper surface velvety to smooth or scaly, pore surface white to pale lemon yellow or straw-coloured, pores minute, tubes free from stalk; stalk leathery on the outside, at first with spongy pith then frequently becoming hollow. Cystidia clavate, fusiform or lageniform. Spores ellipsoid, in mass lemon yellow.

KEY

Flesh not turning blue when cut, upper surface of cap rusty-tawny to chestnut .... *castaneus*
Flesh quickly turning blue when cut, upper surface of cap off-white to pale ochraceous .... *cyanescens*

Gyroporus castaneus (Bull. ex Fr.) Quél. (Fig. 187)
Cap 4–10 cm diam., upper surface velvety to smooth, pore surface unchanged when bruised; stalk 3–9 × 1–3 cm, similar in colour to cap, but usually paler especially at the apex, slightly velvety; flesh white or whitish. Spores 7–11 × 5–6. In woods, mostly with *Quercus*, summer and autumn, uncommon.

Gyroporus cyanescens (Bull. ex Fr.) Quél.
Cap 5–15 cm diam., often with a shaggy margin, upper surface fibrillose to floccose scaly, bruising blue, pore surface bruising blue; stalk 5–10 × 3–8 cm, paler at apex, darker buff below, often with ring-like zones due to cracking. Spores 8–11 × 5–6. On heathy soils with *Calluna*, *Betula* etc., late summer, uncommon.

Hapalopilus

Fruitbodies pileate or resupinate, brightly coloured, orange, pinkish, cinnamon etc., turned violet by KOH, pores round or angular. Hyphae with clamps. No cystidia. Spores ellipsoid to cylindrical, hyaline, smooth, thin-walled, J−.

KEY

Fruitbodies pileate, bracket-like .... *rutilans*
Fruitbodies resupinate .... *salmonicolor*

Hapalopilus rutilans (Pers. ex Fr.) P. Karst (Fig. 188)
Fruitbodies sessile, solitary or imbricate, semicircular, fan-shaped or kidney-shaped, broadly attached, tough, spongy to corky brackets, 2–12 × 2–8 cm, 1–4 cm thick near point of attachment, with smooth, sharp edges, upper surface velvety when young becoming smooth, cinnamon to bright ochraceous, lower surface cinnamon brown or ochraceous, pores 2–4 per mm. Hyphae wide, with conspicuous clamps. Spores narrowly ellipsoid, 3.5–5 × 2–3. On dead wood of deciduous trees, either fallen or standing.

Hapalopilus salmonicolor (Berk. and Curt.) Pouz.
Fruitbodies resupinate, up to 10–12 cm diam., 1 mm thick, firmly attached, soft and orange or pinkish when young, hard and orange brown when old, margin pale orange, byssoid, pores angular, 3–5 per mm. Spores cylindrical to narrowly ellipsoid, 3.5–6 × 2–3. On wood of conifers, especially *Pinus*.

Helicobasidium

Fruitbodies resupinate, membranous or felt-like, loosely attached to substrate. Hyphae without clamps, hyaline or brown. Basidia hyaline to violet, cylindrical, curved, often helically, with 1–3 transverse septa and 2–4 lateral sterigmata. Spores hyaline or pale purple, thin-walled, smooth, J−.

KEY

Spores 10–13 × 6–8 .... *brebissonii*
Spores 13–19 × 4–5 .... *compactum*

Helicobasidium brebissonii (Desm.) Donk (Fig. 189)
Fruitbodies effused, often several cm across, thin, membranous, hymenium smooth or undulating, pinkish violet, pruinose, overlying an ochraceous brown or purplish brown subiculum, with brown, thick-walled hyphae 4–8 diam. Basidia 4–7 wide, 1–3-septate, sterigmata up to 20 long. Spores broadly ellipsoid or somewhat reniform, 10–13 × 6–8, hyaline or pale violet. An economically important parasite of many kinds of plants.

Helicobasidium compactum (Boedijn) Boedijn
Fruitbodies effused, felted to membranous, about 1 mm thick, bordered by a network of brown rhizomorphs, hymenium purplish or rose-coloured, hypochnoid, overlying a brown subiculum with hyphae thick-walled, 5–8 diam. Basidia loosely helically coiled, 5–6 wide, 4-spored. Spores narrowly ellipsoid to clavate, 13–19 × 4–5. On fallen trunks of *Salix triandra*, lumps of plaster etc.

Helicogloea *lagerheimii* Pat. (Fig. 190)
Fruitbodies resupinate, several cm across, softly gelatinous when fresh, waxy when dry, at first reddish brown, becoming greyish and finally blackish, often pruinose. Hyphae without clamps. No cystidia. Probasidia saccate, up to 30 × 12, each one bearing laterally at its base an elongated basidium, the terminal part of which becomes transversely 3-septate; sterigmata 4, lateral, rather short. Spores ellipsoid, hyaline, smooth, thin-walled, 8–14 × 5–7.5, J–, forming secondary spores. On decaying wood of deciduous trees.

Henningsomyces *candidus* (Pers. ex Schleich.) O. Kuntze (Fig. 191)
Fruitbodies solitary or, more commonly, gregarious in dense groups, erect, shortly cylindrical, 0.5–1 mm tall, 0.2–0.4 mm wide, often tapered at base, looking rather like insect eggs, white or very pale cream, with hymenium lining each little cylinder, outer surface smooth or furfuraceous, marginal hairs branched, hyaline, smooth. No subiculum. Hyphae with clamps. Cystidia none. Basidia clavate, 15–20 × 5–7, 4-spored. Spores subspherical, apiculate, 4.5–6 × 4–5, hyaline, smooth, J–. On rotten wood, mainly of conifers.

Hericium
Fruitbodies pileate, often short-stalked and attached laterally, fleshy to tough, in some species richly and repeatedly branched in a coralloid manner, with long, pendent teeth arranged either in irregular clusters or more regularly in rows and then comb-like. Flesh amyloid. Hyphae with clamps. Cystidia none but hyphae with long cells full of oil drops often present in the hymenium. Spores subspherical to ellipsoid, smooth or finely verruculose, J+. On wood, often on dead standing trunks.

KEY

Fruitbodies compact, hedgehog-like .... *erinaceum*
Fruitbodies much branched 1
1.　Pendent teeth in regular rows, spores 3.5–4.5 × 3–3.5 .... *clathroides*
　　Pendent teeth in many irregular clusters, spores 5–7 × 4.5–5.5
　　.. *coralloides*

Hericium clathroides (Pallas ex Fr.) S.F. Gray (Fig. 192)
Fruitbodies white to yellowish brown, up to 20 cm across, richly branched in a coralloid manner, pendent teeth up to 1 cm long, regularly arranged in rows. Spores 3.5–4.5 × 3–3.5, mostly smooth. On dead wood of deciduous trees, most commonly on *Fagus*.

Hericium coralloides (Scop. ex Fr.) S.F. Gray (Fig. 193)
Fruitbodies cream-coloured, up to 35 cm across, richly and repeatedly branched in a coralloid manner, pendent teeth 0.5–2 cm long, arranged in many irregular clusters. Spores 5–7 × 4.5–5.5. On dead standing or fallen trunks of *Abies*.

Hericium erinaceum (Bull. ex Fr.) Pers.
Fruitbodies white when fresh, drying yellowish brown, hedgehog-like, round or broadly ovoid, 5–30 cm long, consisting of a sometimes slightly stalked, swollen protuberance which remains undivided or splits into several broad lobes; pendent teeth 1–3 cm long cover the front and lower surfaces. Hyphae hyaline, often swollen up to 20 wide, with clamps. Basidia 4-spored. Spores ellipsoid, 5–6 × 4–5, smooth or finely verruculose. A wound parasite of deciduous trees; recorded on *Fagus* and *Quercus*.

Herpobasidium
Fruitbodies parasitic on leaves and on fern fronds, resupinate. Hyphae hyaline, thin-walled, smooth, without clamps. No cystidia. Basidia hyaline, cylindrical, narrowly clavate or subulate, with 1–3 transverse septa, sterigmata cylindrical or subulate. Spores ellipsoid, hyaline, thin-walled, smooth, J–, forming secondary spores.

KEY

On ferns .... *filicinum*
On leaves of *Lonicera* .... *deformans*

Herpobasidium deformans Gould
Fruitbodies resupinate, thin, hyphae hyaline, emerging through stomata and spreading over parts of the lower leaf surface. Basidia cylindrical, curved, up to 50 × 5, 3-septate; sterigmata cylindrical, up to 8 long. Spores 9–11 × 5–6. On leaves of *Lonicera*, causing yellow spots on the upper surface.

Herpobasidium filicinum (Rostrup) Lind (Fig. 194)
Fruitbodies forming white or cream patches, 3–4 × 2 mm on the lower surface of fronds. Basidia narrowly clavate or subulate, straight or curved, 40–75 × 6–9, 1-septate, sterigmata subulate, up to 15 long. Spores ellipsoid, often curved, 11–21 × 5–10. On various ferns including bracken, most frequently on *Dryopteris filix-max*, June–July.

***Heterobasidion** annosum* (Fr.) Bref. (Fig. 195)
Root Fomes
Fruitbodies pileate or resupinate; brackets solitary or imbricate, broadly attached, often narrow and shelf-like, but may be up to 30×10–15 cm, 1–3 cm thick, tough and elastic, becoming hard when old, upper surface sulcate-zoned, often lumpy, tomentose becoming smooth, with a white edge and behind this a reddish brown zone, the rest blackish brown, lower surface white or cream, pores round or angular, 3–5 per mm. Hyphae without clamps, skeletal ones abundant, thick-walled, C+. No cystidia. Spores subspherical, 4.5–6×4–5, hyaline, finely but distinctly echinulate, J–, C+. Parasitic on conifers, found commonly at or just above soil level, at the base of trunks and on roots.

***Heterochaetella** dubia* (Bourd. and Galz.) Bourd. and Galz. (Fig. 196)
Fruitbodies resupinate, effused, thin, gelatinous when fresh, waxy when dry, hymenium smooth, finely setose under a lens, white to pale ochraceous. Hyphae narrow, hyaline, with clamps. Cystidia up to 250×4–6, hyaline, cylindrical, thick-walled except towards apex, projecting a long way above the surface of the hymenium. Basidia 8–11×6–9, with longitudinal septa, sterigmata 2–4, up to 20×2. Spores ellipsoid, hyaline, thin-walled, smooth, J–, 5–9×4–5. On rotten stumps and fallen branches of both deciduous trees and conifers.

Heterochaetella brachyspora Luck-Allen
Similar and on the same substrates as *H. dubia* but has spores 4–6×3–5.

Hydnellum
Fruitbodies terrestrial, pileate, centrally stalked, sometimes concrescent, turbinate or shallowly funnel-shaped, tough-fleshed, teeth on lower surface of cap decurrent onto the stalk which expands above to merge with the cap. Cystidia none. Basidia 4-spored. Spores mostly subspherical, brown, tuberculate.

KEY

Stalk covered with very thick, spongy, felt-like tomentum .. *spongiosipes*
Stalk not so 1
1. Cap surface very clearly concentrically zoned, tan, wine and very dark brown .... *concrescens*
    Cap surface not so, if zonate only indistinctly so 2
2. Cap surface at first yellow–ochre soon becoming very dark brown to almost black, flesh firm, hard .... *compactum*
    Cap surface and flesh not so 3
3. Cap surface greyish blue with broad white edge, only very slowly

turning brown .... *caeruleum*
Cap surface orange to orange-brown in centre, margin white
.... *aurantiacum*
Cap surface pinkish brown, bruising black .... *scrobiculatum*
Cap surface becoming dark reddish brown in centre, with broad white
border, often exuding drops of blood-red fluid when young 4
4. Hyphae without clamps .... *ferrugineum*
Hyphae with distinct clamps at some of the septa .... *peckii*

Hydnellum aurantiacum (Batsch. ex Fr.) P. Karst. (Fig. 197)
Fruitbodies solitary or gregarious, turbinate or shallowly funnel-shaped;
cap 2–9 cm diam., 3–8 mm thick, round with broadly crenate, undulating
edge, upper surface velvety at first, becoming smooth, radially wrinkled,
faintly zonate, orange to orange-brown in the middle, paling to whitish
towards the margin, lower surface bearing brownish teeth up to 5 mm long
which are decurrent onto the stalk; stalk 2–9 × 0.5–2 cm, roughly cylindrical
but broadening towards the cap, tapered or swollen at the base, pale orange
to dark orange-brown, stuffed. Flesh of cap pale, of stalk dark orange-
brown. Spores subspherical, 6–7 × 4.5–5, with coarse, blunt tubercles. In
conifer and mixed woods.

Hydnellum caeruleum (Hornem. ex Pers.) P. Karst.
Fruitbodies solitary or gregarious, sometimes concrescent, roughly top-
shaped or very shallowly funnel-shaped; cap irregularly rounded, 3–10 cm
diam., broadly crenate at the edge, upper surface velvety when young,
becoming smooth, concentrically and radially slightly wrinkled, greyish
blue with a broad white margin when young, turning very slowly brownish
or yellowish from the centre, lower surface with decurrent teeth 1–5 mm
long, at first bluish, then white and finally purplish brown; stalk obconical
broadening out towards the cap, 2–5 × 1–2 cm, velvety, yellowish to
orange-brown, stuffed, firmly attached to debris at base. Hyphae with
clamps. Flesh in cap showing different colour zones, e.g. pale orange brown
and greyish-blue, in stalk orange brown or slightly bluish. Spores 5–6 × 3.5–
4.5, with large flattened tubercles. In conifer woods in Scotland, uncommon.

Hydnellum compactum (Pers. ex Fr.) P. Karst.
Fruitbodies solitary or gregarious, top-shaped; cap round with crenate
edge, convex or flat, up to 10 cm diam., upper surface at first velvety, then
rough and pitted, not zonate, whitish to yellow-ochre, becoming dark
brown to almost black, lower surface with decurrent teeth up to 5 mm
long, white to ochre or purplish brown; stalk obconical, 2–4 × 1–3 cm,
concolorous with cap, solid. Flesh very firm to hard, not zoned in cap.
Spores 5.5–6 × 3.5–5, with tubercles sometimes sunken at tip. In deciduous
woods, mostly under *Fagus*.

Hydnellum concrescens (Pers. ex Schw.) Banker (Fig. 198)
Fruitbodies usually in clusters, concrescent, turbinate or shallowly funnel-shaped; cap round with wavy edge, 2–7 cm diam., upper surface very clearly concentrically zoned and radially ribbed, cream to pink and velvety at first, becoming scaly especially near the centre with tan or wine and dark or very dark brown zones, lower surface with teeth up to 3 mm long, white then rose to purplish brown; stalk up to 3 × 1 cm, velvety, rather pale brown, solid. Flesh purplish brown, darkest in stalk. Spores 5–6 × 4–4.5. In conifer and deciduous woods, often half covered by mosses and leaf litter.

Hydnellum ferrugineum (Fr. ex Fr.) P. Karst.
Fruitbodies solitary or in clusters, more or less top-shaped cap 3–10 cm diam., at first convex, pulvinate, later flattening or slightly depressed in the centre, with an uneven surface, velvety to felted, at first whitish to pink, sometimes exuding blood-red drops of fluid, becoming flesh-coloured to rather dark reddish brown, but with wavy margin remaining whitish, lower surface with teeth up to 6 mm long, white to reddish brown; stalk 1–6 × 1–3 cm, concolorous with cap. Flesh reddish or purplish brown with white flecks. Spores 5–6 × 4–5, coarsely tuberculate. Mostly in conifer woods.

Hydnellum peckii Banker apud Peck (Fig. 199)
Fruitbodies solitary or clustered, sometimes concrescent, top-shaped; cap 3–8 cm diam., at first convex then flat or depressed in the middle, edge broadly crenate, upper surface uneven, ridged or pitted, with thick scales, when young velvety and white often exuding blood-red drops, then wine-red, reddish or brownish, often very dark in the centre paling towards the margin which remains white for a long time, lower surface with teeth up to 5 mm long, white becoming purplish brown; stalk 1–6 × 1–3 cm, cylindrical or obconical, concolorous with cap or darker, solid. Flesh zoned, pale pinkish brown. Spores 5–5.5 × 3.5–4, pale brown, tuberculate. Amongst mosses and needle litter in Scottish pine woods.

Hydnellum scrobiculatum (Fr. ex Secr.) P. Karst.
Fruitbodies solitary or clustered and sometimes concrescent, roughly top-shaped; cap 2–6 cm diam., depressed in the centre, edge wavy, upper surface velvety when young, becoming radially wrinkled and rough with raised scales, white at first then pinkish brown, darkest in the middle, bruising black, teeth on lower surface 1–4 mm long, white then purplish brown; stalk 2–3 × 1–1.5 cm, sometimes swollen at base, concolorous with cap, solid. Flesh zoned, pinkish brown. Spores 5.5–7 × 4.5–5. In conifer and mixed woods.

Hydnellum spongiosipes (Peck) Pouzar
Fruitbodies solitary or clustered, sometimes concrescent; cap 2–7 cm diam., convex, flat or depressed in the centre, often with one or two concentric grooves, radially wrinkled, at first yellowish white, turning flesh-

coloured, purplish brown or cinnamon but with margin remaining whitish for a long time, teeth on lower surface up to 6 mm long; stalk 1–9 × 0.5–3 cm, swollen at base, dark reddish brown, covered with very thick, spongy, felt-like tomentum. Spores 6–7 × 4.5–5.5, some tubercles on the surface split and then appearing spine-like. In deciduous woods, often with *Quercus* but also with *Castanea* and *Fagus*.

Hydnum

Fruitbodies terrestrial, fleshy, pileate, stalked; cap convex, flat or slightly depressed in the middle, often with inrolled and sometimes wavy margin, upper surface velvety to suede, lower surface with numerous conical teeth which may be decurrent onto the upper part of the stalk; stalk cylindrical, central or excentric, solid. Hyphae often swollen, with clamps. Spores broadly ellipsoid or subspherical, hyaline, smooth, thin-walled, often with guttules, J–.

KEY

Fruitbodies large, cap yellowish to pale flesh-coloured, stalk often
excentric .... *repandum*
Fruitbodies small, orange brown, stalk central .... *rufescens*

Hydnum repandum L. ex Fr. (Fig. 200)
Hedgehog Fungus
Cap 3–17 cm diam., yellowish to pale flesh-coloured, teeth on lower surface up to 6 mm long, white or pale flesh-coloured; stalk sometimes central but often excentric, 3–7 × 1.5–4 cm, paler than cap, white at the base and, when bruised, yellow. Spores 6.5–9 × 5.5–7. In conifer and deciduous woods.

Hydnum rufescens Fr.
Cap not more than 7 cm diam., orange–brown; stalk usually central, up to 7 × 1.5 cm, white to yellowish orange. Spores 7–8.5 × 5–7. In conifer and deciduous woods.

Hymenochaete

Fruitbodies resupinate, effuso-reflexed or pileate and forming brackets, firmly attached, sometimes free at the margin, membranous, crustose or leathery, hymenium smooth or tuberculate, in some species made velvety by projecting setae. Hyphae yellow or brown, without clamps. Setae present, brown or dark brown, thick-walled, subulate, some immersed, others projecting. No cystidia. Basidia narrowly clavate, hyaline. Spores cylindrical, narrowly ellipsoid or somewhat allantoid, hyaline, thin-walled, smooth, J–.

KEY

Fruitbodies entirely resupinate, effused, not turned up at the edge, often crustose 1

Fruitbodies largely resupinate but turned up at the edge, leathery, sometimes also forming shelf-like brackets on vertical surfaces 2

Fruitbodies mainly pileate, forming brackets 2-4 cm wide ... *rubiginosa*

1. Hymenium cinnamon to rust .... *cinnamomea*

    Hymenium rather pale grey or brownish tinged lilac, cracked to form small polygonal islands .... *corrugata*

    Hymenium dark chocolate brown to sepia, sometimes with deep cracks delimiting square or oblong islands .... *fuliginosa*

2. Hymenium bright red, ageing brownish red .... *mougeotii*

    Hymenium tobacco brown with yellow or whitish margin ..... *tabacina*

Hymenochaete cinnamomea (Pers.) Bres.
Fruitbodies resupinate, effused, up to 1 mm thick, older ones stratified, membranous to crustose, hymenium cinnamon to rust or rusty cinnamon, smooth to rather velvety. Setae 50–130 × 6–8, some immersed, some projecting. Spores oblong-ellipsoid, 5–7.5 × 2–3. On fallen and standing dead trunks and branches of deciduous trees, especially *Corylus*.

Hymenochaete corrugata (Fr.) Lév. (Fig. 201)
Fruitbodies resupinate, effused, crustose, 0.1–0.2 mm thick, cracked to form polygonal islands each 0.5–3 mm diam., hymenium smooth or tuberculate, pale grey or brownish tinged lilac. Setae 50–80 × 6–12, some projecting, some immersed, tips rounded, pale, usually encrusted. Spores 4–5 × 1.5–2. On fallen and standing dead trunks and branches of deciduous trees, especially *Corylus*.

Hymenochaete fuliginosa (Pers.) Bres. (Fig. 202)
Fruitbodies resupinate, effused, 0.5–1 mm thick, not stratified, sometimes with deep cracks delimiting square or oblong islands, hymenium dark chocolate brown to sepia, mostly smooth. Setae 40–80 × 6–8. Spores 5–7 × 2–3. On wood and bark of conifers, especially *Picea*.

Hymenochaete mougeotii (Fr.) Cooke (Fig. 203)
Fruitbodies resupinate, at first small, roughly orbicular, becoming confluent to form larger patches, leathery, up to 0.5 mm thick, with edge turned up 1–4 mm, hymenium bright red when young, becoming somewhat brownish red with age, smooth to tuberculate. Setae 40–80 × 6–9. Spores cylindrical, 6–9 × 2–3. On bark of dead, attached or fallen branches of *Abies*.

Hymenochaete rubiginosa (Dicks. ex Fr.) Lév. (Fig. 204)
Fruitbodies tough and leathery becoming hard, occasionally effuso-reflexed but mainly pileate, brackets sessile or shortly stalked, 2.5–6 × 2–4 cm, 0.5–

1.5 mm thick, imbricate, often oblique or joined laterally, margin wavy or lobed, upper surface velvety then smooth, narrowly concentrically zoned, dark rusty brown to date brown or almost black, lower surface brown with a rust brown border or rust brown all over. Setae 40–70 × 5–9. Spores 4.5–6 × 2.5–3, ellipsoid, often yellowish. Mostly on rotting stumps of *Quercus.*

Hymenochaete tabacina (Schw. ex Fr.) Lév. (Fig. 205)
Fruitbodies leathery, effuso-reflexed, on horizontal surfaces resupinate, at first orbicular, then forming larger patches by confluence, hymenium tobacco brown or greyish brown, tuberculate or with concentric low ridges, margin yellow or whitish, wavy, turned up; on vertical surfaces forming shelf-like, imbricate brackets about 1 cm wide with upper surface orange–brown or greyish brown and somewhat zoned. Setae 70–100 × 7–12. Spores 5–7 × 1.5–2. On dead wood of shrubs and trees, *Salix, Corylus* etc.

Hyphoderma

Fruitbodies resupinate, firmly attached, occasionally orbicular and discrete on bark but mostly effused, membranous, when fresh waxy and soft, mostly white, grey, cream or ochraceous, surface smooth, tuberculate or toothed. Hyphae with clamps except in one species. Cystidia present, sometimes of more than one kind. Basidia fairly large, often constricted, usually containing oil drops; sterigmata often rather long and curved. Spores hyaline, thin-walled, smooth, J−, usually with oil drops or granules. On wood and bark.

KEY

	Cystidia with numerous transverse septa and clamps .... *setigerum*	
	Cystidia not so 1	
1.	Hymenium toothed, teeth 1–5 mm long .... *radula*	
	Hymenium smooth or tuberculate 2	
2.	Cystidia thick-walled, encrusted, fusiform to conical .... *puberum*	
	Cystidia not so 3	
3.	Spores distinctly allantoid, 6–8 × 1.5–2.5 .... *macedonicum*	
	Spores always wider and either not curved or only slightly so 4	
4.	Projecting, long, distinctly capitate cystidia present 5	
	No projecting capitate cystidia 6	
5.	Spores 8–12 × 7–9, hyphae without clamps .... *capitatum*	
	Spores 7–10 × 5–6, hyphae with clamps .... *orphanellum*	
6.	Long projecting cystidia subulate, mostly distinctly pointed at apex .... *pallidum*	
	Projecting cystidia not pointed at apex 7	
7.	Projecting cystidia with swollen base and more or less cylindrical neck often encrusted with lumps of reddish or yellowish matter .... *argillaceum*	

Projecting cystidia not so 8
8. Three kinds of cystidia present, including bladder-like structures each
   with a subequatorial frill of small teeth .... *practermissum*
   Cystidia of one kind, all more or less cylindrical 9
9. Spores 9–12 × 3–4, fruitbodies often with scattered pink patches
   .... *roseocremeum*
   Spores 7–10 × 5–6.5, fruitbodies without pink patches .... *obtusum*

Hyphoderma argillaceum (Bres.) Donk (Fig. 206)
Fruitbodies effused, thin, greyish-white, waxy and soft when fresh and
damp, becoming pale ochraceous, membranous, porose to reticulate under
a lens. Cystidia thin-walled, of two kinds: (1) abundant, up to 200 × 10–20,
projecting a long way, subulate, much swollen at base, upper part more or
less cylindrical, rounded at the apex, often encrusted with lumps of yellowish or reddish matter, (2) only a few, 20–40 × 5–8, capitate and also sometimes encrusted. Spores ellipsoid to subspherical, with oblique basal
apiculus, 6–9 × 4–5.5, packed with small guttules or granules. On wet decaying logs and branches.

Hyphoderma capitatum Erikss. and Strid. (Fig. 207)
Fruitbodies effused, very thin, white or greyish white, drying pale yellowish, reticulate to porose. Hyphae without clamps. Cystidia thin-walled,
projecting, capitate, up to 120 long, 8–14 wide at base, tapering to 4–5,
head 6–8 diam., sometimes encrusted. Basidia 40–50 × 10–12, often
containing oil drops. Spores subspherical, apiculate, 8–12 × 7–9. On decaying wood of conifers.

Hyphoderma macedonicum (Litsch.) Donk (Fig. 208)
Fruitbodies effused, membranous, thin, soft, white and waxy when moist,
drying ochraceous. Hyphae excreting lumps of resinous matter and often
also crystals. Cystidia mostly projecting, subulate, rounded at apex, up to
120 × 6–9, often encrusted with lumps of resinous matter. Short capitate
cystidia present also occasionally. Spores allantoid, 6–8 × 1.5–2.5. On decaying wood of deciduous trees.

Hyphoderma obtusum J. Erikss. (Fig. 209)
Fruitbodies effused, thin, membranous to waxy, watery grey when moist,
drying white or cream, sometimes with pinkish tint. Cystidia cylindrical to
subclavate, sometimes swollen near base, projecting a little way, 50–80 × 8–
12, occasionally with excreted hyaline globule surrounding apex. Spores
broadly ellipsoid, 7–10 × 5–6.5. On coniferous wood only.

Hyphoderma orphanellum (Bourd. and Galz.) Donk (Fig. 210)
Fruitbodies effused, very thin, grey or greyish, waxy to membranous,
smooth. Cystidia numerous, projecting, capitate, 50–80 long, 8–14 wide
near base, tapered to 4–5 below the swollen head which is often encrusted.

Unbranched hyphidia present, their ends coated with crystals. Spores broadly ellipsoid, 7–10 × 5–6, containing oil drops. On rotten wood.

Hyphoderma pallidum (Bres.) Donk (Fig. 211)
Fruitbodies effused, thin, waxy to membranous, whitish to ochraceous or pale reddish brown. Cystidia of three kinds: (1) projecting, subulate to fusiform, with basal clamps, mostly pointed at apex, 50–80 × 6–8, (2) not projecting, capitate, heads encrusted with reddish brown matter, (3) spiny-walled swollen ends of side branches, seen only occasionally. Spores cylindrical, often slightly curved, 7–10 × 2.5–3.5. On decaying decorticated wood, mostly of conifers.

Hyphoderma praetermissum (P. Karst.) Erikss. and Strid. (Fig. 212)
Fruitbodies effused, thin, watery white and waxy when moist, drying cream or ochraceous. Cystidia of three kinds: (1) projecting, cylindrical rounded at apex or capitate, 20–80 × 6–10, sometimes encrusted, (2) embedded gloeocystidia 60–90 × 8–12, with yellowish contents, mostly fusiform, (3) bladder-like structures sometimes called stephanocysts 10–12 wide, each with a subequatorial frill of small teeth. Spores ellipsoid, often flattened or slightly concave on one side, 8–11 × 4–5. On rotten wood and bark.

Hyphoderma puberum (Fr.) Wallr. (Fig. 213)
Fruitbodies effused, up to 0.3 mm thick, waxy, white to ochraceous or clay-coloured. Cystidia (m) abundant, projecting and immersed, mostly fusiform to conical, pointed, thick-walled and encrusted, 70–150 × 12–18. A few immersed gloeocystidia have been seen in a few collections but not all. Spores cylindric-ellipsoid, packed with granules, 6–10 × 3.5–5. On decaying decorticated wood, mostly of deciduous trees.

Hyphoderma radula (Fr.) Donk (Fig. 214)
Colonies on bark orbicular, discrete or confluent, on decorticated wood effused, white at first, then cream to yellow–ochre, with fimbriate margin, hymenium toothed, teeth 1–5 mm long. Cystidia immersed, thin-walled, flexuous to moniliform, 40–60 × 5–7, often only a few present. Spores allantoid, 9–11 × 3–3.5. Common on wood and bark, mostly of deciduous trees.

Hyphoderma roseocremeum (Bres.) Donk (Fig. 215)
Fruitbodies effused, somewhat waxy, white to cream or ochraceous, often with scattered pinkish patches. Cystidia cylindrical, often sinuous, thin-walled, smooth, 60–130 × 6–8. Spores cylindric-ellipsoid, often slightly concave on one side, 9–12 × 3–4. On wood, mostly of deciduous trees.

Hyphoderma setigerum (Fr.) Donk (Fig. 216)
Fruitbodies orbicular and often confluent on bark, effused on wood, waxy to membranous, soft, up to 5 mm thick, surface smooth, tuberculate or with teeth, white at first, becoming ochraceous and cracked when old.

Cystidia usually numerous, projecting, cylindrical or slightly tapered, up to
$200 \times 10\text{–}14$, with many transverse septa and clamps, sometimes encrusted
with crystals. Spores $7\text{–}14 \times 3\text{–}5$, some slightly curved. Common on wood
and bark mostly of deciduous trees.

Hyphodermella *corrugata* (Fr.) Erikss. and Ryv. (Fig. 217)
Fruitbodies resupinate, effused, firmly attached to substrate, waxy, thickening and becoming crustose, white to cream and finally ochraceous, irregularly toothed, teeth short, conical, terminating in sterile tufts of encrusted
hyphae which are stuck together. Hyphae without clamps. No rhizo-
morphs. Basidia clavate, stalked. Spores broadly ellipsoid, $7\text{–}11 \times 4\text{–}7$,
hyaline, smooth, thin-walled, J−, each with usually only one vacuole or oil
drop. On lower surface of fallen dead trunks and branches, mostly of
deciduous trees.

Hypochnella *violacea* (Auersw.) Schroet. (Fig. 218)
Fruitbodies resupinate, effused, thin, dark violet, smooth, soft, mem-
branous. Hyphae without clamps, mostly branched at right angles, some
encrusted, basal ones brownish, smooth. No cystidia. Basidia 4-spored.
Spores thick-walled, violet, smooth, ellipsoid basally or obliquely apiculate,
J+, $6\text{–}8 \times 4\text{–}5$. On decaying wood of deciduous trees.

Hypochniciellum
Fruitbodies resupinate, effused, loosely attached, arachnoid or membranous, soft, white or cream, drying pale ochraceous, hymenium smooth
to slightly merulioid. Hyphae with clamps. Cystidia few or none. Spores
ellipsoid, yellowish, with fairly thick walls, smooth, J− or weakly J+.

KEY

Spores $5\text{–}7 \times 3\text{–}3.5$, cystidia present but few .... *molle*
Spores $3.5\text{–}4.5 \times 2.5\text{–}3$, cystidia absent .... *ovoideum*

Hypochniciellum molle (Fr.) Hjortst (Fig. 219)
Fruitbodies thin, cottony, surface uneven, somewhat merulioid when fresh.
Cystidia present but often only a few, cylindrical, flexuous, hyaline, with
fairly thick walls, $70\text{–}150 \times 6\text{–}10$. Spores $5\text{–}7 \times 3\text{–}3.5$, with granular
contents, weakly J+, walls collapsing when old. On fallen, decaying conifer
wood and on worked timber.

Hypochniciellum ovoideum (Jülich) Hjortst and Ryv.
Fruitbodies with smooth or slightly tuberculate surface. No cystidia. Spores
$3.5\text{–}4.5 \times 2.5\text{–}3$. On debris, especially where piles of branches are lying on
the ground.

Hypochnicium

Fruitbodies resupinate, effused, firmly attached, softly waxy when fresh or membranous, smooth, tuberculate or with small teeth, usually white or cream. Hyphae with clamps. Basidia narrowly clavate, 4-spored, mostly with basal clamps. Cystidia, when present, rather thin walled, smooth, hyaline or yellowish. Spores ellipsoid to spherical, hyaline or yellowish, with thick walls, smooth, verruculose or echinulate, J−, C+.

KEY

	No cystidia	1
	Cystidia present	2
1.	Spores 9–11 × 6–8	*bombycinum*
	Spores 6–7 × 5–5.5	*lundellii*
2.	Cystidia containing numerous oil drops, mostly immersed ...	*analogum*
	Cystidia without oil drops, often some of them protruding	3
3.	Spores with verrucose walls	*punctulatum*
	Spores with smooth walls	4
4.	Spores oblong-ellipsoid, 6–8 × 4–5.5	*geogenium*
	Spores spherical or subspherical, 4–7 diam.	*sphaerosporum*

Hypochnicium analogum (Bourd. and Galz.) J. Erikss. (Fig. 220)
Fruitbodies somewhat waxy, thin, with fruity smell. Cystidia (g) mostly immersed, numerous, cylindrical to fusiform, 190–200 × 6–10, containing many large oil drops. Spores ellipsoid, apiculate, 7.5–10 × 6–8, with verrucose walls. On wood and bark of deciduous trees and occasionally also on conifers.

Hypochnicium bombycinum (Sommerf. ex Fr.) J. Erikss. (Fig. 221)
Fruitbodies often very widely effused, thin or thick, with hyphae sometimes projecting from surface but no cystidia. Basidia containing oil drops. Spores broadly ellipsoid to ovoid or pyriform, smooth, 9–11 × 6–8, sometimes yellowish, containing oil drops or granules. On fallen and still standing trunks and branches, mostly of deciduous trees.

Hypochnicium geogenium (Bres.) J. Erikss. (Fig. 222)
Fruitbodies 0.1–0.3 mm thick, tuberculate or short-toothed, white and waxy when fresh, becoming yellowish white and cracked as they dry out. Cystidia numerous, cylindrical or slightly tapered towards each end, thin-walled, yellowish, 90–150 × 6–8, projecting well above surface of hymenium. Spores oblong-ellipsoid, smooth, 6–8 × 4–5.5, each containing 1–2 oil drops. On decayed wood.

Hypochnicium lundellii (Bourd.) J. Erikss. (Fig. 223)
Fruitbodies thin, slightly bluish white when wet. No cystidia. Spores broadly ellipsoid, 6–7 × 5–5.5, hyaline to pale yellowish, each with one or several oil drops. On fallen branches, twigs and other debris.

Hypochnicium punctulatum (Cooke) J. Erikss. (Fig. 224)
Fruitbodies white, thin. Cystidia numerous, mostly projecting only a short distance, tapering gradually towards rounded apex, thin-walled, occasionally with 1–2 septa, 80–150 × 7–10, slightly yellowish. Spores broadly ellipsoid to subspherical, verruculose to echinulate, 5.5–7 × 4.5–5. Mostly on decaying conifer wood.

Hypochnicium sphaerosporum (v. Höhnel and Litsch.) J. Erikss. (Fig. 225)
Fruitbodies thin, white, somewhat waxy. Cystidia numerous, mostly immersed, cylindrical or tapered gradually towards the rounded apex, 60–150 × 6–12. Spores spherical or subspherical, smooth, 4–7 diam. On decaying wood, mostly of deciduous trees.

Hypochnopsis *mustialaensis* (P. Karst.) P. Karst. (Fig. 226)
Fruitbodies resupinate, effused, very thin and soft, with loose white subiculum, hymenium smooth to minutely tuberculate, sulphur yellow, yellowish green or ochraceous. Hyphae with large clamps. No cystidia. Basidia clavate, mostly 4-spored. Spores ellipsoid, obliquely apiculate, 4–6 × 3.5–4.5, with smooth, rather thick walls, yellowish olive, turning deep violet in KOH solution, J+. On rotten decorticated wood.

Inonotus
Fruitbodies in some species resupinate or effuso-reflexed, but in most pileate forming solitary or imbricate brackets with velvety or smooth upper surface without a thick, hard crust; flesh brown, soft when young, drying hard and brittle, tubes not stratified, whole fungus turning black with KOH. No cystidia. Thick-walled, pointed, brown setae in hymenium and sometimes also in trama or in the upper surface of the cap. Spores hyaline to golden brown, mostly ellipsoid, smooth, J–.

KEY

Fruitbodies mostly bracket-like 1
Fruitbodies resupinate 4
1. Anchor-like brown setae with hooks in upper surface tomentum of
   brackets .... *cuticularis*
   Brackets without such setae 2
2. Brackets small, 1.5–3 × 0.5–2 cm, numerous, with decurrent pore layer
   covering surface of bark between them .... *nodulosus*
   Brackets medium-sized, 6–10 × 2–5 cm, almost always on *Alnus*
   .... *radiatus*
   Brackets large, 10–60 × 5–25 3
3. Upper surface of brackets roughly hairy, felted or bristly, mostly on
   Malus, Fraxinus and *Juglans* .... *hispidus*

Upper surface only finely tomentose, exuding yellowish or brownish
   drops along margin, mostly at base of old trees of *Quercus* .. *dryadeus*
4.   Spores 4–6 × 3–4.5, no distinct conidial state .... *hastifer*
   Spores 7–10 × 3.5–6; conidial state parasitic, especially on *Betula* and
   Ulmus, forming large black, pulvinate tumours .... *obliquus*

Inonotus cuticularis (Bull. ex Fr) P. Karst. (Fig. 227)
Fruitbodies sessile brackets, sometimes solitary but mostly imbricate,
roughly fan-shaped, 5–22 × 3–12 cm, 0.5–3 cm thick, upper surface convex
to flat, margin sharp and curved over, rather pale yellowish brown and
tomentose at first, turning rusty to blackish brown, in part radiately striate
and shaggy from agglutinated hairs, finally almost smooth, lower surface
pale yellowish, almost iridescent when young, then brownish, pores round
or polygonal, 2–4 per mm, flesh reddish or yellowish brown. Setae in
tomentum anchor-shaped, up to 250 × 5–15, with several recurved hooks,
in hymenium conical, sharply pointed, 15–30 × 4–12. Spores pale rusty
brown, 5.5–8 × 4–6. On deciduous trees, especially *Fagus* and *Quercus*.

Inonotus dryadeus (Fr.) Murr. (Fig. 228)
Fruitbodies sessile, solitary or in groups and sometimes confluent, roughly
semicircular brackets 10–60 × 5–25 cm, up to 12 cm thick, or occasionally
large, irregular cushions, upper surface uneven, rather lumpy and pitted,
finely tomentose, cream, whitish or pale grey when young and exuding
yellowish or brownish drops especially along the broadly rounded edge,
becoming rusty or dark brown and finally blackish with a thin crust when
old, lower surface off-white to yellowish brown, darkening with age, pores
round or angular, 3–5 per mm, flesh yellowish to reddish brown. Setae in
hymenium dark brown, swollen at base, often hooked and sharply pointed
at apex, 8–40 × 6–15. Spores pale yellowish, thick-walled, 7–9 × 6.5–8. On
Quercus, mostly at base of old trees; has an unpleasant smell.

Inonotus hastifer Pouz. (Fig. 229)
Fruitbodies resupinate, occasionally nodulose, firmly attached, effused but
not usually more than 10 cm across and 4 mm thick, pale greyish brown
when young, turning cinnamon to rusty brown, often with a silvery sheen,
border pale yellowish, pores rounded to angular, 3–4 per mm. Setae in
trama up to 250 × 8–15, in hymenium few, straight or curved, 12–30 × 5–10.
Spores yellowish brown, thick-walled, 4–6 × 3–4.5. On standing trunks of
deciduous trees, especially *Fagus*.

Inonotus hispidus (Bull. ex Fr.) P. Karst. (Fig. 230)
Fruitbodies mostly solitary brackets, rarely imbricate, sessile, roughly semicircular, up to 30 × 25 cm and 3–10 cm thick, triquetrous, upper surface
convex or flattened, uneven, at first rusty brown, roughly hairy, felted or
bristly, becoming blackish and smooth when old, margin pale yellowish,
lower surface pale ochraceous becoming brown, pores angular, 2–3 per

mm, flesh rusty brown when cut. Setae in hymenium often sparse, or even occasionally absent, subulate to conical, sharply pointed, 15–30×7–11. Spores yellowish to rusty brown, thick-walled, 8–12×7–9. On deciduous trees, especially *Malus*, *Fraxinus* and *Juglans*; has a pleasant smell.

Inonotus nodulosus (Fr.) P. Karst. (Fig. 231)
Fruitbodies tough, corky, composed of a large number of imbricate and sometimes also laterally fused, small brackets, 1.5–3×0.5–2 cm, and knobby protrusions, linked together by a decurrent pore layer which covers the surface of the bark between them, upper surface faintly zonate and radiately striate, velvety and yellowish at first, becoming brownish and smooth, pore layer cream drying cinnamon, pores angular to irregular and lacerate, 3–4 per mm. Setae few in trauma and there up to 70 long, often abundant in hymenium, straight, conical or subulate, sharply pointed, 15–30×5–10. Spores hyaline to pale yellowish, 4.5–6×3.5–4.5. On *Fagus*.

Inonotus obliquus (Pers. ex Fr.) Pilát
Fruitbodies resupinate, developing under the bark which eventually splits open, effused, extending sometimes for more than 1 m, 5–10 mm thick, at first white, becoming brown or dark brown, pores 3–5 per mm, angular to oblong, tubes oblique. Hymenial setae up to 45×5–10. Spores hyaline to pale yellowish, 8–10×5–7.5. Conidial state parasitic, forming conspicuous, irregular, black, pulvinate, cracked tumours up to 20×10 cm, which remain on trunks until the trees die. Setae reddish brown, up to 100×5–10. Conidia ovoid, sometimes 1-septate, olivaceous brown, smooth, 7–10×3.5–6. On deciduous trees, mostly *Betula* and *Ulmus*.

Inonotus radiatus (Sow.ex Fr.) P. Karst. (Fig. 232)
Fruitbodies imbricate, sometimes laterally fused, roughly semicircular brackets, 6–10×2–5 cm, 0.5–2 cm thick, with pore layer decurrent onto the bark, upper surface finely wrinkled radially, at first pale yellowish brown and often remaining this colour at the edge, velvety, soon turning rusty brown, finally almost black and smooth, lower surface whitish to pale brown, pores 2–4 per mm, shining and silvery when young, flesh shiny, zoned, radially fibrillose. Setae in trama occasional, straight or slightly curved, 50–80×4–8, in hymenium often abundant, conical, straight, curved or hooked, 15–30×6–15. Spores hyaline or pale yellowish, 4.5–6×3.5–4.5. On standing dead trunks, most commonly of *Alnus*.

Irpex lacteus (Fr.) Fr. (Fig. 233)
Fruitbodies tough, leathery, mostly pileate or effuso-reflexed, rarely resupinate, often imbricate; brackets narrow, 2–5×0.5–2 cm, usually in rows and often fused laterally; upper surface velvety or hairy, sometimes weakly zonate, white, creamy or greyish, margin sharp, deflexed when dry; lower surface toothed but often also with some angular pores along the edge, cream, teeth round or flattened, up to 5 mm long. Hyphae without

clamps, thick-walled to almost solid skeletal ones 5–9 wide. Cystidia arising from skeletal hyphae, projecting, 50–150 × 4–9, upper part thick-walled, encrusted. Basidia 4-spored. Spores cylindric-ellipsoid, 5–6.5 × 2–3, hyaline, smooth, thin-walled, J–. On dead wood of deciduous trees.

***Irpicodon** pendulus* (Alb. and Schw. ex Fr.) Pouzar (Fig. 234)
Fruitbodies pileate, thin, soft, 1–3 cm diam., shell- or fan-shaped, sometimes lobed, often narrowed into a short stalk, pendulous, sometimes imbricate, upper surface white to creamy ochraceous, when old greyish, at first with adpressed hairs but soon smooth or radially wrinkled, lower surface with irregular rows of shining, white, mostly flattened teeth about 2 mm long or sometimes partly lamellate. Hyphae with clamps. No cystidia. Basidia narrowly clavate, 4-spored. Spores narrowly ellipsoid to allantoid, 4.5–5 × 2–2.5, hyaline, smooth, J+ . On dead and dying trunks of *Pinus sylvestris*, winter.

Ischnoderma

Fruitbodies pileate, fleshy when fresh, hard when dry, sessile, semicircular or fan-shaped brackets, upper surface tomentose when young, becoming smooth, often zonate or with ridges, radially wrinkled when old, lower surface whitish or pale ochraceous, bruising brown, pores round to angular, 4–6 per mm. Hyphae tortuous, thick-walled, with scattered large clamps. No cystidia. Spores cylindrical, slightly curved, hyaline, thin-walled, smooth, J–, 5–6 × 2–2.5.

KEY

Fruitbodies at first dark reddish brown, then with black resinous crust, flesh and tube layers of about equal thickness, mostly on conifers .... *benzoinum*
Fruitbodies ochraceous to ochraceous brown, flesh about twice as thick as tube layer, on deciduous trees .... *resinosum*

Ischnoderma benzoinum (Wahlenb.) P. Karst. (Fig. 235)
Fruitbodies solitary or imbricate, mostly up to 18 × 14 cm, 2 cm thick, upper surface dark reddish brown, with resinous black zones, becoming blackish brown with hard resinous crust, margin pale, flesh and tube layers of about equal thickness, ochraceous brown. Mostly on conifers.

Ischnoderma resinosum (Schrad. ex Fr.) P. Karst.
Fruitbodies mostly solitary, rarely imbricate, up to 15 cm each way, 3 cm thick, upper surface zonate, ochraceous to ochraceous brown, blackish only when quite old, flesh about twice as thick as tube layer, whitish or pale wood colour. On deciduous trees, especially *Fagus*.

Jaapia argillacea Bres. (Fig. 236)

Fruitbodies resupinate, effused, membranous, smooth, soft, whitish drying clay-coloured, projecting cystidia visible under low power dissecting microscope. Hyphae with clamps. Cystidia numerous, almost cylindrical but tapering gradually towards rounded apex, flexuous, hyaline, smooth, 50–200 × 5–8. Basidia clavate, up to 60 × 6–9, 4-spored. Spores fusiform, smooth, 16–25 × 5–7, J–, two ends remaining thin-walled, hyaline and tending to collapse, the main body yellowish, containing oil drops surrounded by a thick wall which stains deeply with cotton blue. Mostly on water-soaked wood by the sides of streams etc.

Junghuhnia nitida (Fr.) Ryv. (Fig. 237)

Fruitbodies resupinate, effused, firmly attached except when old at the edge, 2–3 mm thick, pinkish ochraceous with a pale yellowish or whitish margin; pores round or angular, 5–7 per mm, or sometimes elongated and larger. Hyphae with clamps. Cystidia numerous, arising from skeletal hyphae, projecting slightly from hymenium, up to 200 long, upper half thick-walled, heavily encrusted, 8–12 wide, tapered towards base which bears a clamp. Spores ellipsoid, 4–5 × 2–3, hyaline, smooth J–. On rotten wood, mostly of deciduous trees.

Lachnella

Fruitbodies mostly gregarious, bowl-, shallow cup- or bell-shaped, sessile or with short stalks, the outside covered with long, white, thick-walled, granulate encrusted hairs with dextrinoid walls, the inside lined by smooth hymenium. Hyphae of trama not gelatinized. Cystidia none. Basidia large, clavate, 2–4-spored, often with basal clamps. Spores hyaline, white in mass, smooth, J–.

KEY

Spores 6–8 × 2.5–4 .... *eruciformis*
Spores much larger, always more than 8 long 1
1. Hymenium creamy white, spores 9–14 long .... *villosa*
   Hymenium pale grey, tinged with blue or violet, spores 14–16 long
   .... *alboviolascens*

Lachnella alboviolascens (Alb. and Schw. ex Fr.) Fr. (Fig. 238)

Fruitbodies gregarious, saucer- or shallow cup-shaped, 0.5–1.5 mm diam., without a stalk, attached at one point to the substrate, margin inrolled, very much so when dry, outer surface white due to dense covering of hairs, hymenium lining inner surface pale grey tinged with blue or violet. Hairs cylindrical tapered to a point at the apex, pale brown near base, the upper part hyaline, 200–250 × 5–6. Basidia 60–70 × 10–13. Spores 14–16 × 9–12. On dead branches and woody herbaceous stems.

Lachnella eruciformis (Fr.) W.B. Cooke
Fruitbodies solitary or gregarious, bowl- or bell-shaped, 1–1.5 × 1 mm, covered on the outside with bluish white hairs, hymenium brownish. Hairs 2.5–5 thick, sometimes smooth near tip. Spores tending to be asymmetric, ovoid or allantoid, 6–8 × 2.5–4. On twigs of *Populus*.

Lachnella villosa (Pers. ex Fr.) Gillet (Fig. 239)
Fruitbodies gregarious, sessile or short-stalked, 1–2 mm diam., margin inrolled, outer surface white, covered with adpressed hairs, hymenium lining inner surface creamy white. Hairs cylindrical, tapered to a point, hyaline, 150–200 × 5–8. Basidia 40–60 × 10–11. Spores ellipsoid, 9–14 × 6–9. Common on dead herbaceous stems.

***Laetiporus** sulphureus* (Bull. ex Fr.) Murr. (Fig. 240)
Sulphur Polypore
Fruitbodies pileate, soft-fleshed, usually imbricate, very large, sometimes as much as 50 cm wide, 1–5 cm thick, rather flat or undulating, semicircular or fan-shaped, sulphur yellow to yellowish orange brackets; upper surface suede-like; lower surface yellow; pores round to oblong, 1–4 per mm. Hyphae without clamps, some of them coralloid with numerous short branches. No cystidia. Basidia 4-spored, 10–17 × 5–7. Spores ellipsoid, 5–7.5 × 3.5–5, hyaline, thin-walled, smooth, J–. A similarly coloured chlamydospore state occurs with irregularly tuberous or bracket-like fructifications 3–7 cm wide, 2–5 cm thick, containing, when mature, cinnamon brown powdery masses of thick-walled spores 8–20 × 6–15. On standing, living and dead deciduous trees; especially common on *Quercus*.

***Laetisaria** fuciformis* (McAlpine) Burdsall (Fig. 241)
Fruitbodies resupinate, small, thin, membranous, pinkish when fresh, becoming ochraceous, often with scattered, sterile, pinkish orange, clavarioid or *Isaria*-like bundles of hyphae. Hyphae without clamps. No cystidia but some small, slender hyphidia present. Basidia 40–50 × 6–8, somewhat urn-shaped. Spores ellipsoid, 9–12 × 5–7, hyaline, smooth, thin-walled, J–. On turf grasses, mostly *Festuca* and *Lolium* species, causing red-thread or pink-patch disease.

***Lagarobasidium** detriticum* (Bourd. and Galz.) Jülich (Fig. 242)
Fruitbodies resupinate, effused, loosely attached to the substrate, white, membranous, smooth at first but later with short, conical teeth, without rhizomorphs. Hyphae branched, often at right-angles, from or opposite clamps, frequently with stellate crystals between them. Cystidia clavate or spathulate, rather variable, hyaline, smooth, protruding, 75–100 long, 8–10 thick in the broadest part, with basal clamps. Basidia 4-spored, 16–20 × 4–7. Spores ellipsoid, 4.5–6 × 4–4.5, hyaline, thick-walled, each with one large guttule or oil drop, J–, C+. On decayed wood on the ground, also on debris of dead herbaceous plants and ferns.

Laxitextum *bicolor* (Pers. ex Fr.) Lenz (Fig. 243)
Fruitbodies soft leathery, sometimes imbricate, mostly effuso-reflexed, the brown reflexed part contrasting strongly with the white or creamy hymenium, *Stereum*-like but with very different structure, 1–2 mm thick, firmly attached to substrate; upper surface snuff-brown with adpressed hairs, faintly zoned, hymenium smooth, white to cream or pale ochraceous and cracking when old. Hyphae some hyaline and some brown, with clamps. Cystidia (g) projecting slightly, 50–120 × 6–10, cylindrical or fusiform, often with small terminal bulb, smooth, thin-walled, contents oily, yellowish. Spores oblong ellipsoid, minutely echinulate or verruculose, hyaline, thin-walled, 4–5 × 2–3, J+. On decaying wood of deciduous trees in damp places.

Lazulinospora *cyanea* (Wakef.) Burdsall and Larsen (Fig. 244)
Fruitbodies resupinate, effused, thin, arachnoid or felted, pale to dark blue, sometimes with tinge of green or yellow. Hyphae without clamps. No cystidia. Basidia hyaline, narrowly clavate, 20–40 × 4.5–6. Spores irregularly ellipsoid, depressed on one side, 5–8 × 3–4, blue-green when fresh, clearly blue in KOH, walls verruculose. On conifer wood.

Leccinum

Fruitbodies fleshy, large, each composed of a convex or somewhat flattened cap and a central stalk; cap surface often cracking, the cuticle in a number of species overhanging the edge beyond the tubes to form a short, pendent skirt; pore surface in most species whitish or pale buff, yellow only in *L. crocipodium*; stalk for the most part covered with scales made up of cystidia. Cystidia few in tubes. Spores cylindrical to fusiform, mostly more than 15 long, in mass olivaceous yellow to olivaceous brown.

KEY

Grows under *Carpinus* and *Corylus* .... *carpini*
Grows under *Picea* in mixed woods .... *oxydabile*
Grows under *Pinus sylvestris* amongst *Vaccinium* .... *vulpinum*
Grows under *Salix* .... *salicicola*
Grows under *Populus tremula* 1
Grows mainly under *Quercus* rarely under *Fagus* and *Tilia* 2
Grows under *Betula* 3
1. Cap upper surface orange or rusty orange .... *aurantiacum*
    Cap upper surface greyish brown .... *duriusuculum*
2. Pore surface lemon to chrome yellow .... *crocipodium*
    Pore surface white or off-white .... *quercinum*
3. Pore surface especially when young grey or hazel, cap upper surface
        orange–tawny to burnt sienna .... *versipelle*

     Pore surface always white or pale, cap upper surface other colours
.... 4

4. Cap upper surface always very pale or with bright colours 5
    Cap upper surface dark or with rather dull colours 6
5. Cap upper surface off-white, cream or very pale buff .... *holopus*
    Cap upper surface whitish becoming pale salmon pink or coral,
    blotched when old .... *roseotinctum*
6. Flesh when cut usually remaining white, cap upper surface hazel to
    greyish brown .... *scabrum*
    Flesh when cut reddening, cap upper surface often becoming dark
    brown or blackish brown to almost black 7
7. Cut flesh in stem base bluish green, cap cuticle composed of vertical
    chains of short brown cells .... *variicolor*
    Cut flesh of stem base not bluish green, cells of cap cuticle long
.... *roseofractum*

Leccinum aurantiacum (Bull. ex StAmans) S.F. Gray
Cap 8–20 cm diam., hemispherical to convex, upper surface orange or rusty orange, minutely fibrillose or downy, irregular pendent cuticular skirt up to 3 mm long, pore surface white, bruising wine-coloured, pores minute. Stalk 10–15 × 2–5 cm, cylindrical or slightly swollen at base, white to pale brownish, scales white at first, becoming reddish brown to dark brown. Flesh white to cream, when cut turning sepia except in parts of the cap and base of stalk where it is often more wine-coloured. Spores 13–17 × 4–5. Always grows with *Populus tremula* amongst decaying leaves along grassy verges, July–October.

Leccinum carpini (R. Schulz) Moser ex Reid (Fig. 245)
Cap 3–9 cm diam., convex, upper surface rugulose, cracking, especially when old, buff, olivaceous tawny or sepia, no pendent cuticular skirt, pore surface white or pale buff, bruising blackish. Stalk 7–12 × 1 cm, but often swollen just below the middle or at the base to 2.5–3 cm, upper part buff, darkening downwards, bruising black or blackish, brown or brownish scales tending to form a network in the lower part. Flesh white to pale ochraceous, lead grey to black when cut. Spores 16–19 × 5–6. Under *Carpinus* and *Corylus*, fairly common.

Leccinum crocipodium (Letell.) Watling
Cap 4–15 cm diam., upper surface cinnamon to yellowish or olivaceous brown, somewhat tomentose, cracking, pendent cuticular skirt short, pore surface lemon to chrome yellow, bruising darker, pores very small. Stalk 5–12 × 2–3 cm, broadly fusiform or swollen at base, whitish to pale yellow, with lines of pale yellow to pale cinnamon or buff scales, bruising darker. Flesh pale yellowish, when cut turning first reddish brown then black. Spores 12–18 × 5–7. In deciduous woods under *Quercus*, late summer–autumn.

Leccinum duriusculum (Schulzer) Singer
Cap 6–14 cm diam., hemispherical to convex, upper surface greyish brown, slightly downy to smooth, with irregular pendent cuticular skirt, pore surface dingy white, bruising greyish or olivaceous brown, pores small. Stalk 8–12 × 2–4 cm, whitish near apex, becoming greyish brown towards the base, covered with rather small, brown or blackish scales. Flesh firm, white, when cut turning at first greyish lilac or pinkish then blackish, sometimes bluish green at the base. Spores 13–17 × 5–6. Grows under *Populus tremula*, summer–autumn.

Leccinum holopus (Rostk.) Watling
Cap 5–10 cm diam., convex, upper surface off-white, cream or very pale buff, with a bluish green tint when old, viscid when fresh, no cuticular skirt, pore surface white or pale buff, bruising cinnamon or buff. Stalk 7–12 × 1–3 cm, cylindrical or slightly thicker at base, white or very pale buff, floccose scales white at first then buff or cinnamon. Flesh very soft, white, remaining so when cut or turning slightly pink, cortex of stalk base blue–green. Spores 18–20 × 6–7. Grows amongst *Sphagnum*, usually under *Betula*, autumn.

Leccinum oxydabile (Singer) Singer
Cap 5–15 cm diam., convex, upper surface ochraceous to greyish brown or buff, slightly downy to smooth, stippled with irregular small scales, no cuticular skirt, pore surface white to buff, bruising darker, pores minute. Stalk 6–7 × 2–3.5 cm, cylindrical or slightly swollen towards base, white to pale buff, somewhat ridged, scales brown or greyish above middle, almost black and fibrillose below. Flesh compact, hard, white, reddening slightly in cap, lemon yellow in stalk base. Spores 16–20 × 5–7. Recorded under *Picea* in mixed woods, summer–autumn, rare.

Leccinum quercinum (Pilát) Green and Watling
Cap 5–15 cm diam., hemispherical to convex, upper surface orange brown to reddish rust brown or chestnut, smooth or fibrillose scaly especially at margin, pendent cuticular skirt present, pore surface white or off-white, bruising grey or brownish. Stalk 5–17 × 2.5–5, white or buff, bruising darker, scales whitish to rusty tawny, darkening with age. Flesh white, reddening at first when cut, then turning purplish grey. Spores 12–16 × 4–5. Mostly under *Quercus*, less frequently under *Tilia* and *Fagus*.

Leccinum roseofractum Watling
Cap 7–12 cm diam., convex, upper surface dark umber to blackish brown, greasy at first, then dry and splitting to form small scales near centre, no cuticular skirt; pore surface white or pale ochraceous. Stalk 6–9 × 2–4 cm, white, with greyish brown to black scales which form a diamond-shaped network below the middle. Flesh white, on cutting immediately turning coral then slowly wine to purplish grey. Cells of cap cuticle long, c.f. *L. variicolor*. Spores 15–18 × 5–6. Under *Betula*, autumn.

Leccinum roseotinctum Watling
Cap 7–12 cm diam., upper surface white or whitish, becoming pale salmon pink or coral when old, with rusty blotches, bruising wine to sepia, pendent cuticular skirt irregular; pore surface white or pale ochraceous, bruising brown to sepia. Stalk white, bruising brown to sepia, often bluish green near base, scales floccose, large, at first white then brownish. Flesh white, on cutting turning blackish or blackish purple, bluish green in stalk cortex. Spores 13–18 × 5–6. Under *Betula*, autumn.

Leccinum salicicola Watling
Cap 4–8 cm diam., convex, upper surface cinnamon to brick, or reddish brown, often partly cracked to form areolae; pore surface whitish or pale ochraceous. Stalk 6–8 × 1.5–2.5 cm, scales rather small, brown to black. Flesh white, on cutting turning greyish in cap, wine to blackish purple in stalk. Spores 18–20 × 5.5–6.5. Under *Salix*.

Leccinum scabrum (Bull. ex Fr.) S.F. Gray (Fig. 246)
Brown-Birch Bolete
Cap 5–15 cm diam., convex or somewhat flattened, upper surface hazel to greyish brown, smooth, sticky in wet weather, no cuticular skirt, pore surface white or off-white, bruising ochraceous. Stalk 8–20 × 2–4 cm, white, pale greyish or buff, scales brown or blackish, mostly arranged in roughly parallel lines longitudinally in the upper part. Flesh white usually unchanged, rarely very slightly pinkish when cut, soft, watery. Spores 15–20 × 4–6. Under *Betula*, very common, summer–autumn.

Leccinum variicolor Watling
Cap 5–12 cm diam., hemispherical to convex, upper surface usually rather dark greyish brown to blackish brown, with ochraceous or olivaceous spots, felted to smooth, sometimes viscid, no cuticular skirt, pore surface white or cream, wine-coloured when bruised. Stalk 9–17 × 2–3 cm, white or whitish, with greyish to black scales which tend to form a network towards the base, bruising yellowish green. Flesh becoming reddish when cut except in the base of the stalk where it is dark bluish green. Cap cuticle composed of vertical chains of short brown cells, c.f. *L. roseotinctum*. Spores 13–16 × 5–6. On wet ground under *Betula*, summer–autumn.

Leccinum versipelle (Fr. ex Hök) Snell (Fig. 247)
Cap 5–20 cm diam., convex, upper surface orange-tawny to burnt sienna, somewhat ochraceous towards the margin or when old, downy to smooth and sometimes slightly viscid, pendent cuticular skirt pronounced and irregular; pore surface, especially when young, grey to hazel. Stalk 9–20 × 2–5 cm, mostly obclavate, white to grey, often bruising bluish green at base, with brown or blackish scales. Flesh firm, white, when cut turning wine-coloured, then dark grey, bluish green in base of stalk. Spores 13–16 × 4–5. Under *Betula* in open scrub areas, summer–autumn, fairly common.

Leccinum vulpinum Watling
Cap 3–10 cm diam., upper surface dark brick red to purplish chestnut, pendent cuticular skirt irregular, pore surface whitish or pale cream. Stalk 9–14 × 1–3 cm, white or cream, bruising bluish green at base, scales at first white, turning chestnut. Flesh white, when cut not changing or turning slowly wine-coloured, bluish green in stalk base. Spores 13–15 × 3–4. Amongst *Vaccinium* and *Pinus sylvestris* in Scotland.

Lentaria
Fruitbodies erect, clavarioid, tough, up to 3 cm tall, branched, clustered, without or with ill-defined stalks, often arising from a felt-like subiculum. Hyphae with clamps and swellings, no secondary septa. No cystidia. Basidia 4-spored. Spores narrowly ellipsoid, small, hyaline, smooth, weakly J+.

KEY

Fruitbodies with short side branches, yellowish white or yellowish orange .... *corticola*
Fruitbodies repeatedly branched, remaining white or with pinkish flush .... *delicata*

Lentaria corticola (Quél.) Corner
Fruitbodies 1–3 cm tall, clustered in tufts, with short side branches, not clearly stalked, at the apex not or weakly divided, yellowish white or yellowish orange. Basidia 16–20 × 4. Spores 4–6 × 2.5–3.5. On wood of *Pinus* etc.

Lentaria delicata (Fr.) Corner (Fig. 248)
Fruitbodies 1–3 cm tall, clustered, repeatedly branched, stalk absent or very short, remaining white or becoming flushed with pink. Spores 4.5–6 × 2.5–3. On rotting wood and sawdust.

Lenzites betulina (L. ex Fr.) Fr. (Fig. 249)
Fruitbodies solitary or a few together, pileate, tough, leathery or corky, sessile, flat, hemispherical or fan-shaped; sharp-edged brackets 3–10 × 2–5 cm, 3–20 mm thick; upper surface smooth, hairy or shaggy, mostly zonate, zones whitish, pale brown or greyish brown, frequently covered with algae; lower surface white, yellowish or pale grey, hymenium lamellate, gills thin, radiate, dichotomously forked and sometimes anastomosing, especially near the margin, where there are 10–15 per cm. Spores cylindrical, often slightly curved, 4.5–6 × 2–3, hyaline, smooth, J–. On living and dead deciduous trees, especially *Betula*.

Leptoglossum
Fruitbodies 0.5–5 cm diam., cup-, spatula- or mussel-shaped, often lobed, soft, putrescent, grey, whitish or horn-coloured, attached laterally or by the back, upper or outer surface smooth, hymenium on lower or inner surface covering radiating gills in one species but in the others veins or folds which frequently anastomose, short stalk when present usually lateral. No cystidia. Spores in mass white, smooth, J–. On or associated with mosses.

KEY

Hymenium covering surface of gills .... *tremulum*
Hymenium covering folds or veins 1
1. Fruitbody an inverted, shallow cup .... *retirugum*
   Fruitbodies mussel- or spatula-shaped 2
2. Cap up to 5 cm diam., spores 9–11 × 6–7 .... *lobatum*
   Cap 1–2 cm diam., spores 6–9 × 3.5–5 .... *muscigenum*

Leptoglossum lobatum (Pers. ex Fr.) Ricken
Cap lobed, greyish brown, hymenium with veins anastomosing to form a network towards the margin. On or amongst mosses by sides of streams and in other wet places.

Leptoglossum muscigenum (Bull. ex Fr.) Karst.
Cap hygrophanous, sometimes with concentric furrows, pale greyish brown. Stem lateral, off-white, hairy, 1–4 mm long. On mosses.

Leptoglossum retirugum (Bull. ex Fr.) Ricken (Fig. 250)
Cap off-white or greyish, 0.5–1.5 cm diam. Spores subspherical, up to 10 diam. Attached to mosses.

Leptoglossum tremulum (Schaeff. ex Fr.) Sing.
Cap 1–2 cm diam., greyish brown, margin sulcate when fresh. Stalk grey, up to 4 mm long. Spores 7–8 × 5–7. On *Dicranum* and other large mosses.

Leptoporus *mollis* (Pers. ex Fr.) Pilát (Fig. 251)
Fruitbodies mostly pileate, solitary or imbricate, sometimes confluent, soft, spongy, drying hard, cushion-shaped at first, developing into semicircular, sessile brackets 2–10 × 2–4 cm, triangular in section and 2–4 cm thick, margin thin, inrolled when dry, upper surface softly tomentose, at first pink or pinkish white, later reddish or purplish brown, lower surface white or pinkish, bruising and ageing pink or violet, pores 3–4 per mm. No clamps. Spores allantoid, 5–7.5 × 1.5–2, hyaline, smooth, thin-walled, J–. On dead wood of conifers, especially *Picea*.

Leptosporomyces *galzinii* (Bourd.) Jülich (Fig. 252)
Fruitbodies resupinate, *Athelia*-like, effused, pellicular, smooth, thin, loosely attached to substrate, white or cream, often with a greenish yellow

tint, no rhizomorphs. Hyphae narrow, hyaline, smooth, with clamps. No cystidia. Basidia 4-spored, 8–12 × 3–4. Spores ellipsoid, hyaline, smooth, J–, C–, 3–4 × 1.5–2. On decaying wood, mosses and ferns.

Leucogyrophana

Fruitbodies resupinate, effused, with loose, sparsely branched hyphae in subiculum but subhymenial ones densely packed and richly branched, easily detached from substrate, surface smooth, wrinkled or folded in a merulioid manner or, in one species, with teeth. Hyphae with clamps and sometimes with crystals. No cystidia. Basidia narrowly clavate, 4-spored. Spores ellipsoid, yellowish, smooth, with rather thick walls, J–.

KEY

Fruitbodies developing teeth	....	*pinastri*
Fruitbodies not developing teeth	....	1
1.	Spores 2.5–3 wide	*sororia*
	Spores up to 5 wide	2
2.	Fruitbodies brown, with rhizomorphs	*pulverulenta*
	Fruitbodies yellowish orange to orange	*mollusca*

Leucogyrophana mollusca (Fr.) Pouz. (Fig. 253)
Fruitbodies about 1 mm thick, yellowish orange, orange or brownish orange, waxy, soft, hymenium merulioid, margin paler or whitish, cottony. Basidia 25–40 × 7–10. Spores ellipsoid, 5–7.5 × 4–5. On rotting conifer wood in which sclerotia are often found.

Leucogyrophana pinastri (Fr.) Ginns and Weresub
Fruitbodies membranous to somewhat waxy, yellow, olivaceous yellow or brownish, drying darker, with thin rhizomorphs, hymenium reticulate-poroid at first, then becoming toothed. Basidia 20–40 × 5–8. Spores 5–7 × 3.5–5. On conifer wood.

Leucogyrophana pulverulenta (Fr.) Ginns
Fruitbodies about 2 mm thick, brown, with rhizomorphs, hymenium merulioid. Basidia 25–50 × 6–8. Spores 5–8 × 3–5. On conifer wood.

Leucogyrophana sororia (Burt) Ginns
Fruitbodies pale olivaceous or pale orange, hymenium merulioid. Basidia 15–25 × 4–6. Spores 4–5 × 2.5–3. On conifer wood.

Limonomyces

Fruitbodies resupinate, waxy, smooth, rose-coloured or rosy orange. Hyphae hyaline, with clamps. Basidia hyaline, more or less urn-shaped, with ellipsoid probasidia. Spores ellipsoid, large, hyaline, smooth, thin-walled, J−. Parasitic on grasses and, occasionally, on sedges.

Spores 9–14 × 5–6 .... *roseipellis*
Spores 13–16 × 7–9.5 .... *culmigena*

Limonomyces culmigena (Webster and Reid) Stalpers and Loerakker (Fig. 254)
Fruitbodies rose-coloured. Basidia 50–200 × 7–12, 2-spored. Spores 13–16 × 7–9.5. On the uppermost nodes and internodes of culms of *Dactylis glomerata* during the winter following flowering. Recorded also on *Carex*.

Limonomyces roseipellis Stalpers and Loerakker
Fruitbodies rose to rosy orange when fresh, drying cream. Basidia 30–70 × 7–8.5. Spores 9–14 × 5–6. Parasitic on *Festuca rubra, Lolium perenne* and other grasses.

Lindtneria

Fruitbodies resupinate, effused, softly membranous or cottony, hymenium smooth, merulioid or with large irregular pores, mostly cream, yellowish or ochraceous. Hyphae with clamps. No cystidia. Basidia broad, contents staining deeply with cotton blue. Spores spherical or ellipsoid, hyaline, yellowish or pale brownish, rather thick walled, echinulate or verrucose, J−, C+.

Hymenium poroid, spores spherical, on very rotten wood .. *trachyspora*
Hymenium smooth, spores ellipsoid, on moss .... *leucobryophila*

Lindtneria leucobryophila (P. Henn.) Jülich
Fruitbodies cream to ochraceous. Basidia 4-spored, 40–55 × 7–12. Spores ellipsoid, 7–9 × 6–6.5, yellowish or pale brownish, verrucose to echinulate. On *Leucobryum* and other mosses.

Lindtneria trachyspora (Bourd and Galz.) Pilát (Fig. 255)
Fruitbodies loosely attached to substrate, 1–3 mm thick, with large, irregular pores, yellow to ochraceous, sometimes becoming slightly purplish on handling or with age. Basidia 20–35 × 9–12, containing numerous deeply staining granules. Spores spherical, 6–8 diam., yellowish or brownish, echinulate, spines 1–2 long, some connected by ridges. On very rotten wood.

Litschauerella *clematidis* (Bourd. and Galz.) Erikss and Ryv. (Fig. 256)
Fruitbodies resupinate, effused, firmly attached to substrate, smooth, thin,

waxy, greyish when fresh, drying whitish or cream; projecting cystidia seen easily under dissecting microscope. Hyphae with clamps. Cystidia numerous, conical or cylindrical tapered to a pointed apex, sometimes encircled by narrow hyphae, up to 120 × 10–20, the very thick walls encrusted with small crystals in upper part, smooth lower part stained readily by cotton blue, usually with a number of root-like projections. Basidia often with lateral outgrowth at base, 10–20 × 4–8, 4-spored. Spores subspherical, 5–10 diam., hyaline, verruculose, J–. On lower surface of living stems of *Clematis vitalba*; recorded also on conifer wood.

Loweomyces *wynnei* (Berk. and Br.) Jülich (Fig. 257)
Fruitbodies soft, becoming brittle when dry, pileate, semicircular or fan-shaped, sometimes shortly stalked brackets, imbricate or clustered, sometimes fused together, 2–4, × 1–2 cm, 1–3 mm thick, upper surface smooth or adpressed hairy, saffron yellow, yellowish orange or ochraceous, faintly zonate, margin white or pale, lower surface white or cream, pores angular, occasionally sinuous, 2–4 per mm, flesh next to pores white, the rest ochraceous or orange brown. Effuso-reflexed and resupinate bodies occur also. Hyphae with clamps. Basidia 16–20 × 5–6, 4-spored. Spores ellipsoid, 3–5 × 2.5–3.5, hyaline, smooth, thin-walled, J–. On and mixed with or growing through debris on the ground.

Luellia

Fruitbodies resupinate, effused, firmly attached to substrate, pellicular or membranous, smooth, brown, ochraceous or cream. Hyphae with clamps. No cystidia but sometimes with hyphidia between the basidia. Basidia broadly clavate, 2–4-spored. Spores navicular to fusiform, hyaline, smooth, thin-walled, J–, C–.

KEY

Fruitbodies brown, on wood of conifers .... *recondita*
Fruitbodies cream to ochraceous, on wood of deciduous trees
.... *lembospora*

Luellia lembospora (Bourd.) Jülich
Fruitbodies cream to ochraceous. No hyphidia. Basidia 12–24 × 5–9. Spores navicular, 7–10 × 3–4. On decaying wood of deciduous trees.

Luellia recondita (Jacks.) Larsson and Hjortst (Fig. 258)
Fruitbodies brown. Basal hyphae encrusted with resinous brown matter. Hyphidia present, unbranched. Basidia 20–25 × 5–7. Spores navicular to fusiform, 7–10 × 4–4.5. Mostly on decaying wood of conifers.

Lyomyces sambuci (Pers. ex Fr.) P. Karst. (Fig. 259)
Fruitbodies resupinate, effused, firmly attached to substrate, membranous, somewhat waxy, mostly smooth, occasionally tuberculate, white to pale cream. Hyphae hyaline, with clamps. Cystidia plentiful, projecting only slightly, 40–50 × 4–5, cylindrical, capitate, 1–2-septate with clamps at septa, encrusted with scattered compound crystals and often with a resinous deposit at the apex. Basidia 20–25 × 4. Spores ellipsoid, 5–7 × 3.5–4., hyaline, smooth, with 1–2 guttules, J–, C+. Common on attached branches, especially those of *Sambucus*.

Macrotyphula

Fruitbodies erect, solitary or, more commonly, gregarious, long and thread-like or narrowly clavate, smooth, tough, elastic, not arising from sclerotia, not green with $FeSO_4$, with rather short stalks. Hyphae swollen, with clamps. Basidia large, with basal clamps, 4-spored. Spores often large, hyaline, smooth, ellipsoid or fusiform, J–. On rotting leaves, petioles and branches of deciduous trees.

KEY

Spores not more than 12 long .... *juncea*
Spores more than 12 long 1
1. Spores 12–18 long, fruitbodies tall, narrowly clavate .... *fistulosa*
   Spores up to 24 long, fruitbodies short, contorted ....
   .... *fistulosa* var. *contorta*

Macrotyphula fistulosa (Fr.) Petersen (Fig. 260)
Fruitbodies narrowly clavate, sometimes flattened, 3–30 × 0.3–1 cm, slightly tapered towards the rounded tip, yellowish to reddish or greyish brown, hollow. Basidia 60–80 × 10–12. Spores ellipsoid or broadly fusiform, 12–18 × 6–8. In the var. *contorta* (Holmsk. ex Fr.) Petersen (Fig. 261) the fruitbodies are only 0.5–3 cm tall, much contorted, rather greyish ochraceous. The spores are up to 24 long. Mostly on fallen dead branches of deciduous trees.

Macrotyphula juncea (Fr.) Berthier (Fig. 262)
Fruitbodies thread-like, often bent, 3–12 cm tall, 0.5–1.5 mm thick, pale ochraceous. Basidia 40–50 × 7–8. Spores ellipsoid or almond-shaped, often slightly curved, 7–12 × 3.5–5. On rotting leaves, petioles and twigs of deciduous trees.

Megalocystidium

Fruitbodies resupinate, effused, waxy or membranous, smooth or tuberculate, white, cream, ochraceous or reddish brown. Cystidia (g) subulate, thin-walled with yellowish contents. Hyphidia cylindrical, narrow, mostly unbranched, hyaline. Basidia narrowly clavate, 40–60 × 5–8, longer than in the very similar genus *Gloeocystidiellum*. Spores hyaline, smooth, thin-walled, J+.

KEY

Spores cylindrical to allantoid, 12–20 × 5–7 *leucoxanthum*
Spores ellipsoid 1
1. Spores 6–8 long .... *lactescens*
    Spores 7–12 long .... *luridum*

Megalocystidium lactescens (Berk.) Jülich (Fig. 263)
Fruitbodies up to 1.5 mm thick, white, ochraceous or slightly reddish, smooth. Hyphae without clamps. Cystidia 160–250 × 7–12. Hyphidia 60–80 × 2–2.5. Spores broadly ellipsoid, 6–8 × 4–5. On dead wood.

Megalocystidium leucoxanthum (Bres.) Jülich (Fig. 264)
Fruitbodies up to 1 mm thick, smooth or, more often, tuberculate, cream, ochraceous or pale reddish brown. Hyphae with clamps. Cystidia 70–150 × 7–14. Hyphidia 40–60 × 2.5–3. Spores cylindrical to allantoid, 12–20 × 5–7. On dead branches, especially those of *Alnus*, *Corylus* and *Salix*.

Megalocystidium luridum (Bres.) Jülich (Fig. 265)
Fruitbodies thin, smooth or tuberculate, cream or ochraceous. Hyphae with clamps. Cystidia 50–100 × 6–14. Hyphidia 40–60 × 2.5–3.5. Spores ellipsoid, 7–12 × 4–6. On dead wood of deciduous trees and shrubs.

Meripilus *giganteus* (Pers. ex Fr.) Karst. (Fig. 266)
Fruitbodies pileate, compound, fleshy when fresh, brittle when dry, up to 80 cm diam., made up of a number of overlapping, fan-shaped, lobed caps attached to a common, short, bulbous stalk; individual caps 10–30 cm wide, 1–2 cm thick, upper surface ochraceous brown, zonate, smooth or rather indistinctly scaly, lower surface white, cream or ochraceous, bruising brown, pores 3–5 per mm. Hyphae with numerous septa but no clamps. No cystidia. Spores subspherical, 6–7 × 5–6, hyaline, smooth, J–. Mainly on stumps and roots of deciduous trees

Merismodes
Fruitbodies solitary or gregarious, often closely packed together and forming large colonies, seated on a subiculum, individually bowl-, cup-, tube- or top-shaped, hairy on the outside; hairs long, brown, thick-walled and encrusted with granules except near their thinner-walled, pale, usually swollen tips; hymenium lining cups smooth, pale. Hyphae with clamps. Cystidia none. Spores hyaline or yellowish, smooth, J–.

Fruitbodies flat bowl-shaped, spores 6–9 × 2–2.5 .... *confusus*
Fruitbodies more or less top-shaped, tapered below, spores 8–10 × 4–5
.... *anomalus*
Fruitbodies tube- or deep cup-shaped 1
1. Spores 6–11 × 2–2.5 .... *fasciculatus*
   Spores 5–7.5 × 3–4 .... *ochraceus*

Merismodes anomalus (Pers. ex Fr.) Sing. (Fig. 267)
Fruitbodies usually gregarious, closely packed together forming large colonies, individually cup- to top-shaped, tapered below into a short stalk, ochraceous to cinnamon brown and hairy on the outside, with a white or creamy rim, 0.3–0.5 mm diam.; hairs up to 150 × 2–4; hymenium lining cup cream or pale ochraceous. Basidia 30–40 × 5–7, 4-spored. Spores ellipsoid, some slightly curved, 8–10 × 4–5. On wood and bark of deciduous trees, common.

Merismodes confusus (Bres.) Reid (Fig. 268)
Fruitbodies gregarious, closely packed together, individually flat bowl-shaped, 1–3 mm diam., greyish when young then yellowish brown to dark chestnut brown on the outside, rim fringed with protruding hairs which often end in a bulb; hymenium creamy white. Spores slightly curved, 6–9 × 2–2.5. On dead branches of deciduous trees, including *Betula*.

Merismodes fasciculatus (Schw.) Donk
Fruitbodies solitary or gregarious, deeply cup- or tube-shaped, up to 1 mm diam., brown on the outside; hairs sometimes curled at apex; hymenium cream or yellow. Spores 6–11 × 2–2.5. On wood of deciduous trees.

Merismodes ochraceus (Hoffm. ex Pers.) Reid
Fruitbodies deeply cup- to tube-shaped, becoming rust-coloured on the outside, 1–2 × 0.5 mm; hymenium creamy white. Spores 5–7.5 × 3–4. On dead herbaceous stems.

Meruliopsis
Fruitbodies resupinate or effuso-reflexed, membranous to waxy or soft leathery, hymenium smooth at first but becoming merulioid and sometimes in part poroid. Hyphae without clamps. Cystidia, when present, short and subulate. Basidia narrowly clavate. Spores hyaline, thin-walled, smooth, J–.

Fruitbodies resupinate, with some pores, orange–red to dark purplish
.... *taxicola*
Fruitbodies effuso-reflexed, remaining white for a long time,
   hymenium merulioid to reticulate-poroid .... *corium*

Meruliopsis corium (Fr.) Ginns (Fig. 269)
Fruitbodies most commonly effuso-reflexed, rarely pileate or resupinate, membranous to softly leathery, quite tough, 0.5–1 mm thick, often many cm long, the reflexed parts forming shelf-like brackets up to 2 cm wide, imbricate on vertical surfaces, usually remaining white or whitish but some-times turning ochraceous or brownish when old; upper surface tomentose, faintly zonate, hymenial surface smooth when young, becoming merulioid or reticulate-poroid. Hyphae without clamps, some covered with crystals. No cystidia. Basidia 25–35 × 5–6, 4-spored. Spores cylindrical or narrowly ellipsoid, 5–7 × 2.5–3.5. Mostly on the lower side of fallen dead branches of deciduous trees, very common.

Meruliopsis taxicola (Pers.) Bond.
Fruitbodies resupinate, effused, 1–4 mm thick, waxy when fresh, firmly attached to substrate, orange–red when young, becoming dark purplish, margin white, hymenium partly merulioid with folds up to 1 mm deep, but with definite irregular, angular pores 2–4 per mm especially towards the centre. Cystidia subulate, about same length as basidia. Basidia 15–20 × 4–5. Spores cylindrical and bent to allantoid, 3–6 × 1–1.5. On wood and bark of conifers, especially *Pinus sylvestris*.

Merulius *tremellosus* Fr. (Fig. 270)
Fruitbodies when fresh firmly gelatinous to elastic, horny when dry, usually effuso-reflexed or pileate, occasionally resupinate, 2–5 mm thick; brackets 2–5 cm wide, sometimes imbricate, often in long rows, upper surface off-white or pale grey, hairy, somewhat zonate; hymenium on lower surface or substrate merulioid, i.e. reticulately folded or ridged with irregular alveolae, becoming poroid when old, colour varying from greyish to ochraceous and watery when quite young to yellowish or reddish orange when mature. Hyphae with clamps. Cystidia few, small. Basidia 20–28 × 3–5, 4-spored. Spores allantoid, 3.5–4.5 × 1–1.5, hyaline, thin-walled, smooth, J–. On decaying wood in damp places.

Microstroma *juglandis* (Bereng.) Sacc. (Fig. 271)
Hyphae intercellular, crowded and forming pseudostromata in substomatal cavities. Basidia clustered, erumpent through stomata, hyaline, 20–40 × 6–7, each one forming directly on its surface 2–8 spores, apparently not on sterigmata. Spores hyaline, narrowly ellipsoid, 6–10 × 3–4, budding in a yeast-like manner. On the lower surface of living leaves of *Juglans regia*.

Microstroma album (Desm.) Sacc.
On the lower surface of leaves of *Quercus* spp. Has basidia 20–25 long, and ovate-oblong or somewhat inequilateral spores 5–7 × 3, each with 1 or 2 guttules.

Mucronella *calva* (Alb. and Schw. ex Schw.) Fr. var *aggregata* (Fr.) Quél.
(Fig. 272)
Fruitbodies consisting of closely grouped, downwardly directed, white to
pale yellowish, rather glassy looking, subulate teeth, 0.5–3 mm long,
attached directly to the substrate, often apparently without any subiculum.
No cystidia. Basidia 15–18 × 4–5. Spores ellipsoid, 4–6.5 × 3, hyaline,
smooth, J–. On the lower surface of rotting trunks and branches lying on
the ground.

Mycoacia

Fruitbodies resupinate, firmly attached to the substrate, waxy, with thin
subiculum supporting prominent teeth, 1–3 mm long, which are conical or
cylindrical with pointed tips. Hyphae with or without clamps. Cystidia
sometimes present. Basidia narrowly clavate, closely packed together forming a palisade. Spores small, hyaline, smooth, J–.

KEY

Spores slightly but distinctly curved, mostly 3.5–4.5 × 1.5–2. No
 cystidia .... *aurea*
Spores straight or almost so, 5–6 × 2–3. Cystidia present 1
1. Hyphae in ends of teeth strongly encrusted and cystidium-like,
 fruitbodies becoming blackish brown .... *fuscoatra*
No strongly encrusted hyphae in ends of teeth, fruitbodies yellow to
 ochraceous .... *uda*

Mycoacia aurea (Fr.) Erikss and Ryv. (Fig. 273)
Fruitbodies widely effused, cream when young, yellow–ochre when mature,
margin remaining pale, teeth slender, closely packed together, often
branched at or below the apex. No cystidia. Basidia 12–17 × 4–5. Spores
slightly curved, 3.5–4.5 × 1.5–2. On decaying wood of deciduous trees.

Mycoacia fuscoatra (Fr.) Donk (Fig. 274)
Fruitbodies widely effused, pale cream or yellowish when young, becoming
brown to dark blackish brown when old, teeth often joined together in
small groups, fimbriate at their tips where hyphae are strongly encrusted
and cystidium-like. The true cystidia are sharply pointed, 30–40 × 4–5.
Basidia about 20 × 4–5. Spores cylindric-ellipsoid, 5–6 × 2–2.5. On decaying
wood of deciduous trees.

Mycoacia uda (Fr.) Donk (Fig. 275)
Fruitbodies usually smaller than in the other two species, sulphur or lemon
yellow to ochraceous, teeth mostly only 1–2 mm long, rarely branched.
Cystidia inconspicuous, fusiform, 20–25 × 4. Basidia about 20 × 4–5. Spores
ellipsoid, 5–6 × 2–3. On decaying wood of deciduous trees.

Mycogloea macrospora (Berk. and Br.) McNabb (Fig. 276)
Fruitbodies firmly gelatinous, at first appearing as discrete pustules up to
0.5 mm diam., which become confluent and form rose-coloured masses up
to 2 cm long, drying creamy yellow, thin and waxy. Hyphae hyaline, with
clamps. Basidia cylindrical. 40–60 × 4–5, with 3 transverse septa and a basal
clamp, sterigmata subulate, less than 10 long. Spores ellipsoid, hyaline,
thin-walled, smooth, J–, 10–13 × 4–5.5. On stromata of *Diatrype stigma.*

Myxarium

Fruitbodies pulvinate and made up of solitary pustules or resupinate and
effused, hyaline or pale coloured, gelatinous when fresh but thin and
membranous when dry. In resupinate species the surface is usually smooth
and the edge thin and indistinct. Hyphae hyaline, with clamps. Cystidia
found only in a few species, hyaline. Basidia hyaline, when young stalked,
clavate, with a clamp at the base of the stalk; when mature the basidium has
one or two longitudinal septa and is cut off from its stalk by a clampless
septum. Spores hyaline, smooth, thin-walled, J–. All species saprophytic on
wood.

KEY

Spores frequently more than 10 long, often somewhat curved 1
Spores never more than 10 long, not curved 3
1.   Fruitbodies containing lumps of calcium oxalate crystals ..... *nucleatum*
     Fruitbodies not so 2
2.   Fruitbodies initially pulvinate, resembling lacquer .... *laccatum*
     Fruitbodies resupinate, thin, not resembling lacquer .... *podlachicum*
3.   Spores subglobose, not more than 6 × 5.5, each with a large guttule
     .... *sphaerosporum*
     Spores ellipsoid, frequently more than 6 long 4
4.   Fruitbodies appearing resupinate and smooth but seen under a lens to
     be composed of small, glistening, granule-like pustules . *crystallinum*
     Fruitbodies not so 5
5.   Fruitbodies resupinate, effused .... *subhyalinum*
     Fruitbodies round or lenticular, 0.5 mm diam., or confluent and up to
     5 cm .... *grilletii*

Myxarium crystallinum Reid
Fruitbodies appearing bluish grey, resupinate and effused but seen under
the lens when fresh to be made up of numerous small, glistening, granule-
like, gelatinous pustules which, when dry, form an irregular network.
Basidia 7–8 diam., 2-spored; stalk 15–18 long. Spores mostly broadly ellipsoid and 6–8 × 4–5, occasionally subglobose and about 5 diam. On *Alnus*
and *Betula.*

Myxarium grilletii (Boud.) Reid.
Fruitbodies soft, gelatinous, bluish grey or pale lilac, pruinose, round or lenticular and about 0.5 mm diam., gregarious, or confluent and then forming masses up to 5 cm across. Basidia 2-4-spored, 8-12 × 6-10, sterigmata up to 20 long. Spores 4-10 × 2-5. On wood of deciduous trees.

Myxarium laccatum (Bourd. and Galz.) Reid
Fruitbodies when young pulvinate, 1-2 mm diam., solitary or in small groups, when old tending to become resupinate and effused, softly gelatinous, pale grey to ochraceous brownish, resembling lacquer. Basidia 10-15 × 9-12, 4-spored, sterigmata up to 40 long, stalk up to 25 long. Spores ellipsoid, sometimes curved, 9-15 × 5-9. Usually developing under the bark of both deciduous trees and conifers.

Myxarium nucleatum Wallr. (Fig. 277)
Fruitbodies pulvinate with undulating surface, watery gelatinous, almost hyaline, sometimes tinted rose or violaceous, 1-2 mm diam., each containing a lump of calcium oxalate crystals; sometimes confluent and forming masses up to 4 cm across, when dry appearing as just a brownish film. Basidia 10-16 × 9-11, sterigmata up to 20 long, stalks up to 18 long. Spores cylindrical or somewhat allantoid, 10-18 × 4-6. On wood of deciduous trees, found mostly on *Fagus* and *Fraxinus*.

Myxarium podlachicum (Bres.) Raitviir
Fruitbodies resupinate, waxy to gelatinous, bluish grey to ochraceous brown, or with lilac tints, smooth or granular. Basidia 9-18 × 6-12, 2-4-spored, sterigmata up to 40 long. Spores cylindrical or narrowly ellipsoid, somewhat curved, 7-13 × 4-6.5. On deciduous trees.

Myxarium sphaerosporum (Bourd. and Galz.) Reid (Fig. 278)
Fruitbodies resupinate, waxy to gelatinous, whitish surface undulating when moist, drying tuberculate and yellowish. Basidia 10-20 × 5-10, 2-4-spored, sterigmata up to 30 long. Spores 4-6 × 3.5-5.5, each with a large guttule. On wood of both deciduous trees and conifers.

Myxarium subhyalinum (Pearson) Reid
Fruitbodies resupinate, effused, waxy-gelatinous, dirty white to pale bluish grey or greyish violet, pruinose with tiny crystals. Basidia 7-11 × 6-9, 4-spored, sterigmata 10-20 long, stalk up to 15 long. Spores 5-9 × 3-4. On wood of deciduous trees and conifers.

Oligoporus

Fruitbodies of basidiospore state resupinate or pulvinate, overlying or growing beside a *Ptychogaster* state which produces powdery masses of thick-walled, yellowish or brownish chlamydospores; pores angular, 1-4 per mm. Hyphae with clamps. Spores hyaline, smooth, thin-walled, J−, C+.

KEY

Chlamydospores in mass yellow or yellowish .... *rennyi*
Chlamydospores in mass brown or cinnamon brown .... *ptychogaster*

Oligoporus ptychogaster (F. Ludwig) R. and O. Falck (Fig. 279)
Fruitbodies of basidiospore state overlying or growing beside the *Ptychogaster* chlamydospore state, semicircular or oblong, 1–4 × 1 cm, up to 1.5 cm thick, soft, watery, upper surface flat or convex, white or cream, adpressed hairy, lower surface white drying cream or ochraceous, pores angular, 2–4 per mm, flesh white, soft, cottony. Basidia 12–20 × 5–6. Spores 4.5–5.5 × 2–3.5. *Ptychogaster* state hemispherical or elongated cushions 2–8 cm across, up to 5 cm thick, surface hairy, white to brownish, in section concentrically zoned, yellowish, when ripe full of ellipsoid, thick-walled chlamydospores with rounded or truncate ends, 5–10 × 4–7, in mass brown or cinnamon brown. On stumps of conifers.

Oligoporus rennyi (Berk. and Br.) Donk (Fig. 280)
Fruitbodies resupinate, up to 10 cm long, 1–5 mm thick, white when fresh, often with exuded drops on the surface, becoming creamy ochraceous, pores angular, thin-walled, 1–4 per mm, margin fimbriate, white or made powdery and pale yellow by the production of *Ptychogaster* chlamydospores. Basidia 12–30 × 3–6. Spores narrowly ellipsoid, 4–5.5 × 2–2.5. Chlamydospores broadly ellipsoid, thick-walled, 5–7 × 3.5–5, pale yellow in mass. On very rotten wood of conifers; fruitbodies disintegrate readily or are eaten by insects.

Onnia *tomentosa* (Fr.) P. Karst (Fig. 281)
Fruitbodies often in groups, sometimes concrescent, pileate, centrally or laterally stalked, soft, spongy or corky; cap round, oval or kidney-shaped, 3–10 cm diam., 0.2–1 cm thick in centre, flat or depressed in middle, upper surface finely tomentose, yellowish brown to cinnamon brown, not or very indistinctly zonate, somewhat tuberculate, margin thin, wavy, pale, lower surface greyish brown, pores 2–4 per mm, decurrent; stalk 3–4 × 1–2 cm, broadening towards the cap, yellowish or cinnamon brown. Hyphae without clamps. Setae in and projecting above hymenium, subulate, pointed, dark brown, thick-walled, smooth, 30–75 × 8–16. Basidia 12–15 × 4–5.5. Spores ellipsoid, 4–7 × 3–4, yellowish, smooth, J–, C+. On soil under conifers, parasitic on their roots.

Oxyporus
Fruitbodies on wood, pileate or resupinate, sessile, poroid, flesh white or cream. Cystidia with apical encrustation present in hymenium of all species and often abundant. Hyphae hyaline, without clamps. Spores hyaline or yellowish, smooth, J–.

KEY

Fruitbodies mostly pileate and imbricate, spores spherical or
    subspherical .... *populinus*
Fruitbodies all or mostly resupinate, spores ellipsoid 1
1. Cystidia of two kinds, projecting ones subulate, not encrusted
    .... *corticola*
    Cystidia of one kind, all apically encrusted 2
2. Pores 4–7 per mm, rounded .... *obducens*
    Pores 1–3 per mm, angular, often irpicoid .... *late-marginatus*

Oxyporus corticola (Fr.) Ryv. (Fig. 282)
Fruitbodies resupinate, widely effused, up to 7 mm thick, soft and fleshy
when young, hard and leathery when old, cream to ochraceous, pores 1–4
per mm, thin-walled, angular or irregular, margin up to 1 cm wide, fimbriate. Cystidia in hymenium of two kinds: (1) projecting, subulate, smooth,
25–40 × 6–14, (2) not projecting, 10–20 × 4–7, cylindric-ellipsoid, capped
by a crown of crystals. Spores broadly ellipsoid, 5–6 × 4–4.5, hyaline or
yellowish. On deciduous trees, most commonly *Populus tremula.*

Oxyporus late-marginatus (Dur. and Mont. ex Mont.) Donk
Fruitbodies resupinate or nodulose, 2–8 mm thick, up to 20 cm across,
soft, drying corky, white to grey or brownish, pores 1–3 per mm, occasionally round but mostly angular and quite commonly irpicoid, margin up to
5 mm wide. Cystidia cylindrical to subulate, 20–30 × 4–6, capped by
crystals. Spores ellipsoid, 4.5–6.5 × 3–3.5. On deciduous trees.

Oxyporus obducens (Pers. ex Fr.) Donk
Fruitbodies resupinate, widely effused, up to 1 m across, 1–1.5 cm thick,
white or cream when young, yellowish brown when old, pores round or
almost so, 4–7 per mm, tubes sometimes stratified, margin up to 2 mm
wide. Cystidia broadly ellipsoid, 10–20 × 5–13, capped by crystals. Spores
broadly ellipsoid, 4–5.5 × 3–4. On deciduous trees.

Oxyporus populinus (Schum. ex Fr.) Donk (Fig. 283)
Fruitbodies mostly pileate, often imbricate, sessile, fan-shaped or semicircular, tough, leathery brackets 3–9 × 2–4 cm, 1–4 cm thick, upper surface
adpressed hairy to smooth, whitish grey, greyish buff or tinged ochraceous,
frequently covered by green algae, margin white during active growth,
lower surface white or cream, pores 4–7 per mm, edge sterile, velvety, tubes
stratified. Sometimes resupinate on fallen branches. Cystidia abundant,
clavate, 15–20 × 3–7, thick-walled, apically encrusted. Spores spherical or
subspherical, 4–5.5 diam. On deciduous trees, often in holes or cracks.

Parvobasidium *cretatum* (Bourd. and Galz.) Jülich (Fig. 284)
Fruitbodies resupinate, membranous, smooth, white or cream, firmly
attached to substrate. Hyphae hyaline, with clamps. Cystidia hyaline,

clavate or flask-shaped, 20–30 × 5–9, thin-walled, without oil drops. Basidia 9–12 × 3.5–4, 4-spored. Spores narrowly ellipsoid, 4–4.5 × 2, hyaline, thin-walled, smooth, J–. On rotting stems of ferns.

Paullicorticium

Fruitbodies mostly rather small, very thin, membranous or waxy, often appearing just as a bloom on wet wood. Hyphae narrow, with or without clamps. No cystidia. Basidia hyaline, clavate or more commonly obconical or pyriform with mostly 6 sterigmata. Spores hyaline, navicular or cylindric-ellipsoid, thin-walled, smooth, J–, C–.

KEY

Spores navicular, 6–8 × 2–3 .... *pearsonii*
Spores cylindric-ellipsoid, 7–9 × 3–4 .... *niveo-cremeum*

Paullicorticium niveo-cremeum (Höhn. and Litsch.) Oberw. (Fig. 285)
Fruitbodies effused, filmy, soft, whitish to pale grey. Hyphae with clamps. Basidia clavate, with basal clamps, 20–30 × 5–7.5, mostly 6-spored. Spores cylindric-ellipsoid, 7–9 × 3–4. On the lower surface of dead, fallen, mostly decorticated branches of deciduous trees.

Paullicorticium pearsonii (Bourd. and Galz.) J. Erikss. (Fig. 286)
Fruitbodies so thin as to be almost invisible, appearing as a bloom on dead, wet wood. Hyphae without clamps. Basidia, when fully developed, obconical, without clamps, 10–15 × 5–6, mostly 6-spored. Spores navicular, one side sometimes slightly concave, 6–8 × 2–3. On decaying wood of conifers.

Pellidiscus *pallidus* (Berk. and Br.) Donk (Fig. 287)
Fruitbodies cup-, shell- or disc-shaped, 1–3 mm diam., white or cream, covered on the outside by cystidium-like hairs, hymenium lining cup smooth or wrinkled, becoming yellowish brown. Hyphae without clamps. Spores in mass ochraceous, ellipsoid to almond-shaped, 6–9 × 3.5–5.5, pale honey-coloured by transmitted light, finely punctate, J–. On rotten wood and bark, dead herbaceous stems, leaves etc.

Peniophora

Fruitbodies resupinate, mostly effused, in only a few species discrete and orbicular, various colours, membranous to soft or tough waxy, smooth to tuberculate, firmly attached to substrate but sometimes loosening at the margin. Hyphae with clamps in most species, basal ones often brown. Cystidia always present, sometimes of several different kinds. Dendro-hyphidia present in some species. Basidia narrowly clavate, 4-spored.

Spores often curved, thin-walled, smooth, J−, C−, hyaline, in mass often pale reddish. On attached or fallen dead branches.

KEY

Spores 15–20 × 10–13, on *Alnus* .... *erikssonii*
Spores 9–11 × 6–7, on *Buxus* .... *proxima*
Spores not more than 5 wide 1
1. No conical encrusted cystidia, fruitbodies pale reddish with white pruina, on *Populus tremula* .... *polygonia*
Conical encrusted cystidia present 2
2. Fruitbodies pale or brightly coloured, mostly orange, cream or rose 3
Fruitbodies darker or duller colours, often grey, brownish or violaceous 5
3. Spores 6–7 long, dendrohyphidia present .... *boidinii*
Spores 8 or more long 4
4. On *Carpinus*, spores 9–15 × 3.5–4.5 .... *laeta*
Plurivorous, spores 8–12 × 3.5–5 .... *incarnata*
5. Dendrohyphidia present, spores 9–15 × 3.5–4.5 .... *lycii*
No dendrohyphidia 6
6. Fruitbodies discrete, orbicular, on *Pinus sylvestris* .... *pini*
Fruitbodies elongated, effused 7
7. Margins of fruitbodies loosening from the substrate and often rolling upwards and inwards to expose the dark brown or black lower surface 8
Margins of fruitbodies remaining firmly attached to substrate 9
8. On *Fraxinus*, *Ligustrum* and *Syringa*, margin often black, spores 8–12 × 3–3.5 .... *limitata*
On *Quercus*, spores 10–12 long .... *quercina*
On *Tilia*, spores 7.5–9 long .... *rufomarginata*
9. No cystidia with thin walls and granular or oily contents present, spores 7–10 × 2.5–3.5 .... *cinerea*
Cystidia with thin, smooth walls and granular or oily contents present 10
10. Spores 6–8 × 2.5–3, mostly on *Picea* .... *pithya*
Spores larger 11
11. Spores straight or very slightly curved, 8–11 long .... *nuda*
Spores distinctly curved, 7.5–9 long .... *violaceo-livida*

Peniophora boidinii Reid
Fruitbodies up to 0.13 mm thick, smooth, cream with rosy tint. No horizontal hyphal layer. Dendrohyphidia present. Cystidia (m) conical, thick-walled, encrusted, 20–40 × 7–10. Spores ellipsoid, 6–7 × 3–4. On fallen, dead branches.

Peniophora cinerea (Pers. ex Fr.) Cooke (Fig. 288)
Fruitbodies up to 0.1 mm thick, margin attached, grey, often faintly tinged with violet. Usually no basal horizontal hyphal layer. Hyphae with clamps. Cystidia (m) subulate and smooth when young, becoming conical, thick-walled and encrusted, 15–30 × 7–10, tips sometimes just protruding. Basidia 20–35 × 5–6, 4-spored. Spores slightly curved, 7–10 × 2.5–3.5. On fallen dead branches of deciduous trees, very common.

Peniophora erikssonii Boid. (Fig. 289)
Fruitbodies 0.2–0.3 mm thick, yellow to orange, margin attached. Hyphae with clamps. Cystidia (m) 50–100 × 8–14, conical, thick-walled, encrusted. Cystidia (g) 100–200 × 8–18, smooth, filled with granular protoplasm. Basidia 60–100 × 13–15. Spores ellipsoid, 15–20 × 10–13, yellowish orange in mass. On attached, dead branches of *Alnus*.

Peniophora incarnata (Pers. ex Fr.) P. Karst. (Fig. 290)
Fruitbodies up to 0.3 mm thick, margin attached, orange, fading when dry. Hyphae with clamps. Cystidia (m) conical, thick-walled encrusted part 30–60 × 8–15. Cystidia (g) cylindrical, sometimes with short basal outgrowths, 60–200 × 9–15. Basidia 40–45 × 5–7. Spores oblong-ellipsoid, 8–12 × 3.5–5. Very common on attached and fallen dead branches of deciduous trees and on *Ulex* and *Cytisus*.

Peniophora laeta (Fr.) Donk (Fig. 291)
Fruitbodies about 0.3 mm thick, pale orange drying cream, growing under the bark, pushing it up by means of stout hyphal pegs 1–3 mm long, so that it breaks and rolls back to expose the hymenium. Hyphae with clamps. Cystidia (m) few, conical, encrusted, thick-walled, 45–60 × 12–15. Cystidia (g) abundant, up to 100 × 8–12, cylindrical to clavate, thin-walled, smooth. Basidia 30–45 × 5–7. Spores cylindric-ellipsoid, one side often slightly concave, 9–15 × 3.5–4.5. Common on dead attached branches of *Carpinus*.

Peniophora limitata (Chaillet ex Fr.) Cooke (Fig. 292)
Fruitbodies up to 0.5 mm thick, in section basal layer almost black, surface becoming tuberculate and often cracked, grey, dark greyish brown or brownish violet, margin eventually loosening and rolling upwards and inwards to expose the black base. Hyphae with clamps. Cystidia (m) conical, thick-walled, encrusted, 30–70 × 8–12, basal part brown. Basidia 40–50 × 6–8. Spores allantoid, 8–12 × 3–3.5. On attached and fallen branches of *Fraxinus*, *Ligustrum* and *Syringa*.

Peniophera lycii (Pers.) Höhn, and Litsch. (Fig. 293)
Fruitbodies thin, margin attached, variable in colour but mostly grey to violaceous or bluish and rather pruinose. Hyphae with clamps, often brown. Dendrohyphidia always present, sometimes capping cystidia. Cystidia (m) ellipsoid to pear-shaped 20–35 × 10–25, upper part encrusted.

Cystidia (g) with thin, smooth walls sometimes present, subulate, 40–60 × 5–14. Basidia 20–40 × 5–7. Spores cylindrical to allantoid, 9–13 × 3.5–4.5. Mainly on dead twigs and thin branches of deciduous trees and shrubs, very common.

Peniophora nuda (Fr.) Bres. (Fig. 294)
Fruitbodies thin, pinkish grey or greyish brown, with violaceous tints when wet, margin firmly attached. Hyphae with clamps. Cystidia (m) conical or occasionally almost cylindrical, thick-walled, encrusted part 30–40 × 8–12. Cystidia (g) cylindrical to clavate, smooth, with granular contents, 40–60 × 10–20. Basidia 30–45 × 6–8. Spores cylindrical, slightly curved, 8–11 × 2.5–3.5. On dead, attached or fallen branches of deciduous trees.

Peniophora pini (Schleich. ex Fr.) Boidin (Fig. 295)
Fruitbodies orbicular, 1–4 cm diam., loosening at the edge and finally often only remaining attached in the middle, surface frequently tuberculate, pruinose, reddish violet at first, becoming grey or brownish, 0.1–0.5 mm thick. Hyphae with clamps, often gelatinous, thick-walled, swollen. Cystidia (m) conical, encrusted part 12–18 × 5–9. Cystidia (g) embedded, cylindrical, clavate or ellipsoid, 40–60 × 10–20. Spores allantoid, 7–9 × 2.5–3. On small, attached, dead branches of *Pinus sylvestris*.

Peniophora pithya (Pers.) J. Erikss. (Fig. 296)
Fruitbodies up to 0.15 mm thick, bluish grey to reddish grey, or violaceous, quite dark when wet, only loosening when bark flakes off. Hyphae with clamps. Cystidia (m) conical, encrusted, basal part brown, 40–65 × 12–15. Cystidia (g) cylindrical to subulate, 50–80 × 8–10. Basidia 20–35 × 4–6. Spores 6–8 × 2.5–3, some cylindrical but most allantoid. On wood and bark of conifers, especially *Picea abies*.

Peniophora polygonia (Pers. ex Fr.) Bourd. and Galz.
Fruitbodies 0.1–0.2 mm thick, discrete or becoming confluent but not effused, often emerging through holes in bark and spreading radially to form lobed, orbicular patches with margins firmly attached, pale reddish, covered with white pruina, bruising dark red. Hyphae with clamps. Dendrohyphidia on surface much branched. Cystidia (g) pear-shaped, with long narrow stalks, up to 100 long, expanded part 16–25 wide. Spores cylindrical, slightly curved, 9–12 × 2.5–4. On attached and fallen dead branches of *Populus tremula*.

Peniophora proxima Bres.
Fruitbodies reddish, margin soon free. Hyphae with clamps. Cystidia (m) conical, thick-walled, encrusted, 25–35 × 5–7. Spores ellipsoid, 9–11 × 6–7. On dead branches of *Buxus*.

Peniophora quercina (Pers. ex Fr.) Cooke (Fig. 297)
Fruitbodies 0.2–0.5 mm thick, first appearing as small, sometimes discoid

structures through holes in bark but becoming confluent and eventually widely effused, variously coloured, reddish, violaceous or grey, drying pale ochraceous, margin becoming free and rolling upwards and inwards to expose the dark brown or black lower side. Hyphae with clamps. Cystidia (m) plentiful, conical, thick-walled encrusted part 25–35 × 10–15. Spores allantoid, 10–12 × 3–4. On dead, attached and fallen branches of deciduous trees, most commonly *Quercus*.

Peniophora rufomarginata (Pers.) Litsch. (Fig. 298)
Fruitbodies 0.1–0.5 mm thick, pinkish grey, violaceous or greyish brown, paler when dry, margin when young whitish to pale pinkish grey, loosely attached, rolling upwards and inwards to expose the dark brown or blackish lower side. Hyphae with clamps. Cystidia (m) conical, thick-walled, encrusted part 30–50 × 10–20. Spores 7–9 × 2.5–3.5, allantoid. On dead branches of *Tilia*.

Peniophora violaceo-livida (Sommerf.) Massee (Fig. 299)
Fruitbodies up to 0.2 mm thick, bluish or greyish violet, margin firmly attached to substrate, sometimes white. Hyphae with clamps. Basal layer thin or none. Cystidia (m) with conical, thick-walled encrusted part 15–30 × 5–9. Cystidia (g) often somewhat clavate, 50–70 × 5–12. Spores allantoid, 7.5–9 × 3. On dead, attached or fallen branches of deciduous trees, especially *Populus* and *Salix*.

Perenniporia

Fruitbodies tough and corky to hard and woody, resupinate or pileate and forming brackets, cream or ochraceous when young, pores small, tubes stratified. Hyphae with clamps, skeletal ones yellowish. Basidia short, clavate, 2–4-spored. Spores broadly ellipsoid to subspherical, often somewhat flattened at one end and with a pore at the other, thick-walled, smooth, J–, hyaline to pale yellowish.

KEY

Fruitbodies bracket-like .... *fraxinea*
Fruitbodies resupinate .... *medulla-panis*

Perenniporia fraxinea (Bull. ex Fr.) Ryv. (Fig. 300)
Fruitbodies pileate, solitary or imbricate, sessile, semicircular or fan-shaped brackets up to 17 × 12 cm, 8 cm thick at the point of attachment, triquetrous in section, sometimes concrescent and together measuring up to 60 × 30 cm; upper surface often irregular, undulating, at first velvety and creamy ochraceous, then smooth, greyish or brownish, and finally black with a thin crust; margin rounded; lower surface creamy or the colour of cork, bruising darker, pores 1–5 per mm. Flesh cork-coloured. Basidia 10–

20 × 8–10. Spores 6–7 × 5–6. Mostly near base of trunks of deciduous trees, e.g. *Fraxinus*, uncommon.

Perenniporia medulla-panis (Jacq. ex Fr.) Donk
Fruitbodies resupinate, effused, 20–40 cm across, just over 1 cm thick in the middle, tough, corky, becoming hard when dry, whitish to cream or ochraceous, pores round, 4–5 per mm, sometimes with smooth sterile areas between them. Skeletal hyphae dichotomously branched. Basidia 15–20 × 4–5. Spores 4.5–6.5 × 3–5, hyaline to pale yellowish. On wood of deciduous trees and conifers.

Phaeolus schweinitzii (Fr.) Pat. (Fig. 301)
Fruitbodies pileate, soft and watery when fresh, drying light and fragile, with rather short stalks and often imbricate with several caps from the same stalk, 10–30 × 10–20 cm, 1–3.5 cm thick, almost circular or fan-shaped, flat or sunken in the middle; upper surface at first sulphur yellowish and velvety or hairy, turning dark reddish brown in the centre, shading to yellowish at the edge and becoming almost smooth; lower surface yellowish, bruising or ageing brown or rusty brown, pores decurrent, angular, thin-walled, 0.5–1 mm diam. near edge, 3–4 mm diam. near centre; flesh turning black with KOH; stalk dark brown, about 6 × 4 cm. Hyphae without clamps. Cystidia numerous, seen easily projecting from hymenium, clavate or subulate, often with brown, resinous contents, 50–150 × 7–20. Spores ellipsoid, 5–8 × 3.5–4.5, hyaline or slightly yellowish, smooth, J-. On the ground, arising from roots or stumps of conifers, mostly *Pinus*.

Phanerochaete
Fruitbodies resupinate, effused, often large, membranous or waxy, smooth to tuberculate, loosely or firmly attached, often with fimbriate margin and sometimes with rhizomorphs. Subicular hyphae with rather thick walls, tending to be straight and parallel to the substrate with few branches, contrasting with the subhymenial hyphae which are much branched, thinner walled and often intertwined. Clamps present or not, sometimes several at a septum on basal hyphae. No cystidia at all in two species; in most, however, cystidia are numerous and conspicuous, encrusted or not, but never with oily or granular contents. Basidia narrowly clavate, 4-spored, never with basal clamps. Spores not more than 7 long, ellipsoid, hyaline, smooth, thin-walled, J-, C-.

KEY

No cystidia 1
Cystidia thin-walled, smooth or finely encrusted 2
Cystidia thick-walled, heavily encrusted 4

1. Subiculum very thin .... *avellanea*
    Subiculum up to 0.5 mm thick .... *tuberculata*
2. Cystidia 6–10 wide .... *sordida*
    Cystidia 4–6 wide 3
3. Fruitbodies turning reddish, often with red rhizomorphs and staining
    wood the same colour .... *sanguinea*
    Fruitbodies not so .... *affinis*
4. Cystidia 5–8 wide .... *filamentosa*
    Cystidia 8–15 wide 5
5. Fruitbodies with very conspicuous rhizomorphs, up to 1 mm thick
    .... *leprosa*
    Rhizomorphs, when present, not nearly so thick and conspicuous
    .... *velutina*

Phanerochaete affinis (Burt) Parm. (Fig. 302)
Fruitbodies tuberculate, at first thin and smooth, cracked when dry, yellow-ish flesh-coloured. Cystidia cylindrical to subulate, 40–60 × 4–6, smooth then becoming finely encrusted with small crystals, protruding up to 30. Spores 4.5–6 × 2.5–3. On rotten wood, mostly of deciduous trees.

Phanerochaete avellanea (Bres.) Erikss. and Hjortst (Fig. 303)
Fruitbodies smooth, cream to pale ochraceous, or avellaneous, becoming transversely cracked when dry. Subiculum thin, with hyphae 5–8 wide. No cystidia. Spores rather parallel-sided, oblong-ellipsoid, 5–7 × 2.5–3.5. Mostly on fallen dead branches of bushes and on woody herbaceous stems in spring.

Phanerochaete filamentosa (Berk. and Curt.) Burdsall (Fig. 304)
Fruitbodies up to 0.4 mm thick, loosely attached, smooth, ochraceous, white pruinose, turned violet by KOH, with paler fimbriate margin and yellowish rhizomorphs. Cystidia (m) numerous, protruding, subulate or narrowly conical, heavily encrusted, thick-walled, 60–80 × 5–8. Spores ellipsoid, 4–5 × 2–3, with 1–2 oil drops or guttules. Mostly on decayed wood of deciduous trees.

Phanerochaete leprosa (Bourd. and Galz.) Jülich
Fruitbodies up to 1 mm thick, smooth, cream to ochraceous, with very conspicuous rhizomorphs up to 1 mm thick. Cystidia (m) numerous, conical, thick-walled, heavily encrusted, 50–90 × 8–14. Spores ellipsoid, 4.5–6 × 2–4. On fallen dead branches of deciduous trees.

Phanerochaete sanguinea (Fr.) Pouzar (Fig. 305)
Fruitbodies up to 0.5 mm thick, at first white or cream, turning in part or wholly red or reddish, often with red rhizomorphs and staining the wood underneath the same colour. Cystidia few or many, projecting, subulate or cylindrical tapered towards the apex, rather thin-walled, smooth or slightly

encrusted near the tip, 45–80 × 5–6. Spores cylindrical to ellipsoid, 4.5–6 × 2.5–3. On decaying wood, mostly of conifers.

Phanerochaete sordida (Karst.) Erikss. and Ryv. (Fig. 306)
Fruitbodies up to 0.5 mm thick, smooth, whitish, cream or pale ochraceous, without rhizomorphs. Subicular hyphae rather thick-walled, without clamps, branched at right angles, loosely interwoven. Cystidia protruding, cylindrical or subulate, 70–130 × 6–10, smooth or slightly encrusted. Spores narrowly ellipsoid, 5–7 × 2.5–3. On decaying wood, mostly of deciduous trees, common.

Phanerochaete tuberculata (Karst.) Parm. (Fig. 307)
Fruitbodies often widely effused, up to 0.5 mm thick, tuberculate, white to cream, sometimes with rhizomorphs. No cystidia. Spores 5–6.5 × 3–4. On decaying wood, mostly of deciduous trees, fairly common.

Phanerochaete velutina (DC ex Pers.) P. Karst. (Fig. 308)
Fruitbodies often widely effused, up to 0.5 mm thick, watery grey, whitish, ochraceous or occasionally with reddish tints, margin paler, sometimes with rhizomorphs. Subicular hyphae sometimes up to 10 thick and heavily encrusted. Cystidia (m) abundant, protruding and seen easily under a lens, cylindrical to subulate, thick-walled, heavily encrusted, 50–120 × 8–15. Spores ellipsoid, 5–7 × 2.5–3, apiculus oblique. On decaying wood, mostly of deciduous trees, very common.

Phellinus

Fruitbodies on wood, resupinate, effuso-reflexed or pileate, sessile, perennial. Flesh tough, corky to woody, brown, turning black with KOH. Tubes stratified in many species. No clamps, skeletal hyphae yellowish brown. In most species brown or dark brown, thick-walled setae are present in the hymenium and sometimes also in the margin and arising from subicular hyphae. Cystidia occur in a few species. Spores hyaline or coloured, smooth, J–. Of the 15 species found in Britain some are host-limited, they often cause white rots.

KEY

Fruitbodies pileate. on *Hippophäe* .... *hippophäeicola*
Fruitbodies pileate on *Pinus*, pores 1–2 per mm .... *pini*
Fruitbodies effuso-reflexed, pileate or resupinate 1
Fruitbodies entirely pileate or almost so 3
Fruitbodies entirely resupinate or almost so 8
1. Pores 7–9 per mm, spores 3.5–4.5 × 2.5–3.5, mostly on *Quercus* ... *gilvus*
   Pores 5–6 per mm, spores larger 2

2. Hymenial setae not more than 20 long, almost always on *Prunus*, occasionally on *Crataegus* .... *tuberculosus*

Hymenial setae up to 60 long, almost always on *Salix caprea* .... *conchatus*

3. On *Ribes* and *Euonymus*, no hymenial setae .... *ribis*

On other hosts, hymenial setae present 4

4. Spores 6–9 × 5.5–8 .... *robustus*

Spores always smaller 5

5. Spores 4–6 × 3–4.5, setae often more than 20 long 6

Spores 5–7 × 4.5–6.5, setae not more than 20 long 7

6. On *Alnus* and *Salix* .... *trivialis*

Mostly on *Quercus* .... *torulosus*

7. Tips of setae sharply pointed .... *nigricans*

Tips of setae not sharply pointed .... *igniarius*

8. Spores cylindrical, 6–8 × 2–2.5 .... *ferreus*

Spores ellipsoid to subspherical 9

9. Setae up to 500 long, cylindrical, pointed at tip, arising from margin and subicular hyphae .... *ferruginosus*

Not so, setae shorter and in hymenium only 10

10. Hymenial setae 35–60 long, spores 5–7 long .... *contiguus*

Hymenial setae 10–20 long, spores 4–5 long .... *laevigatus*

Phellinus conchatus (Fr.) Quél. (Fig. 309)

Fruitbodies effuso-reflexed, pileate or resupinate, when pileate mostly imbricate, semicircular or fan-shaped, wavy, sometimes concrescent, 4–14 × 1–4 cm, 0.5–1 cm thick brackets, upper surface when young finely velvety, rust brown but soon becoming smooth, greyish brown to dark blackish brown with a distinct crust, narrowly zonate and radially cracked, margin sharp to rounded, velvety, lower surface when young cinnamon brown, later turning greyish brown, margin paler, pores round, thick-walled, 5–6 per mm, decurrent onto substrate. Resupinate forms up to 10 cm long. Setae 20–60 × 5–11, roughly subulate but often mis-shapen with irregular undulating walls, tips often sharply pointed but sometimes without points. Spores spherical or subspherical, thick-walled, pale yellow-ish, 5–6 × 4.5–5.5. On standing trees, often on branches which are high up, most commonly on *Salix caprea*.

Phellinus contiguus (Fr.) Pat. (Fig. 310)

Fruitbodies resupinate, firmly attached to substrate, spongy to corky when fresh, hard when dry, effused, up to 25 × 5 cm or occasionally larger, about 1 cm thick, smooth, undulating on oblique surfaces, reddish brown to grey-ish brown, pores 2–3 per mm, angular, on vertical surfaces frequently larger, more irregular and often irpicoid at the margin. Hymenial setae numerous, projecting, subulate, bent at base, 35–65 × 6–10. Tramal setae occasional, scattered, narrowly subulate with acutely pointed tips, 50–

120 × 5–12. Spores oblong-ellipsoid, usually with large oil drop, 5–7 × 3–3.5, hyaline, thin-walled. On wood of deciduous trees.

Phellinus ferreus (Pers.) Bourd. and Galz. (Fig. 311)
Fruitbodies resupinate, effused, smooth or undulating, starting off as small round cushions, finally becoming up to 50 cm long and 1.5 cm thick, tough, at first yellowish to rusty brown, then reddish brown to dark brown, margin in young specimens pale cinnamon or yellowish brown, woolly or fluffy, becoming smooth and somewhat grey, pores round to angular, 3–5 per mm, tubes stratified. Setae numerous in hymenium, subulate, 30–50 × 6–10, swollen and rooted at base. Cystidia in hymenium subulate, hyaline, thin-walled, up to 25 × 5–6. Spores cylindrical, hyaline, 6–8 × 2–2.5. On dead attached and fallen branches of deciduous trees, especially *Corylus* and *Quercus*.

Phellinus ferruginosus (Schrad. ex Fr.) Pat. (Fig. 312)
Fruitbodies resupinate, widely effused, 2–10 mm thick, surface smooth or tuberculate, yellowish brown to rust coloured, greyish when old, pores usually 5–6 per mm. Setae in the margin and arising from subicular hyphae straight, cylindrical with pointed tips, up to 500 × 6–10, dark brown, easily seen, those in the trama paler and smaller, less easy to find, those in the hymenium numerous, subulate, rusty brown, 25–60 × 6–8. Spores broadly ellipsoid, hyaline, 4–5 × 3–3.5. Usually on the lower surface of fallen branches and trunks of deciduous trees, common.

Phellinus gilvus (Schw.) Pat.
Fruitbodies mostly effuso-reflexed, more rarely pileate or resupinate, 3–15 × 1–7 cm, up to 1 cm thick, upper surface at first shaggy with matted hairs, later smooth, zonate, ochraceous to rusty yellow, pore surface dark brown, pores 7–9 per mm. Setae in hymenium conical or subulate, 17–30 × 3–6. Spores ellipsoid, 3–4.5 × 2.5–3.5. On wood of deciduous trees, mostly *Quercus*, rare.

Phellinus hippophäeicola Jahn (Fig. 313)
Fruitbodies pileate, brackets broadly attached or growing around branches, 3–6 × 2–5 cm, up to 5 cm thick near point of attachment, with rounded margin, upper surface concentrically undulate, tomentose becoming smooth, at first yellowish brown, then greyish brown, lower surface rust brown or cinnamon brown, pores 5–7 per mm, walls of tubes thin. No setae. Cystidia ampulliform, with slender necks, hyaline, 20–25 × 4–6.5. Spores subspherical, 6–8 × 5–7, hyaline to pale yellowish. On dead standing trunks and attached branches of *Hippophäe rhamnoides*. *P. contiguus* also occurs on this host but has resupinate fruitbodies and numerous setae.

Phellinus igniarius (L. ex Fr.) Quél. (Fig. 314)
Fruitbodies pileate, broadly attached, semicircular or hoof-like brackets 5–

30 × 5–10 cm, 3–18 cm thick; upper surface concentrically sulcate or ridged, smooth, rusty brown at first but soon turning grey and finally black, with crust, becoming cracked, margin rounded, rusty brown, pale grey or cinnamon, velvety; lower surface reddish brown becoming grey or greyish brown, mostly 4–6 pores per mm, tubes in several layers, old tubes permeated by white hyphal felt. Setae in hymenium conical or subulate, tips not sharply pointed, dark reddish brown, 12–20 × 5–9. Spores almost spherical, hyaline, 5–7 × 4.5–6. Parasitic on deciduous trees, especially *Betula, Malus* and *Salix.*

Phellinus laevigatus (Fr.) Bourd. and Galz. (Fig. 315)
Fruitbodies mostly resupinate, widely effused, up to 30 cm long, 2 cm thick, sometimes on vertical surfaces forming a narrow bracket along the upper edge, cinnamon brown or tobacco brown, becoming greyish brown, pores 5–8 per mm. Setae in hymenium 10–20 × 5–8, swollen at base, acutely pointed at apex, dark brown. Spores broadly ellipsoid, hyaline, 4–5 × 3–4. Almost always on dead, fallen trunks and branches of *Betula.*

Phellinus nigricans (Fr.) P. Karst. (Fig. 316)
Fruitbodies pileate, mostly solitary, broadly attached brackets 5–15 × 4–10 cm, 3–6 cm, thick, roughly semicircular, triquetrous in section, with a fairly sharp edge; upper surface flat or convex, smooth, dark grey to black, with narrow concentric sulcate zones, crust present except at margin, becoming cracked, margin usually pale grey; lower surface cinnamon to rusty brown, becoming grey when old, pores 5–6 per mm, not decurrent. Setae numerous in hymenium, subulate, sharply pointed, reddish brown, 15–20 × 5–9. Spores almost spherical, hyaline, 6–7 × 5–6.5, with thick walls. Mostly on dead trunks of *Betula.*

Phellinus pini (Brot. ex Fr.) Ames (Fig. 317)
Fruitbodies solitary, pileate, broadly attached brackets, roughly hoof-shaped, 5–20 × 4–10 cm, up to 12 cm thick where attached, margin often wavy; upper surface sulcate zoned, at first velvety to hairy and dark brown, becoming mostly smooth and dark blackish brown to black, but with some zones remaining hairy; lower surface dark brown or rusty brown, becoming greyish, margin velvety, pores angular, 1–2 per mm. Setae subulate, sharply pointed, broad-based, dark brown, 40–80 × 12–20. Spores subspherical, hyaline or pale yellowish, with thick walls, 4.5–5.5 × 4–5. Chlamydospores in trama dark rusty brown, with very thick walls, 6–7 diam. On trunks of *Pinus sylvestris,* often high up, or on recently felled trees, seldom on other conifers.

Phellinus ribis (Schum. ex Fr.) P. Karst. (Fig. 318)
Fruitbodies almost always pileate, sessile brackets, broadly attached or in part encircling stems, sometimes concrescent, often imbricate, 3–20 × 3–7 cm, 0.5–2 cm thick, soft when fresh, drying leathery or woody; upper

surface flat, sulcate zoned, hairy to spongy at first, becoming more compact and smooth, rusty brown to dark blackish brown, margin yellowish to rust-coloured, in section tomentum seen to be separated from flesh by a black line; lower surface rust brown or cinnamon brown, pores round, 6–7 per mm. No setae. Spores broadly ellipsoid, 3–5 × 2.5–3, pale yellowish. On living plants of *Ribes* spp. and *Euonymus europaeus*, just above soil level.

Phellinus robustus (P. Karst.) Bourd. and Galz. (Fig. 319)
Fruitbodies pileate, broadly attached, cushion- to hoof-shaped or semi-circular brackets 8–30 × 5–20 cm, 5–15 cm thick; upper surface tomentose to smooth, broadly sulcate zoned, rusty to greyish brown or blackish brown when old, with crust becoming cracked; lower surface yellowish brown or rusty brown, pores becoming darker or greyish, 5–6 per mm. Setae in hymenium few or absent, subulate to conical, yellowish brown, 12–20 × 4–6. Cystidia plentiful, hyaline to yellowish brown, up to 100 × 5–8, ampulliform with long necks. Spores almost spherical, pale yellowish, thick-walled, 6–9 × 5.5–8. Mostly parasitic on *Quercus*, but recorded also on trunks of other deciduous trees.

Phellinus torulosus (Pers.) Bourd. and Galz.
Fruitbodies mostly pileate, solitary or imbricate brackets 12–30 × 4–10 cm, 1–3 cm thick, flat, broadly attached, shell-shaped, margin rounded, wavy, felt-like or shaggy, upper surface uneven, orange-brown, rust or brown, lower surface cinnamon, rust or olivaceous brown, pores 5–6 per mm. Setae in hymenium plentiful, subulate, 17–35 × 5–9. Spores broadly ellipsoid, hyaline, 4–6 × 3–4.5. Mostly on *Quercus*.

Phellinus trivialis (Bres.) Kreisel (Fig. 320)
Fruitbodies pileate, solitary or imbricate, broadly attached, roughly semi-circular brackets 10–30 × 5–15 cm, with sharp margin; upper surface dark grey to black, narrowly zonate, sulcate; lower surface ventricose, cinnamon to rust-brown, pores 5–6 per mm, tubes tending to be decurrent onto the bark. Setae plentiful, subulate, pointed, 17–25 × 6–8. Spores subspherical, hyaline, fairly thick-walled, 4.5–6 × 4–4.5. On standing trunks of *Alnus* and *Salix*.

Phellinus tuberculosus (Baumg.) Niemalä (Fig. 321) *P. pomaceus*
Fruitbodies pileate, effuso-reflexed or resupinate, broadly attached brackets solitary or in groups, sometimes imbricate or concrescent, 2–8 × 1–4 cm, 0.5–4 cm thick, dimidiate or hoof-shaped, margin rounded; upper surface finely velvety to smooth, pale to dark grey or greyish brown; lower surface cinnamon brown to dark brown, greyish when old, pores 5–6 per mm, margin pale. Setae fairly plentiful in hymenium, 13–20 × 5–8, conical, pointed. Spores 5.5–7 × 4.5–5, pale yellowish. Mostly on living and dead *Prunus* spp., occasionally on *Crataegus*.

Phellodon
Fruitbodies pileate, erect, terrestrial, tough, stalked, usually in groups and mostly fused together, cap often depressed in the centre, roughly orbicular, the rather wavy edge nearly always white or whitish and contrasting strongly with colours of the rest of the cap, teeth on lower surface subulate, usually turning grey. Hyphae without clamps, not swollen. No cystidia. Basidia 4-spored. Spores spherical or broadly ellipsoid, hyaline, echinulate, white in mass, J−.

KEY

Cap surface mostly clearly zonate 1
Cap surface not clearly zonate 2
1. Cap surface grey to purplish black or olivaceous black .... *niger*
    Cap yellowish brown to dark brown .... *tomentosus*
2. Flesh turned green by KOH .... *melaleucus*
    Flesh not turned green by KOH .... *confluens*

Phellodon confluens (Pers.) Pouzar
Fruitbodies confluent, usually in large groups; cap 2–6 cm diam., flat or slightly depressed in the middle; upper surface at first downy or velvety all over, later becoming roughly pitted centrally, initially whitish, turning cream, grey or dark brown except at the edge; lower surface white becoming grey or tinged with violet, teeth 1–2 mm long, decurrent; stalk 1–3 × 0.5–2 cm, often tapered towards the downy base, white to yellowish or greyish brown; flesh not green with KOH. Spores 3.5–4.5 × 3–4 excluding spines. Under deciduous trees, especially *Quercus*.

Phellodon melaleucus (Fr.) P. Karst. (Fig. 322)
Fruitbodies in groups, fused together, funnel-shaped; cap 1–3.5 cm diam., upper surface velvety at first, becoming radially wrinkled, roughened by little pointed projections in the middle, whitish at first, then grey or bluish grey, greyish or yellowish brown to dark brown except at the edge; teeth on lower surface 1–3 mm long, decurrent; stalk 1–2 cm × 1–5 mm, blackish brown, somewhat fibrillose, flesh green in KOH. Spores 3.5–4.5 × 3–4 excluding spines. In mixed woods and conifer woods.

Phellodon niger (Fr. ex Fr.) P. Karst. (Fig. 323)
Fruitbodies in groups, fused together, somewhat funnel-shaped; cap mostly 3–6 cm diam., upper surface velvety at first, becoming pitted or covered with projecting scales near the middle, more fibrillose towards the margin, at first whitish, then grey to purplish black or olivaceous black, usually distinctly zonate, spines on lower surface 1–3 mm long, bluish grey to grey; stalk 1–5 × 0.5–2 cm, black or grey, tomentose; flesh green in KOH. Spores 3.5–4.5 × 3–3.5 excluding spines. In deciduous and mixed woods.

Phellodon tomentosus (L. ex Fr.) Banker (Fig. 324)
Fruitbodies mostly in groups, fused together, shallowly funnel-shaped; cap 1.5–4 cm diam.; upper surface at first white and velvety, turning yellowish brown to dark brown, zonate, usually wrinkled or ridged, with some pitting or roughening towards the centre; teeth on lower surface 1–2 mm long, white to grey; stalk 0.5–3 cm × 0.5–1 mm, smooth or fibrous, yellowish or reddish brown to dark brown; flesh not green in KOH. Spores 3–3.5 × 2.5–3 excluding spines. In mixed woods and conifer woods.

Phlebia

Fruitbodies resupinate, mostly gelatinous to waxy, with dense texture, often drying horny, surface smooth, tuberculate, radially folded or merulioid to poroid. Hyphae with clamps and often embedded in a gelatinous matrix. Cystidia present in most species. Basidia narrowly clavate, closely packed together forming a dense palisade. Spores hyaline, thin-walled, smooth, J–, C–.

KEY

Surface of fruitbodies radially ridged or folded, mostly reddish orange when young .... *radiata*
Surface of fruitbodies merulioid to poroid .... *rufa*
Surface of fruitbodies smooth or tuberculate 1
1. Spores distinctly curved, allantoid, 4.5–6 × 2–2.5 .... *livida*
    Spores quite straight or very slightly curved 2
2. No cystidia, spores ellipsoid, 5–9 × 2.5–4.5 .... *pallidolivens*
    Subulate, pointed cystidia always present 3
3. Cystidia thick-walled, spores 5.5–7 × 2–2.5 .... *segregata*
    Cystidia thin-walled, spores 6–8 × 2.5–3.5 .... *subochracea*

Phlebia livida (Pers. ex Fr.) Bres. (Fig. 325)
Fruitbodies effused, firmly attached, 0.1–0.4 mm thick, surface at first smooth, then tuberculate, variable in colour, often bluish grey or rosy ochraceous but sometimes flesh-coloured, violaceous or reddish brown, margin sometimes but not always fimbriate. Cystidia usually present but may be difficult to find, subulate, hyaline, thin-walled, 35–50 × 3–4. Spores mostly distinctly curved, allantoid, 4.5–6 × 2–2.5. Masses of crystals are formed which often break out through the hymenium. On decaying wood, mostly of conifers.

Phlebia pallidolivens (Bourd. and Galz.) Parm.
Fruitbodies with smooth hymenium, waxy, cream to ochraceous when fresh, drying yellow–ochre to brown. No cystidia. Spores straight, narrowly ellipsoid, 5–9 × 2.5–4.5. On wood and bark of conifers.

Phlebia radiata Fr. (Fig. 326)
Fruitbodies at first discrete, small and orbicular, then confluent and covering large areas, firmly attached, waxy to gelatinous, up to 1 mm thick, hymenium irregularly radially folded or ridged and often also tuberculate, reddish orange at first, often becoming somewhat greyish–violet or more reddish, margin rather paler and often fimbriate. Cystidia cylindrical to clavate, mostly confined to marginal zone, 50–120 × 8–12, sometimes with yellowish brown contents. Spores slightly curved, 4.5–5 × 1.5–2. On fallen trunks and attached and fallen branches, mostly of deciduous trees.

Phlebia rufa (Fr.) Christ. (Fig. 327)
Fruitbodies soon becoming confluent and widely effused, firmly attached, occasionally effuso-reflexed, surface merulioid to poroid, with pores about 2–3 per mm and up to 2 mm deep, cream, pale ochraceous, yellowish orange, reddish brown or rust brown, darkening with age. Hyphae commonly with some reddish brown clumps of crystals. Cystidia 30–90 × 7–12, mostly clavate, smooth, or with resinous encrustation and containing resinous sap when old. Spores cylindrical to narrowly ellipsoid, often slightly curved, 4.5–6.5 × 2–2.5. On fallen and attached dead branches and on trunks of deciduous trees, especially *Quercus.*

Phlebia segregata (Bourd. and Galz.) Parm. (Fig. 328)
Fruitbodies pale watery grey when fresh, drying cream to ochraceous, hymenial surface smooth. Cystidia plentiful, projecting, subulate, sharply pointed at apex, hyaline, mostly 60–70 × 6–10, becoming thick-walled, sometimes with thin transverse septa and occasionally with apical resinous encrustation. Spores cylindrical, slightly curved, 5.5–7 × 2–2.5. Mostly on decorticated wood of conifers.

Phlebia subochracea (Bres.) Erikss. and Ryv. (Fig. 329)
Fruitbodies onion red or brick red at first, then buff to ochraceous, hymenial surface tuberculate. Cystidia fairly plentiful, projecting, 50–80 × 5–7, subulate tapered to a point, hyaline, thin-walled, smooth. Spores ellipsoid, 6–8 × 2.5–3.5. On decaying wood of deciduous trees.

Phlebiopsis
Fruitbodies resupinate, effused, often large, waxy or gelatinous, surface smooth or tuberculate. Basal hyphae densely compacted, nearly always without clamps. Cystidia (m) hyaline, thick-walled, broadly fusiform to conical, upper part encrusted. Basidia narrowly clavate. Spores hyaline, smooth, thin-walled, J–.

KEY

Fresh fruitbodies tuberculate, on wood and bark of conifers ... *gigantea*

Fruitbodies smooth, on wood and bark of deciduous trees
.... *roumeguerii*

Phlebiopsis gigantea (Fr.) Jülich (Fig. 330)
Fruitbodies up to 0.5 mm thick and 1 m long, white or greyish white, opalescent, to somewhat ochraceous, soft, tuberculate and firmly attached when moist, creamy white, tough, crustose, coming away at the edge and rolling inwards when dry. Cystidia numerous, projecting, 60–100 × 10–20, broadly fusiform or conical, thick-walled, upper part encrusted. Spores 5–7 × 2.5–3, oblong-ellipsoid, with one side flatter than the other. On stumps, fallen trunks and branches of conifers, mostly on *Pinus*.

Phlebiopsis roumeguerii (Bres.) Jülich and Stalpers (Fig. 331)
Fruitbodies cream to ochraceous, smooth. Cystidia conical, thick-walled, encrusted, 70–120 × 12–20. Spores 4.5–6.5 × 2.5–3.5. On dead branches of deciduous trees.

Phleogena faginea (Fr.) Link (Fig. 332)
Fruitbodies erect, 3–6 mm high, with pale brown stalks and almost spherical, off-white to snuff-coloured heads 1–3 mm diam., composed of radiating, branched and twisted hyphae which bear the basidia. Hyphae with clamps. Basidia hyaline, with 3 transverse septa, without any distinct sterigmata, mostly 25–35 × 3–5. Spores borne laterally on basidia, rather pale yellowish brown, spherical or subspherical, smooth, with thick walls, 8–10 × 7–9, J–. On bark of dead branches mostly of deciduous trees and especially *Fagus*.

Physisporinus

Fruitbodies resupinate, annual, widely effused, poroid, soft when fresh, drying waxy and brittle. Hyphae hyaline, without clamps. Basidia clavate. Cystidia, when present, cylindrical, hypha-like, encrusted at apex. Spores spherical or subspherical, hyaline, smooth, J–, often with guttules, apiculate.

KEY

Fruitbodies without cystidia, rapidly turning red on bruising
.... *sanguinolentus*
Fruitbodies often with cystidia, not rapidly turning red on bruising
.... *vitreus*

Physisporinus sanguinolentus (Alb. and Schw. ex Fr.) Pilát (Fig. 333)
Fruitbodies 2–4 mm thick, white or pale cream, bruising rapidly red or reddish, then brown to black, pores round to angular, 3–4 per mm, tubes

1–2 mm deep. No cystidia. Spores 4–6 diam. On very rotten wood and debris on ground.

Physisporinus vitreus (Pers. ex Fr.) P. Karst. (Fig. 334)
Fruitbodies up to 1 cm thick, on vertical surfaces with sterile protuberant areas between pores, creamy white or slightly greyish, not rapidly bruising red but they may turn brown after a time, pores round, 3–5 per mm in fresh specimens, tubes 1–5 mm deep. Subiculum ochraceous or pale brown, filling cavities in substrate. Cystidia cylindrical, 3–6 wide, hyaline, thin-walled, projecting, apically encrusted. Spores 4–5 diam. On stumps and wet wood, spreading onto debris and soil.

Pilacrella *solani* Cohn and Schroet. (Fig. 335)
Fruitbodies gregarious, erect, white or whitish, about 2 mm high, stalked, with flat or slightly convex discoid head about 1 mm diam. Basidia hyaline, with 3 transverse septa, 60–70 × 6–7, sterigmata short. Projecting sterile hyphae present in hymenium. Spores ellipsoid, hyaline, smooth, thin-walled, 14–17 × 7–9. On rotten potatoes (*Solanum tuberosum*).

Piloderma

Fruitbodies resupinate, effused, soft membranous or filmy, smooth, loosely attached to substrate, usually with spongy subiculum and rhizomorphs, hymenium, especially when young, with holes which can be seen easily under a lens. No clamps anywhere. Hyphae in hymenium richly branched, in subiculum straight and sparsely branched, often covered with crystals. No cystidia. Basidia frequently stalked, hyaline or yellowish. Spores ellipsoid to subspherical, smooth, with rather thick walls, yellowish, J–, C–, often with a central oil drop or guttule. On woody debris and litter, mainly of conifers.

KEY

Rhizomorphs and subiculum always bright yellow or yellowish orange
.... *croceum*
Rhizomorphs and subiculum pure white .... *byssinum*

Piloderma byssinum (P. Karst.) Jülich (Fig. 336)
Fruitbodies nearly always pure white, only rarely becoming slightly yellowish, rhizomorphs not always clearly seen. Spores broadly ellipsoid, 3–4.5 × 2.5–3.5, hyaline or yellowish.

Piloderma croceum Erikss. and Hjortst (Fig. 337)
Fruitbodies always bright yellow at first fading slightly with age, rhizomorphs conspicuous, often yellowish orange, thick, sometimes branched. Spores subspherical, mostly 3–3.5 × 2–3.

Piptoporus *betulinus* (Bull. ex Fr.) P. Karst. (Fig. 338)
Birch Polypore or Razor-strop Fungus
Fruitbodies solitary or gregarious, pileate, rather soft and elastic, slightly
convex, semicircular, kidney- or shell-shaped, annual brackets 5–30 × 5–
20 cm, 2–5 cm thick, with inrolled edges, each contracted to a short, fat
knob where attached to the tree; upper surface very smooth, never zonate,
rather pale greyish brown or brown, with a paper-thin cuticle; lower surface
white or cream, brown when quite old, pores round or angular, 3–4 per
mm. Flesh white, rubbery. Hyphae with clamps, skeletal ones in tube layer
as well as in flesh. Spores allantoid, hyaline, thin-walled, smooth, J–, 5–
7 × 1.5–2. Parasitic on trunks of *Betula*, very common.

Pistillaria *pusilla* Fr. (Fig. 339)
Fruitbodies gregarious, erect, delicate, subulate, cylindrical or clavate, 0.5–
1 × 0.1–0.2 mm, white, tomentose towards base, upper part fertile, smooth.
Hyphae 3–6 wide, without clamps, constricted at septa. Basidia clavate,
about 25 × 5–6, with 2 or 4 spores, without basal clamps. Spores ellipsoid
with basal apiculus, 4.5–5.5 × 2.5–3.5, hyaline, smooth, thin-walled, J–. On
fallen debris including *Salix* catkins, rotting leaves etc.

Plicaturopsis *crispa* (Pers. ex Fr.) Reid (Fig. 340)
Fruibodies mostly pileate, gregarious and often imbricate, when fresh soft
and elastic, semicircular, fan- or shell-shaped brackets 2–4 cm wide, sessile
or tapered to a short stalk, with pale margin lobed or undulating, often
rolled downwards and inwards; upper surface weakly zonate, finely velvety
with adpressed hairs, pale ochraceous to reddish brown; lower surface
whitish or pale greyish, with hymenium covering radial folds or veins which
branch and anastomose. Cystidia none. Basidia clavate 4-spored. Hyphae
with large looped clamps. Spores hyaline, white in mass, allantoid, about 3–
4.5 × 1, smooth, with 2 guttules, J+. On dead standing or fallen trunks and
branches, mainly of *Fagus* but also of *Alnus* and *Corylus*.

Podoscypha *multizonata* (Berk. and Br.) Pat. (Fig. 341)
Fruitbodies erect, compound, forming rosettes about 20 cm diam., individually roughly fan-shaped, thin, tough leathery, up to 7 × 5 cm, overlapping, with wavy or indented margin, stalks joined below into a common
base; upper surface smooth, zonate, at first flesh-coloured or rosy
ochraceous with dark reddish brown bands, drying pale orange–brown or
blackish; lower surface with smooth hymenium rosy ochraceous, yellowish
brown or greyish. Hyphae with clamps, skeletal ones hyaline. Basidia
narrowly clavate. Cystidia in hymenium hyaline, cylindrical, sinuous, sometimes swollen at base, up to 120 × 12, thin-walled, smooth. Spores broadly
ellipsoid to subspherical, 4.5–6.5 × 4–5, hyaline, thin-walled, smooth, J–.
Growing on the ground in deciduous woods, probably as a rule from
buried roots or wood.

Polyporus
Fruitbodies pileate, composed of cap and stalk, fleshy or tough when fresh, hard or fragile when dry; cap upper surface smooth or scaly, mostly yellowish brown or brown, lower surface white or cream, pores round or angular, tubes often decurrent; stalk central or lateral. Flesh pale. Binding hyphae mostly dichotomously branched. No cystidia. Spores hyaline, smooth, thin-walled. Almost always on wood which may be buried in soil.

KEY

     Cap with large scales darker than surface arranged roughly concentrically, spores 10–16 long 1
     Cap without such scales, spores not more than 11 long 2
1. Cap very large, up to 70 cm diam., scales without dark, pointed tips
     .... *squamosus*
     Cap 2–12 cm diam., scales with dark pointed tips .... *tuberaster*
2. Stalk pale, spores slightly curved, 5–7 × 2–2.5 3
     Stalk always dark brown to black at least in part, spores straight, larger 4
3. Pores 2–3 per mm .... *brumalis*
     Pores 5–7 per mm .... *ciliatus*
4. Cap large, up to 25 cm diam., pores 6–8 per mm, generative hyphae without clamps .... *badius*
     Cap small to medium and not more than 10 cm diam., pores mostly 4–6 per mm, generative hyphae with clamps 5
5. Stalk with upper part ochraceous, lower part black .... *varius*
     Stalk very dark brown or black all the way up 6
6. Cap small, 1–3 cm diam., spores cylindrical, 8–11 × 2.5–3.5
     .... *varius* var. *nummularius*
     Cap medium, 2–10 cm diam., spores oblong-ellipsoid, 6–9 × 3–4
     .... *melanopus*

Polyporus badius (Pers. ex S.F. Gray) Schw. (Fig. 342)
Fruitbodies erect, solitary or gregarious, sometimes with two or more arising from a common stalk. Cap round or kidney-shaped, often lobed or with wavy margin, when young convex, then flat to funnel-shaped, 5–25 cm diam., 1–4 mm thick; upper surface smooth, rather glossy, finely wrinkled radially when old, ochraceous or bay brown, shading to dark chestnut brown in the middle, margin often pale; lower surface white or cream, ochraceous when old, pores round, 6–8 per mm, tubes decurrent. Stalk central or lateral, 2–8 × 0.5–1.5 cm, velvety and dark brown to blackish brown, black and longitudinally wrinkled when old. Generative hyphae without clamps. Spores ellipsoid or oblong-ellipsoid, 6–9 × 3–4. On standing or fallen trunks and branches of deciduous trees, including *Alnus*, *Populus* and *Salix*.

Polyporus brumalis Pers. ex Fr. (Fig. 343)
Winter Polypore
Fruitbodies erect, solitary or gregarious, centrally stalked. Cap round, flat to somewhat umbilicate, 2–8 cm diam., 2–4 mm thick, thin margin involute when dry, surface smooth, edge ciliate only when young, greyish brown to umber or blackish brown, lower surface white to cream, pores round or angular, thin-walled, 2–3 per mm, becoming larger and more elongated towards the centre. Stalk 2–6 × 0.3–0.8 cm, at first floccose and ochraceous, becoming smooth and pale brown or greyish brown. Generative hyphae with clamps. Spores cylindrical to allantoid, 5–7 × 2–2.5. Common in winter on dead fallen branches of deciduous trees. Distinguished easily by its pale stalk and large pores.

Polyporus ciliatus Fr. (Fig. 344)
Fruitbodies erect, solitary or gregarious, centrally stalked. Cap round, flat to umbilicate, 3–10 cm diam., 2–7 mm thick, thin margin involute when dry; upper surface smooth or very finely adpressed squamulose, edge ciliate when young, ochraceous or pale brown at first, becoming umber to dark brown or blackish brown; lower surface pale cream to straw-coloured, pores 5–7 per mm. Stalk 2–7 × 0.5–1 cm, finely tomentose, pale ochraceous to greyish brown, sometimes spotted or streaked. Generative hyphae with clamps. Spores cylindrical to allantoid, 5–7 × 2. On stumps and dead branches of mainly deciduous trees lying on the ground, May–August. Distinguished in the field by the pale stalk and very small pores.

Polyporus melanopus Fr. (Fig. 345)
Fruitbodies erect, solitary or gregarious, centrally or excentrically stalked. Cap 2–10 cm diam., 2–5 mm thick, at first convex, becoming depressed in the centre and somewhat funnel-shaped, round, with a wavy or lobed, thin, deflexed margin, upper surface smooth, finely velvety or floccose and radially fibrillose, pale ochraceous or yellowish brown when young, more reddish or purplish brown and wrinkled when old, lower surface white to cream, or straw-coloured, pores 4–7 per mm. Stalk 1.5–5.5 × 0.3–1.5 cm, velvety and dark brown when young, becoming black and longitudinally wrinkled when old, flesh white. Generative hyphae with clamps. Spores ellipsoid or oblong-ellipsoid, 6–9 × 3–4. On dead wood buried in the soil.

Polyporus squamosus Fr. (Fig. 346)
Dryad's Saddle
Fruitbodies very large and conspicuous, usually laterally stalked and attached to standing trees in a bracket-like manner, often imbricate, sometimes with several arising from a common base. Cap semicircular, broadly fan-, saddle- or kidney-shaped, fleshy when fresh, hard when dry, 5–70 cm wide, 1–5 cm thick, flat or convex, margin involute when dry; upper surface rather pale yellowish brown or ochraceous, with roughly concentric zones

of large, brown to dark brown adpressed scales; lower surface white to cream, pores 1–3 per mm, becoming lacerate. Stalk 3–10 × 2–6 cm, basal part blackish brown to black. Spores oblong-ellipsoid, 11–16 × 4–6. Common on living and dead deciduous trees.

Polyporus tuberaster Pers. ex Fr. (Fig. 347)
P. lentus
Fruitbodies solitary or in small groups, on horizontal substrates centrally stalked, on vertical ones laterally stalked. Cap round or kidney-shaped, convex or somewhat funnel-shaped, 2–12 cm diam., 0.5–1.5 cm thick, fleshy when fresh; upper surface pale brown or straw-coloured, with roughly concentric zones of brown scales dark and pointed at their tips; lower surface white or cream, pores angular or elongated, 1–2 mm diam. Stalk 3–8 × 0.5–1.5 cm, straight or curved, whitish or pale ochraceous, finely hairy. On branches of deciduous trees and shrubs. We have found this species a number of times on *Ulex*.

Polyporus varius Pers. ex Fr. (Fig. 348)
Fruitbodies erect, solitary or a few together, centrally or laterally stalked, sometimes with two or more arising from a common base. Cap round, fan- or kidney-shaped, convex to funnel-shaped, 1–10 cm diam, 0.3–1 cm thick, margin sometimes lobed, thin, involute when dry, upper surface smooth, with fine radial lines, pale yellowish brown, then more reddish brown or ochraceous, lower surface white, cream or straw-coloured, pores 4–6 per mm, angular, tubes decurrent. Stalk 1–6 × 0.3–1 cm, upper part ochraceous, lower part black. Spores cylindrical, 8–11 × 2.5–3.5. On attached and fallen branches of deciduous trees. In var. *nummularius* Bull. ex Fr. the cap is only 1–3 cm diam. and the central stalk black all the way up. Found mostly on twigs and slender branches.

Porphyrellus *pseudoscaber* (Secr.) Sing. (Fig. 349)
Fruitbodies pileate, terrestrial, centrally stalked, poroid. Cap 6–17 cm diam., convex, fleshy; upper surface greyish brown, hazel or, when old, blackish brown, margin paler, velvety, lower surface vinaceous buff, bruising bluish green, pores large, angular. Stalk 7–18 × 1–4 cm, cylindrical to obclavate, greyish brown to blackish brown, velvety or smooth. Flesh whitish to buff, when cut turning purplish grey then olivaceous grey. Cystidia fusiform or lageniform. Spores 12–16 × 5–6, in mass purplish brown. In deciduous and mixed woods, later summer–autumn, uncommon.

Porpomyces *mucidus* (Pers. ex Fr.) Jülich (Fig. 350)
Fruitbodies resupinate, effused, poroid, up to 3 mm thick, soft when fresh, drying brittle, loosely attached, with or without rhizomorphs, white or cream at first, then ochraceous or saffron yellow, pores polygonal, 3–5 per mm, thin-walled. Hyphae with clamps, some thin-walled and some thick-walled. No cystidia. Basidia clavate, 8–12 × 4–5.5. Spores broadly ellipsoid,

hyaline, thin-walled, smooth, J−, 2.5–3.5 × 2–2.5, with small oil-drops. On rotten wood. Superficially rather like *Trechispora mollusca* which has larger pores and spiny spores.

Postia

Fruitbodies pileate and forming brackets, effuso-reflexed or resupinate, poroid, when fresh soft-fleshed, at first often white but turning cream or other colours. Hyphae with clamps, hyaline, metachromatic with cresyl blue. Cystidia present in a few species. Spores hyaline, thin-walled, smooth, maximum size 4–6 × 2–3, mostly narrower and often slightly curved, J− in most species. On wood.

KEY

Fruitbodies mostly pileate 1
Fruitbodies mostly resupinate to effuso-reflexed, when reflexed brackets not more than 2 cm wide, never bluish 7
Fruitbodies entirely resupinate, with numerous cystidia in hymenium .... *sericeomollis*
1. Fruitbodies forming rosettes .... *floriformis*
   Fruitbodies pendent .... *ceriflua*
   Fruitbodies bracket-like, neither pendent nor forming rosettes 2
2. Upper surface of cap tinted blue or bluish 3
   No blue or bluish tinting of cap 4
3. Blue colour often pronounced, spores 1.5–2 wide, on wood of conifers .... *caesia*
   Blue tinting slight, spores 1–1.5 wide, on deciduous trees and shrubs .... *subcaesia*
4. Cystidia present in hymenium, cap surface zonate, with grey and ochraceous zones .... *balsamea*
   No cystidia, no zonation 5
5. Spores not curved, 1.5–2.5 wide .... *stiptica*
   Spores often slightly curved, 1–1.5 wide 6
6. Taste distinctly bitter .... *lactea*
   Taste very mild .... *tephroleuca*
7. Cystidia (g) present in hymenium .... *leucomallela*
   No cystidia present .... *undosa*

Postia balsamea (Peck) Jülich
Fruitbodies tough and leathery when fresh, brittle when dry, mostly pileate but also occasionally resupinate or effuso-reflexed; brackets sessile, flat, often imbricate, 2–10 × 1–5 cm, 3–5 mm thick, margin thin, sharp, upper surface finely velvety and white when young, becoming almost smooth or slightly wrinkled and zonate with greyish and ochraceous zones, finally

brownish, lower surface white when fresh, sometimes bruising pink, becoming cream to pale reddish brown, pores angular, 3–5 per mm, more irregular when decurrent onto the substrate. Hyphae with clamps. Cystidia in hymenium, near tube mouth, often subulate, up to 30×4–7, sometimes crowned by crystals. Spores ellipsoid, 4–6×2–3, J–. Mostly on gymnosperms.

Postia caesia (Schrad. ex Fr.) P. Karst. (Fig. 351)
Fruitbodies mostly pileate, soft when fresh, solitary, laterally fused or imbricate, semicircular or fan-shaped brackets, 3–6×1–4 cm, 0.2–1 cm thick, sessile, often with a narrow point of attachment, margin undulating, sharp, upper surface velvety and white at first, becoming smooth and usually blue or bluish grey, lower surface white to grey or bluish, pores round to angular, dentate, 3–5 per mm. Hyphae with clamps. No cystidia. Spores cylindrical, slightly curved, 4–5×1.5–2. with oil drops. On fallen trunks, branches and stumps mostly of conifers.

Postia ceriflua (Berk. and Curt.) Jülich
Fruitbodies mostly pileate, occasionally resupinate, soft when fresh, flat, round or semicircular, 1–4 cm diam., 2–6 mm thick, pendent, substipitate or with a dorsal, umbonate point of attachment, margin sharp, inrolled when dry, upper surface when young finely velvety and white, becoming smooth and cream or pale grey; lower surface white or cream, drying straw-coloured or ochraceous, pores 2–5 per mm. Hyphae with clamps. No cystidia. Spores oblong-ellipsoid, 3.5–4.5×2–2.5, J–. On dead wood of deciduous trees and conifers.

Postia floriformis (Quél. in Bres.) Jülich
Fruitbodies pileate, soft, shortly and laterally stalked, flat, fan-shaped brackets about 3 cm wide, 1–2 mm thick, often arising from a common base and forming rosettes, margin thin, markedly incurved when dry, upper surface finely velvety to smooth, not or weakly zonate, white to cream or greyish, lower surface white to cream, pores angular, 5–6 per mm, tubes decurrent onto stalk. Hyphae with clamps. Spores oblong-ellipsoid, 3.5–4.5×2–2.5, J–. On conifer wood, frequently on buried roots and stumps.

Postia lactea (Fr.) P. Karst.
Fruitbodies pileate, solitary, soft and fleshy when fresh, with a bitter taste, sessile, semicircular or somewhat elongated brackets 4–8×2–7 cm, up to 2 cm thick, upper surface finely adpressed floccose and white when young, becoming smooth or with radiating fibres and grey or greyish brown when old, not or very weakly zonate, lower surface white or cream, pores angular, thin-walled, 3–5 per mm, tubes slightly decurrent onto substrate. Hyphae with clamps. No cystidia. Spores narrowly cylindrical to allantoid, 4–6×1–1.5. Mostly on wood of deciduous trees.

Postia leucomallela (Murrill) Jülich (Fig. 352)
Fruitbodies pileate to resupinate, sometimes imbricate or fused laterally, soft, very bitter tasting, sessile brackets 2–8 × 0.5–2 cm, with wavy margin, upper surface white to ochraceous or slightly brownish, velvety to smooth, lower surface white to cream or ochraceous, pores 2–4 per mm, tubes decurrent onto substrate. Hyphae with clamps. Cystidia (g) in hymenium at the base of the tubes, cylindrical to clavate, with yellowish contents, scarcely protruding, 15–35 × 4–8. Spores narrowly cylindrical to allantoid, 4.5–6 × 1–1.5, J–. On dead wood of conifers.

Postia scriceomollis (Romell) Jülich
Fruibodies resupinate, 2–4 mm thick, small and discrete or widely effused, soft and fleshy, drying leathery, white to cream, pores round or angular, 3–4 per mm, margin narrow, white, coming away from the substrate and becoming inrolled. Hyphae with clamps. Cystidia abundant in hymenium, cylindrical to conical, hyaline, 10–30 × 4–8, often becoming thick-walled and crowned by clusters of tetrahedral crystals. Spores oblong-ellipsoid, 4–5.5 × 2–3, J–. On rotten wood of conifers.

Postia stiptica (Pers. ex Fr.) Jülich (Fig. 353)
Fruitbodies pileate, solitary or imbricate, but only in small numbers, semi-circular or kidney-shaped, flat, soft, very bitter tasting, usually sessile brackets 3–8 × 2–6 cm, 0.3–2.5 cm thick, upper surface milky white, cream when dry, tuberculate, lower surface white to cream, pores round or angular, mostly 3–4 per mm but a few sometimes larger and up to 1–4 mm diam., tubes in damp weather often exuding whitish droplets. Hyphae with clamps. No cystidia. Spores oblong-ellipsoid to cylindrical, 3.5–5 × 1.5–2.5, J–. Common on wood of conifers.

Postia subcaesia (David) Jülich
Fruitbodies pileate, solitary or imbricate, often in rows, soft, sessile, frequently centrally and narrowly attached, fleshy, semicircular or fan-shaped brackets 3–8 × 2–4 cm, 2–3 cm thick, upper surface velvety to smooth or wrinkled, white to creamy ochraceous, faintly tinged blue or bluish green, weakly zonate, margin white, lower surface often tinged bluish grey, pores 4–6 per mm. Hyphae with clamps. No cystidia. Spores allantoid, weakly J+, 4–5 × 1–1.5. On dead and fallen branches of deciduous trees and shrubs; we have found this species several times in different localities on *Ulex.*

Postia tephroleuca (Fr.) Jülich
Fruitbodies pileate, solitary or occasionally imbricate, soft, fleshy, mild tasting, sessile brackets, 4–10 × 2–5 cm, up to 5 cm thick at point of attach-ment, triquetrous in section; upper surface whitish to pale brown; lower surface white, pores 4–5 per mm. Hyphae with clamps. No cystidia. Spores

narrowly cylindrical to allantoid, 4–5 × 1–1.5. On dead wood of deciduous trees.

Postia undosa (Peck) Jülich
Fruitbodies resupinate to effuso-reflexed, rarely pileate and then only with small brackets not more than 2 cm wide, white to cream, pores angular, 1–2 per mm, sometimes labyrinthine. Hyphae with clamps. Spores narrowly cylindrical to allantoid, 4.5–6 × 1–1.5. On dead wood of conifers.

Protodontia *subgelatinosa* (P. Karst.) Pilát (Fig. 354)
Fruitbodies resupinate, bluish grey to purplish, up to about 2 cm², with very thin subiculum, seen under a lens to be densely covered by small warts or subulate teeth up to 0.5 mm long and 0.1 mm wide, which have fertile bases and sterile, often fimbriate tips. Basidia stalked, broadly ellipsoid to subspherical, 7–12 × 6–9, with cruciately arranged longitudinal septa and sterigmata 10–20 long; stalks 10–15 × 1–1.5, each with a clamp at its base. Spores ellipsoid, 5–8 × 3–4.5., hyaline, thin-walled, J–. On rotten wood, mostly of deciduous trees, including *Alnus, Betula, Fagus, Fraxinus* and *Quercus,* December–May.

Pseudocraterellus *sinuosus* (Fr.) Reid (Fig. 355)
Fruitbodies pileate, erect, solitary or gregarious, sometimes with several arising from one stalk, soft, thin, fibrous, irregularly funnel-shaped, with wavy and lobed margin; cap 1–5 cm diam.; upper surface greyish brown, undulating; lower surface at first smooth and greyish, becoming radially veined or wrinkled and pale ochraceous or yellowish, veins decurrent; stalk 2–10 × 0.3–0.8 cm, becoming hollow, upper part cream, lower part grey or yellowish grey. Hyphae without clamps, up to 17 wide, constricted at septa. Basidia clavate, 90–120 × 8–11, with 4–5 sterigmata. Spores broadly ellipsoid to ovoid, 10–13 × 7–9, hyaline to pale ochraceous, smooth, J–. Amongst leaf litter under deciduous trees.

Pseudohydnum *gelatinosum* (Scop. ex Fr.) P. Karst. (Fig. 356)
Fruitbodies pileate, fan- or shell-shaped brackets 2–7 cm broad, attached to wood by thick, short, usually lateral stalks, gelatinous, translucent, creamy white at first but sometimes becoming grey or greyish brown, darkening with age; sterile upper surface often wrinkled, furfuraceous; lower fertile surface densely covered with soft teeth 1–4 mm long. Hyphae with clamps, Basidia stalked, 10–16 × 7–12, with longitudinal septa, sterigmata 2–4, up to 20 long, stalks fairly long, each with a clamp at its base. Spores spherical or subspherical, with protuberant hilum, hyaline, smooth, thin-walled, J–, often with guttules, 5–8.5 diam. On rotten wood, mostly but not always of conifers; it has been recorded on *Quercus* and *Ulmus* as well as *Abies, Picea* and *Pinus,* August–November.

***Pseudomerulius** aureus* (Fr.) Jülich (Fig. 357)
Fruitbodies resupinate or effuso-reflexed, waxy when fresh, brittle when dry, at first discrete, orbicular, 1–2 cm diam., often becoming confluent and forming larger patches, 1–2 mm thick, yellow, yellowish orange or bright yellowish brown, margin paler yellowish, velvety, tending to turn up, hymenium phlebioid to merulioid, with the folds sometimes forming pores 1–3 mm wide. Hyphae anastomosing and with widely bowed clamps. No cystidia. Basidia clavate, 15–20 × 4–6, 4-spored, with basal clamps. Spores cylindrical, slightly curved, 4–5 × 1.5–2, smooth, yellowish, brown in mass, J–, C+. On decaying, decorticated wood of conifers.

Pseudotomentella

Fruitbodies resupinate, effused, soft, membranous, some species with rhizomorphs. Hyphae yellowish or brownish, without clamps. No cystidia. Basidia stalked when mature, 4-spored. Spores yellowish or brown, subspherical, verrucose, warts often dichotomously notched.

KEY

Fruitbodies ochraceous to pale orange brown, margin white, with
    slender rhizomorphs .... *mucidula*
Fruitbodies rust-coloured, brown or bluish black, no rhizomorphs
    .... *tristis*

Pseudotomentella mucidula (P. Karst.) Svrček (Fig. 358)
Fruitbodies loosely attached, soft membranous to spongy, about 0.5 mm thick, surface smooth to tuberculate, ochraceous to pale orange brown, margin white or pale, with slender rhizomorphs not more than 20 thick. Basidia 30–50 × 7–9. Spores subspherical, yellowish, verrucose to echinulate, 6–9 diam. excluding warts, 7–11 with, some warts dichotomously notched. On rotten wood and worked timber.

Pseudotomentella tristis (P. Karst.) M.J. Larsen
Fruitbodies membranous, up to 2 mm thick, smooth or wrinkled, rust-coloured, brown or bluish black, without rhizomorphs. Hyphae encrusted. Basidia 40–70 × 8–12. Spores subspherical, brown, verrucose, 7–11 diam. On rotten wood.

***Pteridomyces** galzinii* (Bres. in Bourd. and Galz.) Jülich (Fig. 359)
Fruitbodies thin, resupinate, membranous, cream, covered with very small, conical, sterile teeth. Hyphae hyaline, with clamps. No cystidia. Basidia clavate, 10–15 × 4–5. Spores ellipsoid, 5–8 × 3–4, hyaline, smooth, thin-walled, J–. On rotting fern fronds.

Pterula

Fruitbodies erect, very slender, often needle-like, solitary or in groups, unbranched to richly branched near the base, the branches vertical and bristle-like. Hyphae with clamps, not swollen. Cystidia present in some species. Basidia with 2 or 4 spores. Spores hyaline, smooth, J−, without any guttules or oil drops.

KEY

Spores more than 10 long 1
Spores less than 10 long 2
1. Basidia 2-spored, on grasses and other plants in swampy or wet places
    .... *gracilis*
    Basidia 4-spored, on *Carex pendula* .... *caricis-pendulae*
2. Cystidia present, on *Juncus* .... *debilis*
    No cystidia, not on *Juncus* 3
3. Spores 3–3.5 wide, on conifer twigs and needles .... *multifida*
    Spores 3.5–5 wide, on soil .... *subulata*

Pterula caricis-pendulae Corner

Fruitbodies mostly solitary, not branched, up to 15 × 0.1–0.2 mm, needle-shaped, white, without a sclerotium, stalk 1–4 mm long, brown. Cystidia in hymenium clavate, thin-walled, up to 45 × 10–20, on stalk conical, 30–65 × 5–18. Basidia 30–40 × 9–10, 4-spored. Spores narrowly ellipsoid to somewhat fusiform, 12–15 × 4.5–5. On dead leaves of *Carex pendula*.

Pterula debilis Corner

Fruitbodies fairly tough, solitary or in groups, up to 5 cm tall, loosely branched, branches not anastomosing, at first white, becoming pale brown from the base up, side-branches 0.1–0.3 mm thick, themselves scarcely branched, stalk 5–25 × 0.2–0.4 mm. Cystidia fusiform, 30–40 × 5–8. Basidia 4-spored, 20–25 × 6–7. Spores 7–9.5 × 3.5–4. On *Juncus* in swampy places.

Pterula gracilis (Desm. and Berk.) Corner (Fig.360)

Fruitbodies solitary or in groups, mostly unbranched, needle-shaped, 1–10 × 0.1–0.3 mm, at first white, then dirty flesh-coloured to pale brownish, very finely hairy. Cystidia subulate, thin-walled, 25–50 × 5–8. Basidia 2-spored, 20–30 × 6–10. Spores ellipsoid, 10–16 × 4.5–7. Common in wet, swampy places on dead leaves and stems of grasses, *Carex, Equisetum, Eupatorium, Iris, Juncus, Typha* etc.

Pterula multifida E.P. Fries ex Fr. (Fig. 361)

Fruitbodies in dense groups, 1–6 cm tall, richly branched from the base, bushy, the side branches not anastomosing, less than 1 mm thick, at first whitish, pale grey, pale yellowish or flesh-coloured, when old ochraceous or brownish, stalk up to 12 × 0.5–1.5 mm. No cystidia. Basidia 25–35 × 6–8,

4-spored. Spores 5.5–8 × 3–3.5, ellipsoid to pyriform. On needles and twigs of conifers, rarely on other twigs.

Pterula subulata Fr.
Fruitbodies in dense groups, 1–6 cm tall, richly branched from the base, the side-branches less than 1 mm thick, below flattened and up to 1.5–2.5 mm in the middle, anastomosing, pale greyish brown, becoming flesh-coloured just below the white tips, stalk buried, up to 10 × 1–3 mm. No cystidia. Basidia 30–50 × 8–9, 4-spored. Spores 6–8 × 3.5–5, ellipsoid or pyriform. On the ground in woods, parks, flower-beds etc., mostly on buried plant remains.

Pulcherricium *caeruleum* (Schrad. ex Fr.) Parm. (Fig. 362)
Fruitbodies resupinate, discrete and orbicular at first but soon becoming confluent and widely effused, firmly attached except at the edge, soft, membranous to waxy, up to 0.5 mm thick, surface smooth or tuberculate, dark blue when fresh with narrow, pale margin. Hyphae with clamps. Dendrohyphidia mixed with basidia, also some structures that are intermediate in form between the two. Basidia narrowly clavate, hyaline or blue, 40–60 × 5–7, 4-spored. Spores ellipsoid, 7–12 × 4–7, hyaline or pale bluish, thin-walled, smooth, J–. On dead branches of deciduous trees.

Pycnoporus *cinnabarinus* (Jacq. ex Fr.) P. Karst. (Fig. 363)
Cinnabar Polypore
Fruitbodies pileate, tough, broadly attached, flat, semicircular or fan-shaped brackets 3–10 × 2–8 cm, 0.5–2 cm thick; upper surface slightly velvety at first, soon smooth, sometimes tuberculate, wrinkled or zonately furrowed, bright cinnabar or reddish orange when fresh, paling with age; lower surface also reddish orange but darker, pores angular, 2–3 per mm; flesh turned black by KOH. Hyphae with clamps. No cystidia. Basidia 12–15 × 4–5. Spores cylindrical, slightly curved, 5–6 × 2–2.5, hyaline, smooth, thin-walled, J–. On dead wood of deciduous trees, summer–autumn, uncommon.

Ramaria
Fruitbodies erect, branched like coral, mostly growing on soil, only a few species on wood, brittle or tough. Clamps in most species. No cystidia. Basidia mostly 4-spored, with basal clamps. Spores ellipsoid to oblong-ellipsoid, fusiform in one species, yellowish, ochraceous or brownish, J–, warted, echinulate or striate, ornamentation C+.

KEY

Fruitbodies turning greenish when bruised .... *abietina*
Fruitbodies not greenish when bruised 1

1. Tips of branches bright pink, red or wine-coloured .... *botrytis*
   Tips of branches not so 2
2. Spores never more than 7 long 3
   Spores more than 7 long 4
3. Fruitbodies pale, bruising reddish, spore walls finely echinulate
   .... *myceliosa*
   Fruitbodies pale yellow–ochre, not bruising reddish, spore walls
   finely verruculose .... *gracilis*
4. Spores fusiform, 13–20 × 5–8 .... *broomei*
   Spores ellipsoid, never more than 16 × 6.5 5
5. Spore walls echinulate, spines up to 1 long 6
   Spore walls verrucose 7
6. Hymenium on one side of branches only, spores 6–9 long ...... *flaccida*
   Hymenium on both sides of branches, spores 6–11 long ...... *eumorpha*
7. Fruitbodies distinctly yellow 8
   Fruitbodies not so 9
8. Fruitbodies golden or yolk yellow, warts on spore walls in distinct
   longitudinal rows .... *aurea*
   Fruitbodies sulphur yellow, warts not in rows .... *flava*
9. Growing on wood, which may be buried 10
   Growing on soil 11
10. Fruitbodies pinkish ochraceous to ochraceous brown, darkening
    from base up but tips remaining pale, whitish or greenish .. *apiculata*
    Fruitbodies pale yellow–ochre to ochraceous brown bruising reddish
    brown, tips not pale .... *stricta*
11. Branches pinkish ochraceous to pinkish orange, with lemon yellow
    tips, spores 9–15 × 5–6 .... *formosa*
    Branches creamy ochraceous to milky coffee, spores 9–16 × 4.5–6.5
    .... *pallida*
    Branches other colours, spores not more than 13 long 12
12. Branches brown or beige often tinted lilac, warts on spore walls in
    rows .... *fumigata*
    Branches other colours, warts not in rows 13
13. Branches reddish brown to brown .... *condensata*
    Branches flesh-coloured to rosy ochraceous or ochraceous ...... *suecica*

Ramaria abietina (Pers. ex Fr.) Quél (Fig. 364)
Fruitbodies in groups or rows, seldom solitary, arising from a white
mycelial mat, richly and often dichotomously branched, 3–7 cm tall and 3–
5 cm wide, stalk 5–15 × 4–8 mm, branches rounded or flattened, 2–5 mm
thick, tips with 2–4 points, dirty ochraceous to somewhat olivaceous, more
greenish when old and always becoming greenish when bruised. Hyphae
with clamps. Basidia mostly 4-spored, with basal clamps, 40–60 × 6–7.5.
Spores ellipsoid to pip-shaped, yellowish, echinulate, 7–10 × 3.5–5, spines
up to 1 long. Amongst needle litter on the ground in conifer woods.

Ramaria apiculata (Fr.) Donk

Fruitbodies solitary or in clumps, up to 7 cm tall, stalk 3–4 mm thick, richly branched from or near the base, branches long, somewhat flattened, terminating in 2 or 3 long points, at first pinkish ochraceous or pale yellowish cream, becoming darker to ochraceous brown from the base up, the tips remaining pale, whitish or sometimes greenish. Hyphae with clamps. Basidia 35–45 × 7–8, 4-spored, with basal clamps. Spores ellipsoid, 7–10 × 3.5–5, finely verruculose. On wood, bark and litter of conifers, mainly *Pinus*.

Ramaria aurea (Schaeff. ex Fr.) Quél. (Fig. 365)

Fruitbodies solitary or in clumps, 5–12 cm tall, 8–20 cm wide, stalk 2–4 × 2–5 cm, white at the base, lemon yellow above, richly branched, branches short, the lower ones up to 2 cm thick, the terminal ones usually with bifurcate tips, golden or yolk yellow, slightly tinted rose. Hyphae without clamps. Basidia 4-spored, without basal clamps, 40–60 × 8–14. Spores ellipsoid, 8–15 × 4–6, finely verruculose, warts tending to be arranged longitudinally in rows, yellow in mass. On the ground in deciduous and mixed woods, most commonly under *Fagus*.

Ramaria botrytis (Fr.) Ricken

Fruitbodies solitary or gregarious, 8–15 cm tall, 7–20 cm wide, stalk white or yellowish, 3–4 × 2–6 cm, richly branched, the lower branches few and up to 3 cm thick, white to cream or ochraceous, the terminal ones each ending in 2–4 sharp points which are bright pink, red or wine colour. Hyphae with clamps. Basidia 4-spored, with basal clamps, 50–70 × 8–12. Spores narrowly ellipsoid, 14–20 × 4.5–6, pale yellow, longitudinally striate. In deciduous woods, late summer–autumn.

Ramaria broomei (Cotton and Wakef.) Petersen

Fruitbodies solitary or gregarious, 4–8 cm tall, 2–4 cm wide, stalk up to 4 × 1 cm, cylindrical, at first white, bruising pinkish, when old blackish brown, richly branched, branches up to 4 mm thick, orange–ochraceous, rapidly turning brown on bruising, then black, tips flattened, dark orange. Hyphae with clamps. Basidia 50–80 × 8–9, 2-spored. Spores broadly fusiform, 13–20 × 5–8, echinulate. Mostly in woods, under conifers.

Ramaria condensata (Fr.) Quél.

Fruitbodies 5–10 cm tall, 3–5 cm wide, stalk 1–2 × 0.5–1 cm, pale brown, richly branched, branches brown or rather reddish brown, erect, parallel, pointed. Basidia 30–40 × 4–5. Spores ellipsoid, 8–12 × 4–5, verrucose. In conifer and deciduous woods.

Ramaria eumorpha (P. Karst.) Corner

Fruitbodies solitary or in clumps, 5–8 cm tall, 4–6 cm wide, stalk distinct, 1–3 × 0.5–1.5 cm, white or pale ochraceous, with white felt or white or yellowish rhizomorphs at base, richly branched, branches erect, cylindrical,

pointed, the lower, main branches up to 2.5 mm thick, yellow–ochre turning ochraceous brown when old, terminal branches each ending in 2–3
sharp, yellowish points 1–2 mm long. Hyphae with clamps. Basidia 40–
50 × 8–10, mostly 4-spored, with basal clamps. Spores ellipsoid, 6–11 × 3.5–
5, distinctly echinulate with spines 0.5–1 long. Mostly in conifer woods.

Ramaria flaccida (Fr.) Bourdot
Fruitbodies usually in rows or forming fairy rings, 4–6 cm tall, 3–4 cm
wide, flaccid, stalk about 1.5 × 0.4 cm, ochraceous, whitish towards base
where there are cream coloured rhizomorphs, richly branched, branches 1–
3 mm thick, pale ochraceous, becoming yellowish brown, tips paler with 2–
3 points. Hyphae with clamps. Basidia 30–60 × 5–8, 4-spored, with basal
clamps. Spores ellipsoid, 6–9 × 3–5, echinulate, spines nearly 1 long. Mostly
in conifer woods.

Ramaria flava (Schaeff. ex Fr.) Quél
Fruitbodies solitary or in groups, sometimes forming rings, 10–20 cm tall,
7–15 cm wide, stalk short but distinct, 5–8 × 4–7 cm, lemon yellow, white
at base, richly branched but with relatively few main branches, pale yellow
below but mostly sulphur yellow, turning ochraceous when old, tips blunt
or with 2–3 points. Clamps mostly confined to hyphae in outer part of stalk.
Basidia 50–60 × 8–10, with basal clamps. Spores ellipsoid, 10–16 × 4–6, pale
ochraceous, verrucose, warts scattered. Mostly in deciduous woods.

Ramaria formosa (Fr.) Quél.
Fruitbodies solitary or in groups, chalky brittle when dry, 8–30 cm tall, 7–
15 cm wide, stalk 3–6 × 2–6 cm, with white base, richly branched, with
main branches 1.5–2 cm thick, pinkish ochraceous or pinkish orange, with
lemon yellow tips. Hyphae with clamps. Basidia 40–60 × 7–9, 4-spored,
with basal clamps. Spores oblong-ellipsoid, 9–15 × 5–6, verrucose. Mostly
in deciduous woods, especially under *Fagus*.

Ramaria fumigata (Peck) Corner
Fruitbodies solitary or gregarious, 6–13 cm tall, 5–12 cm wide, stalk white
or whitish tinted violaceous, 2–5 cm thick, richly branched, with the 2–4
main branches dividing repeatedly, violet when young, then brown or beige
tinted lilac, the 2 or 3 points at the tips often remaining purplish. Hyphae
with clamps. Basidia 30–70 × 6–10, 4-spored, with basal clamps. Spores
narrowly ellipsoid, 8–13 × 4–6, verruculose with warts often in rows. In
deciduous woods, mostly under *Fagus* and *Quercus*.

Ramaria gracilis (Fr.) Quél
Fruitbodies solitary or gregarious, 3–6 cm tall, 2–5 cm wide, stalk pale, up
to 1.5 × 0.4 cm, with white mycelial felt and rhizomorphs at base, fairly
richly branched, branches always slender, not more than 1–2 mm thick,
pale yellow–ochre, sometimes with flesh tints, whitish towards tips. Hyphae

with clamps. Basidia 30–45 × 5–7, 4-spored, with basal clamps. Spores ellipsoid, 4.5–7 × 3.5–4.5, hyaline, finely verruculose. In conifer forests, often growing on pieces of wood.

Ramaria myceliosa (Peck) Corner
Fruitbodies 2–5 cm tall, 1–3 cm wide, stalk up to 2 × 0.4 cm, yellow–ochre with white mycelial felt and rhizomorphs at base, branches up to 3 mm thick, somewhat flattened, at first pale cream, becoming olive-ochraceous, turning reddish when bruised. Hyphae with clamps. Basidia 30–35 × 4–4.5, 4-spored. Spores 4–6 × 2.5–3.5, finely echinulate. In pine woods.

Ramaria pallida (Schaeff. ex Schulzer) Ricken
Fruitbodies solitary or in groups, 4–15 cm tall and wide, stalk 3–8 × 2–4 cm, pale ochraceous brown, white at base, richly branched, with main branches 1–1.5 cm thick, pale creamy ochraceous to milky coffee colour, sometimes faintly tinted violet, more noticeably so at the tips. Hyphae and basidia without clamps. Basidia 4-spored, 50–70 × 8–12. Spores ellipsoid, 9–16 × 4.5–6.5, finely verruculose. In deciduous and mixed woods.

Ramaria stricta (Fr.) Quél
Fruitbodies solitary or gregarious, 4–10 cm tall, 4–6 cm wide, stalk 1–6 × 0.5–1 cm, pale, with white mycelial felt and rhizomorphs at base, richly branched, branches 1–5 mm thick, erect, more or less parallel, pale yellow–ochre to ochraceous brown or cinnamon brown, bruising reddish brown. Hyphae and basidia with clamps. Basidia 25–40 × 8–9, 4-spored. Spores ellipsoid, 8–10 × 4–5, verruculose. Mycelial felt contains thick-walled skeletal hyphae. On rotten wood, mostly of *Fagus*.

Ramaria suecica (Fr.) Donk
Fruitbodies drying chalky brittle, 3–10 cm tall, stalk 0.5–2 cm thick, with white felt and rhizomorphs at base, with few main branches below and with brush-like branched tips, pale flesh colour to rosy ochraceous or ochraceous, ends sometimes whitish. Hyphae with clamps. Basidia 30–50 × 5–8, 4-spored. Spores narrowly ellipsoid, 8–11 × 4–5, verruculose. In conifer woods.

Ramariopsis
Fruitbodies erect, stalked, branched several times, often dichotomously, sometimes antler-like, branches cylindrical or flattened, with points or rounded at apex, brittle or fairly tough, various colours. Hyphae more or less swollen, with clamps. Basidia mostly 4-spored. Spores broadly ellipsoid or subspherical, small, maximum length 5.5, hyaline, finely echinulate or warted, J–, each with a large oil-drop or vacuole.

KEY

Fruitbodies up to 12 cm tall, richly branched, white, cream or pale brown .... *kunzei*

Fruitbodies shorter, sparsely branched, other colours 1

1. Fruitbodies up to 5 cm tall, golden to yellowish orange .... *crocea*

Fruitbodies up to 3.5 cm tall, pale straw to pale yellow–ochre .... *tenuiramosa*

Fruitbodies 1–2 cm tall, violet or lilac, at least at the tips .... *pulchella*

Ramariopsis crocea (Pers. ex Fr.) Corner

Fruitbodies solitary or in small groups, up to 5 cm tall, golden yellow to yellowish orange, loosely branched, the branches divided dichotomously 2–4 times, cylindrical, 1–2 mm thick, bent, not brittle, tapered to a point or blunt at apex, stalk distinct, up to 10 × 1–2 mm, upper part smooth, yellow-ochre to orange, lower part white, felted. Basidia 15–25 × 4–5, 4-spored. Spores subspherical, 2.5–4.5 × 2–4, finely echinulate. Amongst grass in woods, has been recorded under *Juniperus* and *Rhododendron*.

Ramariopsis kunzei (Fr.) Corner

Fruitbodies solitary, gregarious or clustered, 3–12 cm tall, tough, richly and closely branched, pure white to cream or pale brownish, rarely tinted pinkish, stalk up to 2.5 × 0.5 cm, downy or felted, lower main branches 2–5 mm thick, divided many times, upper ones branched dichotomously, 1–2 mm thick, somewhat flattened, tapered to a point or rounded at apex. Basidia 25–45 × 6–7. Spores broadly ellipsoid to subspherical, finely echinulate, 3.5–5.5 × 2.5–4.5. On the soil, rarely on rotten wood, in woods and pastures.

Ramariopsis pulchella (Boud.) Corner (Fig. 366)

Fruitbodies solitary or clustered, rather soft, waxy, violet or lilac, 1–2 cm tall, slim, 1–3 times dichotomously branched, antler-like, with most of the violet colour in the spine-like tips, stalk short, 1–2 mm thick, white, yellow or reddish yellow, white tomentose at base. Basidia 20–30 × 5–7. Spores subspherical, finely verruculose, 3–4.5 × 2.5–3.5. On ground in woods.

Ramariopsis tenuiramosa Corner

Fruitbodies solitary or gregarious, up to 3.5 cm tall, pale straw-coloured or pale yellow–ochre, sparingly and loosely branched, stalk 3–10 × 1 mm, smooth or somewhat tomentose at base, lower branches divided 2–3 times, cylindrical, 0.5–1 mm thick, upper branching dichotomous. Basidia 30–40 × 5–7. Spores 3.5–5 × 3–4, finely verruculose. Amongst moss, grass, pine needles, heather etc.

Resinicium *bicolor* (Alb. and Schw. ex Fr.) Parm. (Fig. 367)

Fruitbodies resupinate, widely effused, firmly attached to substrate, thin, waxy to membranous, cream or pale ochraceous, often overgrown by green algae, surface toothed, teeth usually small and conical but sometimes longer. Hyphae with clamps. Cystidia mostly 10–20 long, with basal clamps, of two distinct kinds: (1) capitate, each with a thin apical bladder 15–20 diam., containing oily or resinaceous matter, (2) tapered and capped by star-like clusters of crystals. Basidia 4-spored, up to 25 × 5–6. Spores oblong, often slightly curved, hyaline, smooth, 5–8 × 2.5–3.5, J–, C–. On decorticated wood, mostly of conifers, e.g. *Picea*.

Rigidoporus *ulmarius* (Sow. ex Fr.) Imazeki (Fig. 368)
Fruitbodies pileate, broadly attached, roughly semicircular but often malformed brackets 10–50 × 7–20 cm, 3–8 cm thick, tough and corky fresh, woody hard when dry, with thick, rounded margin; upper surface slightly tomentose at first, soon smooth, concentrically ridged and sometimes lumpy, cream to dirty wood-colour; lower surface orange or pinkish brown, drying buff, pores 5–8 per mm, round or angular, tubes stratified, about 5 mm long, pinkish brown to brown tube layers separated by bands of pale flesh. Cystidia in hymenium subulate, ventricose, 15–25 × 5–9. Basidia 10–15 × 6–7. Spores spherical or almost so, 6–7.5 diam., hyaline to pale yellowish, smooth, J–. At the base of trunks of deciduous trees, especially *Ulmus*, common.

Sarcodon
Fruitbodies terrestrial, erect, solitary or in small groups, pileate, centrally stalked, stalks sometimes concrescent, hymenium on lower surface of cap toothed, flesh brittle, firm or rather soft, not corky. No cystidia. Basidia 4-spored. Spores brownish, irregular in outline, tuberculate.

KEY

Large scales on cap erect, in concentric circles, hyphae with clamps .... *imbricatus*
Scales on cap not so, hyphae without clamps 1
1. Flesh in cap pale purple or bluish violet .... *fuligineo-violaceus*
   Flesh in cap other colours 2
2. Scales often reddish brown, flesh in cap soon reddish .... *scabrosus*
   Scales often greyish brown, flesh in cap mostly yellowish grey ... *regalis*

Sarcodon fuligineo-violaceus (Kalchbr. apud Fr.) Pat.
Cap up to 13 cm diam., convex, flat or somewhat sunken in the middle; upper surface at first downy, then with separated and slightly raised scales, at first yellowish brown, then dark brown or slightly bluish, finally blue–grey; lower surface becoming purplish brown, teeth up to 4 mm long; flesh pale purple, then bluish grey or bluish violet. Stalk solid, 3–6 × 1–4 cm, cylindrical, tomentose to smooth, paler than or concolorous with cap, flesh

reddish, greyish green in base. Hyphae without clamps. Spores 5.5–6.5 × 4–5, brownish, tuberculate. In conifer woods.

Sarcodon imbricatus (L. ex Fr.) P.Karst.
Cap round or somewhat irregular, 5–30 cm diam., at first convex then flattening and becoming depressed in the middle, with inrolled margin, when old sometimes funnel-shaped, upper surface velvety, with large, erect, often dark brown or blackish brown scales arranged more or less in concentric circles and contrasting with the paler cap, lower surface grey to purplish brown, teeth up to 1 cm long, decurrent. Stalk solid, 4–8 × 2–4 cm, cylindrical or slightly swollen at base, velvety, at first whitish, turning brownish. Flesh mostly whitish, but may be brown in base. Hyphae with clamps. Basidia 4-spored, 30–40 × 7–8. Spores subspherical, 6.5–8 × 5–6, coarsely tuberculate, brownish. In pine woods.

Sarcodon regalis Maas Geesteranus
Cap up to 10 cm diam., slightly convex to flat; upper surface yellowish brown, with greyish brown or dark brown scales in the middle which lie flat or have slightly raised tips; lower surface becoming purplish brown, teeth up to 4 mm long, flesh whitish to pale yellowish grey, tinted wine red or mottled violet below the surface. Stalk 5–6 × 1.5–2 cm, concrescent or branched, solid, cylindrical or tapered towards the base, velvety or tomentose, upper part pale orange brown, lower part pale, dirty violet; flesh in the upper part pale reddish brown, in the lower part greyish green. Hyphae without clamps. Spores 5.5–6 × 4–5, brownish, tuberculate. In conifer and mixed woods.

Sarcodon scabrosus (Fr.) P.Karst. (Fig. 369)
Cap up to 14 cm diam., slightly convex, often sunken in the middle, upper surface dirty yellowish, at first velvety or tomentose, then cracked becoming tesselated, finally with large, separate, reddish to dark brown scales, flesh whitish, then soon reddish or occasionally yellowish, lower surface for a long time pale yellowish, finally purplish brown, teeth up to 1 cm long. Stalk 3–10 × 1–3 cm, solid, cylindrical or tapered towards base, tomentose to fibrous scaly, upper part brownish, lower part greyish brown to grey–green, flesh in lower part grey–green. Hyphae without clamps. Spores brownish, tuberculate, 6–7.5 × 4–5. In deciduous and mixed woods.

Sarcodontia *setosa* (Pers.) Donk (Fig. 370)
Fruitbodies resupinate, effused, firmly attached to substrate, waxy, with numerous crowded, acutely pointed, conical teeth 0.5–2 cm long, sulphur yellow, with teeth sometimes turning reddish. Hyphae with clamps. No cystidia. Basidia narrowly clavate, 20–40 × 4–5, 4-spored. Spores broadly ellipsoid to subspherical, with basal apiculus, 4.5–6 × 3.5–4, hyaline, smooth, J–, rather thick-walled, each with one oil drop. Parasitic on

Rosaceae, especially *Malus*, rarely on other trees and shrubs, usually in knot-holes or under rotten bark.

***Schizopora** paradoxa* (Schrad. ex Fr.) Donk (Fig. 371)
Fruitbodies resupinate, effused, firmly attached to substrate, sometimes with protruding fertile nodules on vertical surfaces, but no true pilei, white to cream or ochraceous, darkening with age, 1–5 mm thick, soft but tough, brittle when dry, hymenium poroid to irpicoid, pores at first rounded to angular, 1–3 per mm but soon becoming irregular, sometimes labyrinthine, walls splitting and often elongating to form teeth. Hyphae with clamps, hyphal ends often encrusted with granular crystals. Capitate cystidioles present, heads often covered with resinous matter. Basidia 12–20 × 3–5. Spores ellipsoid, 5–6 × 3.5–4, hyaline, smooth, thin-walled, J–, usually each with one oil drop. On dead attached and fallen branches mostly of deciduous trees, a very common species.

***Scopuloides** rimosa* (Cooke) Jülich (Fig. 372)
Fruitbodies resupinate, effused, firmly attached to substrate, thin, gelatinous to waxy, greyish and rather transparent, sometimes with lilac tinge, teeth short, often rounded, pale. Hyphae with short cells, without clamps. Basidia hyaline to brownish, full of granules or small guttules, walled, encrusted, 40–70 × 7–10, projecting above hymenium; in the teeth there are cylindrical, septate, encrusted ones 50–120 × 8–15. Basidia 10–15 × 4–5. Spores cylindrical to narrowly ellipsoid, some slightly curved, 3.5–4 × 1.5–2, hyaline, smooth J–. On rotten wood of deciduous trees, common on fallen branches of *Fagus*.

***Scotomyces** subviolaceus* (Peck) Jülich (Fig. 373)
Fruitbodies resupinate, loosely attached to substrate, several cm diam., less than 1 mm thick, softly membranous to somewhat gelatinous, greyish olive to dark purplish grey, paler when dry. Hyphae hyaline to brown, with clamps. Basidia hyaline to brownish full of granules or small guttules, clavate, 20–35 × 6–9, each terminating in 4 subulate sterigmata up to 16 long. Spores hyaline or brownish, packed with granules or guttules, J–, ellipsoid, 7–9 × 4–6, forming secondary spores. On rotten wood and other debris.

***Scytinostroma** ochroleucum* (Bres. and Torr.) Donk (Fig. 374)
Fruitbodies resupinate, effused, membranous soft to leathery, smooth, cream, pale ochraceous or pale cinnamon brown. Hyphae without clamps, skeletal ones yellow, thick-walled. Cystidia abundant, with resinous-oily contents, tubular, sinuous, sometimes tapered to a point, 50–200 × 6–10. Basidia clavate, 4-spored, 40–70 × 6–10. Spores broadly ellipsoid, 10–15 × 5–8, hyaline, smooth, J–. On wood of both deciduous trees and conifers, uncommon.

Sebacina

Fruitbodies resupinate, effused, waxy, gelatinous or leathery. Hyphae hyaline, without clamps. Basidia hyaline, without clamps, with longitudinal septa and 2-4 sterigmata. Spores hyaline, thin-walled, smooth, J-, forming secondary spores.

KEY

Spores 20-35 long .... *calospora*
Spores not more than 18 long 1
1.   Fruitbodies thick, leathery, often encrusting parts of plants .. *incrustans*
     Fruitbodies thin, gelatinous, on soil and humus .... *epigaea*

Sebacina calospora (Bourd. and Galz.) Bourd. and Galz.
Fruitbodies resupinate, greyish or whitish, waxy to gelatinous. Basidia subspherical, 8-11 diam., sterigmata up to 10 long. Spores narrowly spindle-shaped, sometimes curved, 20-35 × 3-4. On wood, e.g. *Fagus*, and on herbaceous plants.

Sebacina epigaea (Berk and Br.) Neuh.
Fruitbodies resupinate, effused, gelatinous, thin, whitish or pale grey, when dry scarcely visible. Basidia 15-22 × 10-12, sterigmata up to 100 × 3. Spores 7-13 × 5-9. On soil and humus.

Sebacina incrustans (Pers. ex Fr.) Tul. (Fig. 375)
Fruitbodies forming irregular crusts, sometimes *Thelephora*-like, rather thick, lightly attached to substrate, cartilaginous or leathery, dirty greyish white to cream, ochraceous or reddish brown. Basidia ellipsoid, 15-25 × 10-12, sterigmata up to 90 × 3. Spores ellipsoid, 10-18 × 8-10, with granular contents. On soil, debris, or encrusting bases of living herbaceous plants including grasses or those of woody plants.

Serpula

Fruitbodies resupinate or effuso-reflexed, fleshy or waxy, with merulioid to reticulate-poroid or labyrinthine hymenium. Rhizomorphs contain skeletal hyphae. Hyphae with clamps. Cystidioles sometimes present. Basidia 4-spored, with basal clamps. Spores broadly ellipsoid, often flattened on one side, smooth, yellowish to brownish or rust-coloured in mass, with rather thick, double walls and an apical germ pore, J-, C+.

KEY

Fruitbodies on conifer trunks and branches in woods, mostly 2-5 cm
     diam. .... *himantioides*
Fruitbodies found only on worked timber in buildings, often very
     large, causes dry rot .... *lacrimans*

Serpula himantioides(Fr.) P.Karst.
Fruitbodies mostly 2–5 cm across, up to 2 mm thick, almost smooth and lilac when quite young, soon becoming merulioid to reticulate-poroid, yellowish, olivaceous, cinnamon or dark brown with a narrow white or whitish border, subhymenium not or scarcely gelatinous, rhizomorphs greyish, usually not more than 1 mm thick. Basidia 30–70×6–9. Spores almost ellipsoid, often flattened on one side, yellowish, 9–12×5–7. On conifer wood especially on lower side of large logs, causes brown rot.

Serpula lacrimans(Wulf ex Fr.) Schroet. (Fig. 376)
Dry-rot Fungus
Fruitbodies resupinate or effuso-reflexed, fleshy, often covering very large areas, occasionally forming brackets on vertical surfaces which are up to 2 cm thick and may project as much as 10 cm, hymenium mostly merulioid, sometimes reticulate-poroid or labyrinthine, sulphur yellow to dark rust brown, with a thick, white or slightly yellowish margin and rhizomorphs. Subhymenium thick, often gelatinous. Basidia 30–70×6–8. Spores 9–12×5–9, rust brown in mass. Mostly on worked coniferous timber in buildings. Wood attacked by this species cracks both along and across the grain, breaking into cubical chunks. Hyphae of the fungus permeate the wood in all directions. Under damp conditions the mycelium forms on the surface of the wood a fluffy white growth with patches of yellow here and there and on drying this may turn into a greyish skin with occasional touches of pale purple. The external mycelium gives rise to thick grey rhizomorphs and these find their way through brickwork and plaster, mortar and stone walls, along pipes and over the surface of concrete, often travelling long distances before finding more wood. During its action on wood a great deal of water is formed by the fungus and this is carried by the rhizomorphs enabling them to attack wood that is only slightly moist. Droplets of water are seen on the mycelium as well as on the fruitbodies; the specific epithet of the fungus '*lacrimans*' means weeping.

Sistotrema
Fruitbodies in most species resupinate, effused, pellicular, membranous or waxy, usually white or whitish, in one species pileate and fan-shaped, hymenium smooth, tuberculate, toothed or reticulate-poroid. Hyphae often with oily contents, sometimes swollen at the septa, generally with clamps. Cystidia in only one species. Basidia rather short, urniform, nearly always with 6–8 sterigmata and a basal clamp. Spores hyaline, thin-walled, smooth, J–, C–.

KEY

Fruitbodies pileate, fan-shaped, stalked .... *confluens*
Fruitbodies resupinate, effused 1

1. Hymenium toothed .... *brinkmannii*
    Hymenium smooth or slightly tuberculate 2
2. Cystidia present, spores 5–7 long .... *coroniferum*
    Cystidia absent, spores 3–6 long 3
3. Spores subspherical or broadly ellipsoid, 3–5 wide .... *diademiferum*
    Spores narrower 4
4. Spores 1.5–2 wide .... *oblongisporum*
    Spores 2–2.5 wide .... *commune*

Sistotrema brinkmannii (Bres.) J. Erikss. (Fig. 377)
Fruitbodies resupinate, effused, becoming tuberculate to toothed, teeth conical or cylindrical, rounded at apex, up to 1 mm long, membranous to waxy, soft and white when fresh, brittle and pale cream to ochraceous when dry, crystals often present in hymenium. Basidia 10–20 × 5–8, urniform, with 6–8 sterigmata. Spores mostly 4–5 × 2–2.5, slightly curved. On very rotten wood and debris.

Sistotrema commune J. Erikss.
Fruitbodies resupinate, effused, smooth or tuberculate, loosely attached, thin, soft, membranous, whitish. Basidia 20–30 × 5–7, with 6–8 sterigmata. Spores narrowly ellipsoid or pyriform, 4–6 × 2–3. On dead wood and bark and overgrowing mosses.

Sistotrema confluens Pers. ex Fr. (Fig. 378)
Fruitbodies often concrescent, mostly pileate, fan- to funnel-shaped, with incised margin, 1–2 cm wide, tapered gradually into a stalk about 1 cm long and 2 mm thick, white to cream and soft, bruising and ageing yellow or brownish, brittle when dry, upper surface tomentose, smooth or concentrically wrinkled, lower surface reticulate-poroid to irpicoid or toothed with teeth 1–2 mm long, stalk base blackish. Basidia 12–20 × 5–7, with 6–8 sterigmata. Spores 4–6 × 2–2.5, slightly curved. Amongst mosses and leaf litter in woods.

Sistotrema coroniferum (Höhn. and Litsch.) Donk (Fig. 379)
Fruitbodies resupinate, effused, loosely attached, smooth, thin, membranous, white. Cystidia (g) present but often difficult to find, cylindrical or slightly tapered, flexuous, 50–90 × 6–9, thin-walled, hyaline or yellowish, containing granules or oil drops. Basidia 15–25 × 4–7, with 6–8 sterigmata. Spores 5–7 × 2–3, oblong-ellipsoid, slightly curved. On wet, decaying wood of both deciduous trees and conifers.

Sistotrema diademiferum (Bourd. and Galz.) Donk. (Fig. 380)
Fruitbodies resupinate, thin, effused, firmly attached to substrate, whitish or slightly greyish, pale yellow when dry, smooth, waxy. Basidia 10–30 × 5–8, with 6–8 sterigmata. Spores broadly ellipsoid to subspherical, 3–6 × 3–5. On fallen branches of both deciduous trees and conifers.

Sistotrema oblongisporum Christ. and Hauerslev (Fig. 381)
Fruitbodies resupinate, effused, firmly attached to substrate, smooth, thin, whitish grey, waxy. Basidia 10–25 × 5–7, with 6–8 sterigmata. Spores 4.5–6 × 1.5–2, mostly curved. On fallen branches of deciduous trees.

Skeletocutis

Fruitbodies pileate, effuso-reflexed or resupinate, sessile, white, ochraceous or brown, poroid. Hyphae of the dissepiments, especially in the region of the pore mouth, strongly encrusted with crystalline or amorphous matter. Cystidioles hyaline, subulate to fusiform, thin-walled, smooth. Basidia 4-spored. Spores mostly curved, hyaline, smooth, J–, 3–5 × 0.5–1.5.

KEY

Fruitbodies resupinate .... *tschulymica*
Fruitbodies pileate or effuso-reflexed 1
1. Pores 3–4 per mm .... *amorpha*
    Pores 6–8 per mm .... *nivea*

Skeletocutis amorpha (Fr.) Kotl. and Pouz.
Fruitbodies leathery when fresh, brittle when dry, mostly effuso-reflexed but sometimes pileate or resupinate, broadly but loosely attached to substrate, brackets up to 10 × 1–6 cm, 2–5 mm thick, solitary, fused laterally or imbricate, margin wavy, fimbriate; upper surface tomentose to smooth, white or greyish white; lower surface and resupinate parts at first white, soon becoming orange–pink or salmon pink, with white margin, pores thin-walled, angular, 3–4 per mm. Flesh 2-layered, upper layer white, fibrous, containing skeletal hyphae, lower one yellowish, gelatinous to horny. Hyphae with clamps, ends sometimes encrusted. Cystidioles fusiform, 10–15 × 3–5. Basidia 10–15 × 4–5. Spores allantoid, 3–5 × 1–1.5. On dead wood and bark of conifers.

Skeletocutis nivea (Jungh.) Keller (Fig. 382)
Fruitbodies pileate to effuso-reflexed, corky tough to hard, roughly semicircular brackets 1–6 × 1–3 cm, 2–4 mm thick, sometimes fused laterally, upper surface at first white or cream and finely tomentose, becoming smooth and brown or dark brown with a white edge when old, lower surface white or cream, bruising brownish, pores very small, 6–8 per mm. Hyphae with clamps. Cystidioles few, fusiform. Basidia 8–10 × 3–4. Spores slightly curved, 3–4 × 0.5–1. On dead branches of deciduous trees.

Skeletocutis tschulymica (Pilát) Keller
Fruitbodies resupinate, effused, up to 1 cm thick, when fresh soft gelatinous, when dry brittle, white at first, bruising and ageing ochraceous brown, pores 2–4 per mm, angular or slightly elongated, margin white,

1–2 mm wide. Hyphae with clamps. Cystidioles subulate, 8–14 × 2–3. Basidia 8–12 × 4–5. Spores often slightly curved, 3.5–5 × 1–1.5. Mostly on wood of conifers, rarely on that of deciduous trees.

Sparassis

Fruitbodies mostly large, fleshy to elastic, pulvinate or roughly hemi-spherical, resembling sponges or cauliflowers, with thick stalks and numerous ribbon-like, flat, wavy, crisped or crinkled branches and lobes which sometimes anastomose, they bear the hymenium mainly on the lower surface and are usually cream, straw or pale ochraceous with edges tinged brown when old. Basidia narrowly clavate, with clamps, 4-spored. Spores ellipsoid or ovoid, hyaline, cream in mass, smooth, J–.

KEY

    Fruitbodies not more than 8 cm tall, spores 5–9 long .... *simplex*
    Fruitbodies up to 20–25 cm tall, spores 4.5–6 long 1
1. In coniferous woods, mostly with *Pinus*, branches and lobes crisped or crinkled .... *crispa*
    In deciduous woods, mostly with *Quercus*, branches and lobes not crisped or crinkled .... *laminosa*

Sparassis crispa (Wulf.) Fr. (Fig. 383)
Cauliflower or Brain Fungus
Fruitbodies up to 20 cm tall and 10–40 cm wide, branches and lobes crisped or crinkled. Spores ovoid, 4.5–6 × 3–4.5. On the ground in conifer woods, mostly under *Pinus* spp. and parasitic on their roots.

Sparassis laminosa Fr.
Fruitbodies up to 25 cm tall, 30–40 cm wide, branches erect, densely crowded, leaf-like, not much crisped or crinkled, straw yellow. Spores ovoid, 4.5–6 × 3–4.5. In deciduous woods, mostly associated with *Quercus*.

Sparassis simplex D.A. Reid
Fruitbodies not more than 8 cm tall and 9 cm wide, somewhat flat fan-shaped, divided many times into flat segments, creamy ochraceous with pale cream edges, basal part resupinate, fertile. Spores ellipsoid, 5–9 × 3.5–5. On debris of conifers.

Spongipellis

Fruitbodies pileate or effuso-reflexed, white to ochraceous, soft and fleshy or leathery when fresh, hard when dry, hymenium poroid, irpicoid or hydnoid; flesh white to cream, 2-layered, upper part soft, lower part firmer, fibrous, tough. Hyphae with clamps. No cystidia. Basidia slender, clavate,

4-spored. Spores broadly ellipsoid to spherical, hyaline, smooth, with rather thick walls, J–.

KEY

> Hymenium toothed, irpicoid or hydnoid .... *pachyodon*
> Hymenium poroid 1
> 1. Pores round, entire, 1–4 per mm .... *spumeus*
>   Pores irregular, lacerate, 1–3 mm wide .... *delectans*

Spongipellis delectans (Peck) Murrill
Fruitbodies pileate, solitary or imbricate, soft and fleshy when fresh, hard and brittle when dry, flat, semicircular, broadly attached brackets 4–14 × 4–12 cm, 0.5–3 cm thick; upper surface finely tomentose and white to cream when young, becoming more or less smooth and pale ochraceous when old; lower surface concolorous with upper surface, pores large and irregular, angular to labyrinthine, lacerate, sometimes irpicoid when quite old, tubes up to 1.5 cm long. Spores broadly ellipsoid, 5.5–8 × 4.5–6. On deciduous trees, e.g. *Fagus, Populus* and *Tilia*.

Spongipellis pachyodon (Pers.) Kotl. and Pouz. (Fig. 384)
Fruitbodies pileate to effuso-reflexed, solitary or imbricate, leathery when fresh, hard when dry, broadly attached, roughly semicircular brackets, sometimes with lobed margins, measuring about 5 cm each way, up to 1.5 cm thick; upper surface at first white and finely tomentose, becoming ochraceous and smooth; lower surface white to ochraceous, irpicoid to hydnoid with teeth up to 1 cm long. Spores spherical or subspherical, 5–7 diam., with large oil drops. On deciduous trees, e.g. *Acer, Fagus* and *Quercus*.

Spongipellis spumeus (Sow. ex Fr.) Pat. (Fig. 385)
Fruitbodies pileate, usually solitary, soft and fleshy fresh, hard when dry, flat and broadly attached or dimidiate more narrowly attached brackets 5–25 × 4–10 cm, 2–10 cm thick at base; upper surface finely tomentose or hairy, fresh white to cream, dry ochraceous; lower surface white to pale ochraceous, pores round, entire, 1–4 per mm, tubes up to 1.5 cm long. Spores broadly ellipsoid to subspherical, 6–9 × 5–7. On standing deciduous trees.

Steccherinum
Fruitbodies tough, resupinate or effuso-reflexed, sometimes imbricate, hymenium toothed. Hyphae with clamps, skeletal ones thick-walled. Cystidia abundant in and projecting from the ends of teeth, thick-walled, encrusted towards the rounded tips. Basidia small, 4-spored. Spores ellipsoid, small, hyaline, smooth, thin-walled, J–, C–.

Fruitbodies resupinate, pinkish grey or greyish violet, margin fringed .... *fimbriatum*
Fruitbodies effuso-reflexed, cream, ochraceous or pale orange
.... *ochraceum*

Steccherinum fimbriatum (Pers. ex Fr.) J. Erikss. (Fig. 386)
Fruitbodies resupinate, loosely attached to substrate, effused, pale pinkish grey or greyish violet, drying ochraceous brownish, clearly fringed margin with rhizomorphs, teeth up to 0.5 mm long, conical but splayed out at tip. Cystidia abundant in teeth and protruding from their ends, strongly encrusted, 8–10 thick. Basidia clavate, 15–20 × 4–4.5, with basal clamps. Spores ellipsoid, 3–4 × 2–2.5. On dead wood and bark, mostly of deciduous trees.

Steccherinum ochraceum (Pers. ex Fr.) S.F. Grey (Fig. 387)
Fruitbodies effuso-reflexed or resupinate, cream to ochraceous imbricate, the reflexed parts standing out 0.5–1 cm, with velvety and sometimes zonate upper surface, teeth 0.5–3 mm long. Cystidia abundant especially in the teeth but also between them, upper part strongly encrusted, 7–10 thick. Basidia 15–20 × 4–5. Spores ellipsoid, 3–4.5 × 2–2.5. On dead wood mostly of deciduous trees and common on *Fagus*, occasionally also on *Lupinus arboreus*, *Ulex europaeus* etc.

Stereopsis *vitellina* (Plowr.) Reid (Fig. 388)
Fruitbodies solitary or concrescent, erect, short-stalked, up to 3 cm tall, leathery, thin, spathulate or fan-shaped, with deeply incised or crenate margin, upper surface yellow when fresh, drying creamy ochraceous, lower surface yellow or orange-yellow, smooth or radially wrinkled to merulioid. Hyphae without clamps. No cystidia. Basidia narrowly clavate, 4-spored, 30–40 × 4–5. Spores ellipsoid, 3–4 × 2–3, hyaline, smooth, thin-walled, J–. On the ground, sometimes attached to pieces of wood, uncommon.

Stereum

Fruitbodies effuso-reflexed, pileate or resupinate, sessile or shortly stalked, leathery becoming woody, hymenium smooth or tuberculate. Skeletal hyphae present, their ends often simulating cystidia, no clamps. Basidia narrowly clavate. Spores hyaline, thin-walled, smooth, J+.

Hymenium bleeding red when rubbed 1
Hymenium not bleeding when rubbed 3
1. Fruitbodies on conifer wood .... *sanguinolentum*

Fruitbodies on deciduous wood 2
2. Fruitbodies almost entirely resupinate .... *rugosum*
    Fruitbodies effuso-reflexed, brackets standing out 1–2 cm, brown
    with white margin, mostly on *Quercus* .... *gausapatum*
3. Fruitbodies pileate, stalked, often on *Alnus* .... *subtomentosum*
    Fruitbodies effuso-reflexed 4
4. Bracket parts standing out 1–3 cm, hymenium bright yellow or
    orange–yellow when fresh, spores mostly 5.5–7 long .... *hirsutum*
    Bracket parts not standing out more than 0.5 cm, hymenium greyish
    yellow, spores mostly 7–9 long .... *ochraceo-flavum*

Stereum gausapatum (Fr.) Fr.
Fruitbodies effuso-reflexed, reflexed part forming brackets 1–2 mm thick,
standing out 1–2 cm, often fused laterally or imbricate, upper surface hairy,
ochraceous brown, greyish brown or rusty from hairs, with a white,
undulating margin, lower surface and resupinate areas chestnut or greyish
brown, bleeding red when rubbed, margin white. Spores 6–10 × 3–4.5. On
dead branches of deciduous trees, most commonly on *Quercus.*

Stereum hirsutum (Willd. ex Fr.) S.F. Gray (Fig. 389)
Hairy Stereum
Fruitbodies effuso-reflexed, pileate or resupinate, brackets standing out 1–
3 cm, fused laterally or imbricate, upper surface yellow or ochraceous when
fresh and covered with whitish grey shaggy hairs, somewhat zonate, fading
to greyish when old, hymenial surface bright yellow or orange–yellow,
greyish brown when old. In section a dark line is seen below the tomentum.
Spores 5.5–7 × 2.5–3. On stumps, logs and fallen branches of deciduous
trees, very common.

Stereum ochraceo-flavum (Schw.) Ellis
Fruitbodies effuso-reflexed or forming small brackets which stand out only
0.5 cm and are up to 1 cm long, often concrescent and forming long rows,
pale ochraceous, covered with greyish white shaggy hairs, hymenial surface
smooth or tuberculate, greyish yellow. Spores mostly 7–9 × 2–3. Usually on
thin twigs or small branches of deciduous trees.

Stereum rugosum (Pers. ex Fr.) Fr. (Fig. 390)
Common Bleeding Stereum
Fruitbodies almost entirely resupinate, sometimes coming away at the edge
and protruding a few mm when old, up to 3.5 mm thick, hymenial surface
smooth or uneven, ochraceous to pinkish buff, greyish when old, bleeding
red when rubbed. Spores 7–12 × 3.5–4.5. On dead attached and fallen
branches of deciduous trees.

Stereum sanguinolentum (Alb. and Schw. ex Fr.) Fr.
Fruitbodies effuso-reflexed, the reflexed part forming brackets standing out

up to 1.5 cm, 0.5 mm thick, with wavy or crenate pale edges, often in long rows and imbricate, upper surface hairy, greyish brown or ochraceous cinnamon, lower surface and resupinate parts smooth or tuberculate, greyish buff or brownish, frequently tinted violet, bleeding red when rubbed. Spores 6.5–9 × 2–3. Common on coniferous wood and bark.

Stereum subtomentosum Pouzar
Fruitbodies mostly pileate and shortly stalked, rarely effuso-reflexed, semicircular or fan-shaped, leathery brackets 3–7 × 3–5 cm, sometimes fused laterally, often imbricate, upper surface shortly tomentose to smooth, undulating, concentrically zonate, ochraceous to reddish brown, becoming grey or greyish with whitish margin, hymenial surface smooth or tuberculate, yellow to pinkish ochraceous or greyish yellow. Spores 5.5–7 × 2–3. On dead wood of deciduous trees especially *Alnus* and sometimes also *Salix* growing in alder carrs.

Stigmatolemma *poriaeforme* (Pers. ex Fr.) Cooke (Fig. 391)
Fruitbodies sessile, grey or brownish cups 1–2 mm tall, closely packed together on a greyish stroma, cups hairy on the outside and covered with rosette-like crystals, hymenium lining them pale grey. Hyphae of trama partly gelatinized. No cystidia. Basidia 18–25 × 5–8, with 2–4 straight sterigmata 5–6 long. Spores spherical or subspherical, 5–7 diam., hyaline, smooth, each with a large guttule, J–. On wood and bark, uncommon.

Stilbum *vulgare* Tode ex Mérat (Fig. 392)
Fruitbodies erect, stalked, 1–2 mm tall, resembling coremia with hyphae held together in a gel, white or yellowish, fertile head spherical to ovoid, stalk sterile, smooth. Basidia pear-shaped or narrowly ellipsoid, with one transverse septum, with or without short sterigmata. Spores ellipsoid, hyaline, smooth, thin-walled, mostly 8 × 5–6. On rotten wood, uncommon.

Strobilomyces *floccopus* (Vahl ex Fr.) P.Karsten (Fig. 393)
Fruitbodies pileate, centrally stalked. Cap 6–13 cm diam., convex, upper surface grey or greyish brown, covered with large floccose, woolly, overlapping scales which protrude beyond the margin making it look shaggy, pore surface whitish, then grey or somewhat olivaceous, reddening when bruised, pores large, angular. Stalk 7–12 × 1–2 cm, cylindrical, grey or greyish brown, covered with large scales up to the peronate ring, above this paler, smooth or slightly sulcate. Flesh at first white, when cut turning gradually wine-coloured, then brown. Cystidia fusiform or lageniform. Spores 10–12 × 9–11, almost spherical, with reticulate walls, purplish black in mass. In woods, early autumn, uncommon.

Stromatoscypha *fimbriata* (Pers. ex Fr.) Donk (Fig. 394)
Fruitbodies resupinate, effused, loosely attached to substrate, membranous, about 1 mm thick, white to cream, appearing poroid but in fact made up of

separate, crowded cups each 0.1–0.3 mm diam., embedded in a soft, cottony or fibrous subiculum. Hyphae with clamps. No cystidia. Basidia 20–25 × 6–7, 4-spored. Spores ellipsoid, 4.5–5.5 × 2.5–3, hyaline, smooth, J–, with guttules. On the lower surface of fallen trunks and branches of deciduous trees, especially *Fagus*.

Stypella vermiformis (Berk.) Reid (Fig. 395)
Fruitbodies resupinate, effused, whitish or very pale ochraceous, glistening, thin, waxy, with crowded teeth which are 0.1–0.2 mm long or occasionally longer. Hyphae hyaline, with clamps. Cystidia (g), which are formed in the main axis of each tooth, cylindrical, flexuous, hyaline, up to 200 × 5–12. Basidia broadly ellipsoid to subspherical, 7–12 × 7–9, with longitudinal septa and 2–4 sterigmata up to 20 long. Spores subspherical or broadly ellipsoid, with protuberant hilum, 4–6 × 3.5–4.5, hyaline, smooth, with guttules. On rotten wood of conifers.

Subulicystidium longisporum (Pat.) Parm. (Fig. 396)
Fruitbodies resupinate, effused, loosely attached to substrate, arachnoid or membranous, white or whitish, with protuberant cystidia. Hyphae with clamps at all septa. Cystidia abundant, 50–100 × 3–4, subulate, encrusted, with the encrustations arranged spirally in rows. Basidia 10–15 × 4–5, 4-spored, with basal clamps. Spores narrowly fusiform to sigmoid, 10–20 × 2–3, hyaline, smooth, J–. On fallen dead wood and other debris.

Suillus

Fruitbodies terrestrial, erect, fleshy spongy, composed of a more or less convex or flattened cap and a central stalk; cap with upper surface usually glutinous or viscid when moist, tubes frequently short and sometimes decurrent, stalk with ring and/or glands. Cystidia in tubes fasciculate, often encrusted. Spores in mass olivaceous, ochraceous or snuff brown, mostly under 12 long.

KEY

    Growing under *Taxus* .... *tridentinus*
    Under *Larix*, stalk always with ring or ring zone 1
    Under other conifers, mainly *Pinus*, stalk with or without ring 4
1. Cap upper surface orange to burnt sienna .... *tridentinus*
    Cap surface other colours 2
2. Pale margin of cap always with remains of veil attached to it .... *nueschii*
    Margin of cap without veil remains 3
3. Cap upper surface lemon to golden yellow, pore surface lemon to
      sulphur yellow .... *grevillei*
    Cap upper surface creamy white to ochraceous or olivaceous grey,

pore surface greyish or olivaceous grey .... *aeruginascens*
4.   Stalk with ring 5
      Stalk without ring 7
5.   Cap upper surface purplish chestnut, ring at first large and floppy,
      under *Pinus* spp. .... *luteus*
      Cap upper surface never chestnut, ring smaller 6
6.   Cap upper surface buff to ochraceous with rusty tawny streaks and
      brown scales, under *Pseudotsuga* .... *amabilis*
      Cap upper surface greyish lemon yellow, or pale straw colour, under
      Pinus amongst *Sphagnum* .... *flavidus*
7.   Pore surface lemon yellow and exuding milky droplets when fresh ... 8
      Pore surface sand-colour to brownish olive or olivaceous buff 9
8.   Cap upper surface burnt sienna or reddish brown with yellowish
      tinge, margin not clearly inrolled, stalk with granule-like glands at
      apex which exude milky juice .... *granulatus*
      Cap upper surface reddish buff, marbled by a rather darker purplish
      brown network, margin clearly inrolled .... *fluryi*
9.   Cap upper surface ochraceous or yellowish brown, scaly ...... *variegatus*
      Cap orange–buff or cinnamon with pale margin, not scaly .... *bovinus*

Suillus aeruginascens (Secret.) Snell
Cap 4–10 cm diam., convex or somewhat flattened, upper surface pale,
creamy white, ochraceous or olivaceous grey to hazel, viscid, with separable
cuticle, pore surface greyish or olivaceous grey, bruising darker and more
greenish, pores broad, angular. Stalk 5–8 × 1–1.5 cm, viscid, pale yellowish
or greenish grey with ring which is at first white then brown. Flesh soft,
whitish to straw colour, when cut faintly bluish green. Spores 10–14 × 3–5.
Always under *Larix*, widespread but not common.

Suillus amabilis (Peck) Singer
Cap 5–18 cm diam., upper surface buff to ochraceous, with rusty tawny
streaks and brown scales, pore surface ochraceous, bruising rusty tawny,
pores large, angular, tubes adnate to decurrent. Stalk 6–18 × 1–4 cm,
obclavate, with thin membranous ring, pale straw to lemon yellow above
ring and without glands, rust colour below. Flesh pale yellowish cinnamon
turning greenish in base of stalk. Spores 8–12 × 4–5.5. Under *Pseudotsuga*.

Suillus bovinus (Fr.) O. Kuntze
Cap 4–10 cm diam., convex to somewhat conical or flattened, upper surface
buff, orange–buff or cinnamon, with white or very pale margin, viscid,
sometimes wrinkled, pore surface olivaceous buff to ochraceous, pores
large, angular, divided into smaller pores below the surface, radially
elongated near the stalk where the tubes are often decurrent. Stalk 4–
6 × 0.5–1 cm, without a ring, same colour as cap or slightly yellower and
paler, smooth. Flesh yellowish or slightly pinkish, sometimes becoming

faintly blue when cut. Spores 8–11 × 3–5. In conifer woods, mostly under *Pinus sylvestris*, late autumn, common.

Suillus flavidus (Fr.) Singer
Cap 3–7 cm diam., often umbonate, upper surface greyish lemon yellow or pale straw colour, viscid, pore surface yellow, pores large, angular, divided into smaller pores below the surface, tubes decurrent. Stalk 5–8 × 0.5 cm, straw colour to buff, with pale glands and a gelatinous, collapsing, whitish to tawny ring. Flesh pale yellow, becoming somewhat wine colour when cut. Spores 8–10 × 4. Mostly amongst *Sphagnum* under *Pinus sylvestris*, uncommon except in some Scottish pine woods, late summer–autumn.

Suillus fluryi Huijsman
Cap 4–10 cm diam., convex with margin inrolled, upper surface reddish buff, marbled by a rather darker, purplish brown network, pore surface lemon yellow or yellowish, pores irregular, rather small. Stalk 3–8 × 1–2 cm, cylindrical with abruptly tapered base, rather pale yellowish with purplish brown flecks formed by coalescing glands. Flesh pale yellow, when cut turning somewhat wine colour under cuticle and in base of stalk. Spores 8–11 × 3.5–4. In conifer woods, under *Pinus sylvestris*, autumn.

Suillus granulatus (L. ex Fr.) O. Kuntze (Fig. 397)
Cap 4–10 cm diam., convex or flattened, upper surface burnt sienna or reddish brown, often with yellowish tinge, viscid, smooth and shiny when dry, pore surface lemon yellow, when fresh exuding milky droplets which are formed also by granule-like glands at the stalk apex, pores small, tubes often slightly decurrent. Stalk 4–8 × 0.5–1 cm, without a ring, pale yellow, with numerous granule-like glands, often wine colour or brownish at base especially when old. Flesh pale yellow, not changing colour when cut. Spores 8–10 × 3–4. In conifer woods, common under various species of *Pinus*, including *P. nigra* and *P. sylvestris*, autumn.

Suillus grevillei (Klotsch) Singer (Fig. 398)
Cap 3–12 cm diam., convex to somewhat conical or umbonate, upper surface lemon to golden yellow, sometimes tinged with burnt sienna especially in the middle, glutinous, shiny when dry, pore surface lemon to sulphur yellow, bruising burnt sienna, pores angular, small, tubes often slightly decurrent. Stalk 5–10 × 1.5–2 cm, with whitish or pale yellow ring which darkens when coated with spores, yellow above ring, cinnamon below it. Flesh yellow, scarcely changing when cut as a rule but occasionally becoming wine coloured or bluish in stalk. Spores 8–11 × 3–4. Always under *Larix* and sometimes called the Larch Bolete, August–October, common.

Suillus luteus (Fr.) S.F. Gray (Fig. 399)
Slippery Jack
Cap 5–12 cm diam., convex, upper surface rather purplish chestnut, slimy when wet, shiny when dry, sometimes streaked radially with darker lines, pore surface lemon yellow at first, becoming brownish or reddish brown when old, pores small, round. Stalk 5–10 × 2–3.5 cm, whitish or pale yellowish, with wine-coloured glands, brownish below ring when old, ring large, floppy, white or cream, turning brown. Flesh yellowish white. Spores 7–11 × 3–4. Very common in pine woods on sandy soil, often growing amongst grass along the sides of rides, autumn.

Suillus nueschii Singer
Cap 3–6 cm diam., upper surface greyish yellow or dull ochraceous, becoming somewhat blotched as the yellowish gluten dries, paler at the margin where the veil persists, pore surface greyish yellow to ochraceous. Stalk 2–4 × 1–1.5 cm, pale yellowish with peronate, woolly ring. Flesh pale yellow, turning bluish or wine colour when cut. Spores 7–10 × 3–5. Under *Larix*, rare.

Suillus tridentinus (Bres.) Singer
Cap 4–12 cm diam., convex, upper surface orange to burnt sienna, or reddish brown, with slightly darker adpressed scales which are somewhat fibrillose, covered with gluten, pore surface yellowish orange to burnt sienna, pores broad, angular, tubes often slightly decurrent. Stalk 4–8 × 1.5–2 cm, whitish or pale above yellowish ring, yellowish or orange below it, with reddish brown net or striations. Flesh lemon yellow to apricot. Spores 10–14 × 4–5. Mostly under *Larix* but recorded also under *Taxus*.

Suillus variegatus (Fr.) O. Kuntze
Cap 5–15 cm diam., hemispherical to convex, upper surface ochraceous or yellowish brown, slightly sticky or greasy in damp weather, felt-like when dry and covered with soft, small, darker scales, pore surface sand coloured to brownish olive, bluish when bruised, pores small, unequal. Stalk 5–10 × 1.5–2 cm, cylindrical to obclavate, pale ochraceous, lower part darker, sometimes reddish brown or olivaceous, without a ring. Flesh pale yellow or slightly orange, sometimes with bluish patches when cut. Spores 8–11 × 3–4. In conifer woods, usually under *Pinus*, August–October.

Thanatephorus
Including *Uthatobasidium* in the Key
Fruitbodies resupinate, cobwebby, pellicular or membranous, smooth. Hyphae hyaline or brownish, without clamps. Basidia hyaline, cylindrical, barrel-shaped or clavate, about the same width as the subtending hyphae. Spores hyaline or yellowish, smooth, J–, often forming secondary spores. Parasitic on herbaceous plants and shrubs or saprophytic on wood.

KEY

Spores fusiform up to 20 long .... *Uthatobasidium fusisporum*
Spores other shapes, not more than 15 long 1
1. On orchids .... *orchidicola*
    On *Plantago* .... *langlei-regis*
    On other plants or plant remains 2
2. White or cream; parasitic on many different plants including some of
    economic importance .... *cucumeris*
    Ochraceous; saprophytic on rotten wood and plant debris
    .... *Uthatobasidium ochraceum*

Thanatephorus cucumeris (Frank) Donk (Fig. 400)
Corticium solani
Fruitbodies resupinate, cobwebby or membranous, white or cream. Basidia
cylindrical, 12–20 × 8–12, 2–6-spored. Spores ellipsoid, 8–14 × 5–8.
Hymenium formed as a thin sheet or collar on stems and leaves at soil level
and on soil particles. The fungus can survive in the soil for a long time in the
form of brown sclerotia. Parasitic on herbaceous plants and shrubs and of
economic importance on potato (*Solanum tuberosum*), beans (*Phaseolus*), beet
(*Beta*), cabbage (*Brassica*), tomato (*Lycopersicon*), wheat (*Triticum*) and turf
grasses. It causes damping off, stem canker etc. and is primarily a soil-borne
pathogen.

Thanatephorus langlei-regis D.A. Reid
Fruitbodies membranous, pale yellow–ochre. Basidia cylindrical to clavate,
15–30 × 6–9. Spores ovoid, pyriform or ellipsoid, 8–15 × 5–7.5. On leaves of
Plantago.

Thanatephorus orchidicola Warcup and Talbot
Fruitbodies resupinate, cobwebby or floccose, white to brownish. Basidia
broadly cylindrical, 15–20 × 9–11, 4-spored. Sterigmata up to 12 long.
Spores pale brownish, 9–12 × 7–10, commonly producing secondary
spores. On orchids.

Thelephora

Fruitbodies mostly terrestrial, soft leathery, fan-shaped, antler-like or with
resupinate part and narrow outgrowths or branches. Hyphae with clamps
except in one species. Basidia narrowly clavate, mostly 4-spored. Spores
brown or yellowish, echinulate, often somewhat angular or irregular in
outline, J–.

KEY

Fruitbodies with a strong, unpleasant, rather garlicky smell .... *palmata*
Smell none or not unpleasant 1

1. Fruitbodies broad fan-shaped or funnel-shaped 2
    Fruitbodies clavarioid or with resupinate part and relatively narrow branches or outgrowths 4
2. Fruitbodies black when moist, a rare, little known species .... *cuticularis*
    Fruitbodies white or brownish, uncommon .... *mollissima*
    Fruitbodies reddish to chocolate brown or blackish brown 3
3. Fruitbodies mostly stalked and somewhat funnel-shaped ... *caryophyllea*
    Fruitbodies splayed-out fans forming rosettes close to the ground .... *terrestris*
4. Fruitbodies clavarioid, branches reddish brown or purplish brown with white tips 5
    Fruitbodies with some part resupinate, other colours 6
5. Branches flat, spatula-shaped or narrow fan-shaped .... *anthocephala*
    Branches cylindrical or subulate .... *anthocephala* var. *clavularis*
6. Fruitbodies crust-like, with short outgrowths, dark grey or blackish .... *atra*
    Fruitbodies long remaining white, slowly turning brown from base, branches with dentate or penicillately fringed tips .... *penicillata*

Thelephora anthocephala (Bull. ex Fr.) Pers. (Fig. 401)
Fruitbodies clavarioid, tufted, erect, up to 6 cm tall, short-stalked, branches 0.5–1.5 cm wide, divided many times, flattened, spatula-shaped or narrow fan-shaped, fairly thick, deeply divided, reddish brown to purplish brown with white tips, hymenium on lower side smooth, dark brown or greyish violet. Basidia 50–80 × 8–11, usually 4-spored. Spores subspherical to angular, brown, 8–10 × 6–8.5, spines 0.5–2 long. On the ground in woods, not strong smelling. The var. *clavularis* (Fr.) Quél. has amphigenous hymenium on branches which are cylindrical to subulate and 0.5–3 mm thick.

Thelephora atra Weinm.
Fruitbodies crust-like with cylindrical or laterally compressed outgrowths up to 2 × 0.5–1.5 cm, dark grey or blackish with paler yellowish green tips, hymenium smooth, covering all parts. Basidia 50–100 × 8–12. Spores subspherical or somewhat angular, 9–13 × 8–11, spines 1–2 long. On soil and amongst mosses, uncommon.

Thelephora caryophyllea (Schaeff.) Fr.
Fruitbodies 1.5–4 cm tall, 1–5 cm wide, usually shortly stalked, roughly funnel-shaped, made up of many broadly fan-shaped, imbricate segments forming the funnel, these again deeply divided into wedges which are toothed or laciniate; upper surface purplish brown, somewhat zonate; lower, hymenial surface smooth, brown or purplish brown with paler margin. Basidia 50–80 × 9–12. Spores angular-ellipsoid or irregular, brown, 7–8.5 × 5–7, warted to echinulate, spines 1–1.5 long. On sandy soil in conifer woods.

Thelephora cuticularis Berk.
Fruitbodies pileate or somewhat effuso-reflexed, sessile, flat, fan-shaped, 2–4 × 2–3 cm, 1–2 mm thick, finely hairy, when moist black, drying pale brown or cinnamon brown, lower, hymenial surface smooth, black when moist, purplish brown when dry. Hyphae without clamps. Spores ellipsoid to kidney-shaped, brown. On mossy bark and twigs.

Thelephora mollissima Pers. ex Fr.
Fruitbodies effuso-reflexed to pileate, sessile or short-stalked, fan-shaped, 2–6 cm tall, soft, upper surface white at first, becoming brownish, lower, hymenial surface smooth or tuberculate, brown. Basidia 50–80 × 9–12, 4-spored. Spores yellowish brown, angular or irregular, 9–12 × 7–9, spines up to 1 long. On soil in conifer woods.

Thelephora palmata (Scop.) Fr. (Fig. 402)
Fruitbodies 2–7 cm tall, 1–4 cm wide, bushy, usually short-stalked, main branches dichotomously branched several times, branches cylindrical or laterally flattened, at the apex subulate, flattened or toothed, purplish brown to blackish brown, hymenium smooth, amphigenous or all on lower surface of branches, brown or blackish brown. Basidia 80–100 × 10–12. Spores angular or irregular, brown, 8–11 × 7–9, spines up to 1.5 long. On soil in conifer woods. Has a strong, unpleasant, somewhat garlicky smell.

Thelephora penicillata (Pers.) Fr.
Fruitbodies with an indistinct resupinate part 3–15 cm wide and numerous erect, spatula-shaped or narrow fan-shaped branches which are dentate or penicillately fringed at their tips, long remaining white but slowly turning brown from the base up, drying yellowish cream. Basidia 30–70 × 8–10, 4-spored. Spores brown, angular or irregular, 7–10 × 5–8, with spines up to 1.5 long. Overgrowing and encrusting small twigs and leaf litter in woods.

Thelephora terrestris Pers. ex Fr. (Fig. 403)
Earth Fan
Fruitbodies gregarious and sometimes in large colonies, broad fan-shaped, up to 6 cm wide, often splayed out and forming rosettes close to the ground, 1–3 mm thick, tough but soft; upper surface reddish to chocolate brown or blackish brown, paler at the dentate or fringed margin, weakly concentrically zoned and covered with radiating fibres; lower, hymenial surface wrinkled, clay brown. Basidia 50–90 × 8–12. Spores broadly ellipsoid to angular, brown, 8–10 × 5–7, with spines about 0.5 long. On sandy soil of pine woods and heaths, often overgrowing twigs, leaf litter and mosses, common.

Tomentella

Fruitbodies resupinate, effused, arachnoid or membranous, rather soft, smooth, granular, tuberculate or with teeth, various colours. Hyphae hyaline or brown, clamps at nearly every septum, skeletal hyphae, when present, yellowish brown, hyphae in basal layer often uniting to form thick, rope-like strands. Cystidia in only a few species. Basidia hyaline or golden brown, clavate, mostly 4-spored. Spores yellowish to brown, spherical or irregular in outline, sometimes lobed or sinuate, echinulate or, more rarely, verrucose, J–, seldom blue in KOH. Mostly on rotten wood and debris lying on the ground.

KEY

Hymenial surface toothed, spores verrucose 1
Hymenial surface smooth or granular, spores echinulate 2
1. Spores 4–6 diam. .... *calcicola*
   Spores 7–10 diam. .... *crinalis*
2. With projecting cystidia 3
   Without cystidia 6
3. Cystidia golden brown, hyphal strands up to 200 thick, extending beyond edge as brown rhizomorphs .... *pilosa*
   Cystidia hyaline, hyphal strands absent or very thin 4
4. Cystidia cylindrical to subulate, 40–60 long, fruitbodies pinkish buff to pinkish brown .... *subtestacea*
   Cystidia often clavate or capitate, up to 90 or 130 long, fruitbodies other colours 5
5. Fruitbodies ochraceous brown to greyish brown .... *subclavigera*
   Fruitbodies greyish green to olivaceous brown .... *viridula*
6. Fruitbodies growing only on *Cladium mariscus* .... *cladii*
   Fruitbodies not on this host 7
7. Basidia swollen up to 20 towards base .... *terrestris*
   Basidia not so 8
8. Fruitbodies dark brown with white or whitish border .... *albomarginata*
   Fruitbodies reddish brown to blackish brown with broad sulphur yellow or ochraceous border .... *ellisii*
   Fruitbodies lacking such contrasting borders 9
9. Hyphae often with distinct swellings, torulose, vesicular or ampulliform 10
   Hyphae not so 11
10. Fruitbodies brick red .... *lateritia*
    Fruitbodies ochraceous brown or reddish brown .... *sublilacina*
11. Spores regular in outline, spherical or subspherical 12
    Spores irregular in outline, lobed, angular or sinuate 15
12. Spores 5–6 diam. .... *cinerascens*
    Spores more than 7 diam. 13
13. Spores with short spines .... *asperula*
    Spores with spines up to 3 long 14

14. Spores 8–10 diam. .... *bryophila*
    Spores 10–12 diam. .... *bresadolae*
15. Basal layer with few or no hyphal strands 16
    Basal layer with hyphal strands 17
16. Spores minutely but closely echinulate .... *coerulea*
    Spores with scattered spines 1–2 long .... *puberula*
17. Hyphal strands 35–40 thick 18
    Hyphal strands up to 70–90 thick 20
18. Basidia 8–14 thick .... *rhodophaea*
    Basidia 6–9 thick 19
19. Fruitbodies cinnabar red to reddish brown .... *punicea*
    Fruitbodies blackish brown, olivaceous brown or purplish brown
    .... *botryoides*
20. Fruitbodies pale creamy buff to ochraceous .... *avellanea*
    Fruitbodies bright ferruginous or darker, mostly rust brown or dark
    brown 21
21. Spores subspherical, spines often dichotomously notched .... *italica*
    Spores elongated, lobed or sinuate 22
22. Spores minutely echinulate .... *rubiginosa*
    Spores with spines 1.5–2 long .... *ferruginea*

Tomentella albomarginata (Bourd. and Galz.) Christiansen
Fruitbodies densely membranous, smooth, closely adnate, dark brown with broad white or cream, fibrillose margin. Hyphal strands up to 25 thick. Hyphae yellowish brown. Basidia 30–60 × 8–12. Spores irregular in outline, often somewhat flattened on one side, 7–10 × 6–9, brownish, echinulate. On dead branches of *Betula* and other deciduous trees on the ground.

Tomentella asperula (P. Karst.) Höhn. and Litsch.
Fruitbodies membranous, surface granular or warted, greyish brown, olivaceous brown or cinnamon brown, subiculum often whitish or yellowish. Hyphal strands up to 80 thick. Basidia 40–60 × 8–10, 2–4-spored, sterigmata up to 8 long. Spores spherical or subspherical, 7–10 diam., smoky brown, shortly echinulate. On rotten wood of both deciduous trees and conifers.

Tomentella avellanea (Burt.) Bourd. and Galz. (Fig. 404)
Fruitbodies membranous, soft, smooth, pale creamy buff or ochraceous, with darker subiculum. Hyphal strands up to 75 thick. Basidia 50–70 × 8–14, sterigmata curved, up to 9 long. Spores irregularly oblong, subspherical or angular, 8–10 × 5–7, clear yellowish brown, shortly echinulate. On twigs and other debris on the ground, has been found on rotting stems of *Filipendula ulmaria*.

Tomentella botryoides (Schm.) Bourd. and Galz.
Fruitbodies membranous, surface granular or warted, always dark coloured,

blackish brown, olivaceous brown or purplish brown, with a paler brownish subiculum and pale yellowish brown margin. Hyphal strands up to 40 thick. Hyphae yellowish brown to brown, ends often capped with blackish purple exudate. Basidia 40–70 × 7–9, 4-spored, curved sterigmata 7–8 long. Spores often irregular in outline, 5–7 × 6–8, brown with hyaline apiculus, closely and shortly echinulate. On very rotten wood and on leaf litter.

Tomentella bresadolae (Brinkm. ex Bres.) Höhn. and Litsch.
Fruitbodies membranous, smooth, rarely finely granular, rather thick, reddish brown, with concolorous subiculum. No hyphal strands. Basidia 50–70 × 10–14, hyaline or pale brownish, curved sterigmata 8–15 long. Spores spherical, 10–12 diam., brown, echinulate, spines up to 3 long. On rotten wood of both deciduous trees and conifers.

Tomentella bryophila (Pers.) M.J. Larsen (Fig. 405)
Fruitbodies membranous, smooth, rust brown or reddish orange, subiculum darker than hymenium. No hyphal strands. Hyphae yellowish brown. Basidia 40–60 × 8–12, 4-spored. Spores spherical or subspherical, 8–10 diam., echinulate, with spines up to 3 long. Overgrowing mosses on rotten wood.

Tomentella calcicola (Bourd. and Galz.) M.J. Larsen
Fruitbodies arachnoid to thinly membranous, densely hydnoid, with teeth 1–2 mm long, cylindrical, rounded or somewhat pointed at apex, golden brown to reddish brown with darker subiculum. Hyphal strands up to 40 thick. Narrow skeletal hyphae present. Basidia 30–40 × 6–8, 4-spored. Spores spherical or subspherical, 5–6 × 4–5, mostly 4–5 diam., golden brown, verrucose. On decaying wood.

Tomentella cinerascens (P. Karst.) Höhn. and Litsch.
Fruitbodies membranous, granular or warted, seldom smooth, grey, greyish brown or ochraceous brown, subiculum paler, pale ochraceous or whitish cream. Hyphal strands up to 80 thick. Basidia 30–40 × 5–8. Spores spherical or subspherical, 5–6 diam., spines rather short. On dead fallen branches of deciduous trees.

Tomentella cladii Wakef. (Fig. 406)
Fruitbodies very thin, membranous, smooth or granular, chestnut brown, subiculum scanty, darker than hymenium. No hyphal strands. Basidia hyaline, 35–40 × 7–9, with four short sterigmata. Spores subspherical to broadly ellipsoid, sometimes flattened on one side, 7–9 × 5–6, brownish, densely but minutely echinulate. On dead stems of *Cladium mariscus*.

Tomentella coerulea (Bres.) Höhn. and Litsch.
Fruitbodies membranous, thin, granular or warted, ochraceous or claycoloured, sometimes with a rosy tinge, with pale subiculum. No hyphal strands. Basidia 50–60 × 7–9. Spores subspherical or slightly angular or

sinuate, 6–8.5 × 5–7, pale brown with hyaline apiculus, minutely but closely echinulate. On rotten wood and bark.

Tomentella crinalis (Fr.) M.J. Larsen
Fruitbodies loosely attached, membranous, rust brown to dark brown, covered with bluntly conical teeth up to 3 mm long. Hyphal strands up to 25 thick. Hyphae yellowish brown. Basidia 40–60 × 7–10, with four curved sterigmata up to 6 long. Spores spherical or subspherical, 7–10 diam., brown, coarsely verrucose, warts sometimes dichotomously notched. On rotten wood of stumps and fallen branches.

Tomentella ellisii (Sacc.) Jülich and Stalpers
Fruitbodies membranous, smooth, reddish brown to purplish or blackish brown, subiculum forming broad margin, pale sulphur yellow or ochraceous. Hyphal strands up to 80 thick. Hyphae sometimes with swellings. Basidia 30–40 × 8–9. Spores spherical or irregularly subspherical, 7–9 diam. or 6–7 × 8–9, verrucose to echinulate. On rotten wood on the ground.

Tomentella ferruginea (Pers.) Pat. (Fig. 407)
Fruitbodies arachnoid or membranous, bright ferruginous, greenish or olive blackish hymenium, reddish or yellowish brown subiculum. Hyphal strands up to 90 thick. Basidia 30–45 × 7–8. Spores irregular in outline, lobed or sometimes sinuate, 8–9 × 6–7, brown, with small hyaline apiculus, echinulate, spines 1.5–2 long. On ground, rotten wood, debris etc.

Tomentella italica (Sacc.) M.J. Larsen
Fruitbodies membranous, smooth or, rarely, warted, brown to dark brown, subiculum concolorous or darker. Hyphal strands up to 80 thick. Basidia 40–50 × 7–8. Spores subspherical, irregular in outline, 7–9 diam., brownish, verrucose to echinulate, spines often dichotomously notched. On soil and rotten wood.

Tomentella lateritia Pat.
Fruitbodies membranous, granular or warted, brick red with paler subiculum. No hyphal strands. Hyphae with hyaline, vesicular cells up to 20 wide in subiculum and subhymenium. Basidia 50–60 × 7–8. Spores spherical or irregular in outline, 7–9 diam., echinulate. On rotten wood.

Tomentella pilosa (Burt.) Bourd. and Galz. (Fig. 408)
Fruitbodies loosely attached, membranous, smooth or granular, yellowish brown to olivaceous brown, margin often paler. Hyphal strands up to 200 thick, often extending beyond the edge as brown rhizomorphs. Cystidia clavate, golden brown, smooth, somewhat thicker walled towards the base, 60–120 × 12, with granular contents, sometimes septate with clamps. Basidia 40–60 × 6–10. Spores roughly spherical or sinuate, often flattened on one side, 8–10 × 7–9, yellowish brown, scattered spines 1.5–2 long. On rotten wood and debris, sometimes overgrowing mosses.

Tomentella puberula Bourd. and Galz.
Fruitbodies loosely attached, membranous, rather thick, granular or warted, ochraceous or ochraceous brown, with paler subiculum. Hyphal strands very few. Basidia 40–50 × 6–10, 2–4-spored, hyaline but often with reddish brown granular contents even when young. Spores irregular in outline, somewhat angular to sinuate, 5–8 × 5–7, brown, with scattered spines 1–2 long. On dead wood and rotting leaves.

Tomentella punicea (Alb. and Schw. ex Fr.) Schroet.
Fruitbodies firmly attached to substrate, membranous, smooth or warted, when fresh cinnabar red, becoming darker to brownish red, subiculum concolorous. Hyphal strands up to 35 thick. Basidia 30–50 × 6–9, with 2–4 curved sterigmata 5–7 long. Spores subspherical, irregular in outline, 7–10 × 5–9, reddish to brown, spines crowded, 1–1.5 long. On rotten wood and debris on the ground.

Tomentella rhodophaea Höhn. and Litsch.
Fruitbodies membranous, smooth, ochraceous to yellowish brown or brownish grey, subiculum concolorous, margin sometimes pale pink when fresh. Hyphal strands up to 40 thick. Basidia 40–60 × 8–14, 4-spored. Spores subspherical, often irregular in outline, 7–10 diam., yellowish brown, echinulate. On rotten wood.

Tomentella rubiginosa (Bres.) R. Maire
Fruitbodies soft, arachnoid to membranous, surface granular or warted, yellowish rust brown, reddish rust or dark rust brown, sometimes slightly greenish, with yellowish brown subiculum. Hyphal strands up to 70 thick. Basidia 30–50 × 6–9, with four short sterigmata. Spores irregular in outline, sinuate to angular, 6–9 × 5–7, yellowish brown, minutely echinulate. On dead fallen twigs, branches and leaves.

Tomentella subclavigera Litsch.
Fruitbodies arachnoid to softly membranous, in small patches, smooth to granular, minutely hairy under lens, pale ochraceous brown to greyish brown, margin thin, slightly paler. No hyphal strands. Cystidia projecting, hyaline, cylindrical to clavate, 80–130 long, 2–5 wide at base, 6–12 at apex. Basidia 20–35 × 6–9, sterigmata 6–7 long. Spores subspherical or broadly ellipsoid, flattened on one side, 6–9 × 6–7, pale yellowish brown, with lateral hyaline apiculus, densely and minutely echinulate. On rotten wood.

Tomentella sublilacina (Ellis and Holway) Wakef.
Fruitbodies membranous, smooth, ochraceous brown to reddish brown, subiculum concolorous or darker. No hyphal strands. Hyphae yellowish brown, those in subhymenium torulose, those in the subiculum sometimes swollen and ampulliform. Basidia 50–60 × 8–12, 4-spored. Spores irregular in outline, lobed, 8–11 diam., echinulate, spines 1–2 long. On dead, rotting wood and on old rotten polypores.

Tomentella subtestacea (Bourd. and Galz.) Svrček. (Fig. 409)
Fruitbodies loosely attached, membranous, granular or smooth, pinkish buff to pinkish brown, margin arachnoid, pale. Hyphal strands up to 15 thick. Cystidia hyaline, thin-walled, cylindrical to subulate, with rounded apex, 40–60 × 4–6, protruding up to 30. Basidia 40–50 × 5–8, some with reddish yellow contents. Spores subspherical or irregular in outline, sometimes lobed, 7–9 × 6–8, echinulate, with spines 1–1.5 long. On decaying wood and old polypores.

Tomentella terrestris (Berk. and Br.) M.J. Larsen (Fig. 410)
Fruitbodies membranous, smooth or somewhat tuberculate, dark brown to blackish brown, tinted wine red, subiculum paler. Hyphal strands up to 50 thick. Basidia 60–90 × 10–20, mostly swollen towards the base. Spores subspherical and mostly irregular in outline, 6–8 diam., yellowish brown, echinulate. On very rotten wood and on soil.

Tomentella viridula (Bourd. and Galz.) Svrček
Fruitbodies softly membranous, thin, smooth or granular, greyish green to olivaceous brown, margin paler. No hyphal strands. Cystidia clavate or capitate, hyaline or pale, 50–90 × 2–5, swollen at apex to 7–10, protruding about 45. Basidia 40–60 × 5–9, sterigmata 5–7 long. Spores irregular in outline, sometimes lobed, 6–8 diam., or 7–9 × 6.5–7, yellowish brown, shortly echinulate. On rotten wood and on soil.

Tomentellastrum
Fruitbodies resupinate, effused, arachnoid to membranous. Hyphae yellowish to brown, mostly without clamps, in one species an occasional clamp is seen in the subhymenium. No cystidia. Basidia clavate, 4-spored, hyaline to yellowish. Spores mostly 8–14 diam., brown, echinulate, J–.

KEY

Fruitbodies greyish brown, basidia 10–15 wide .... *fuscocinereum*
Fruitbodies ochraceous to pale brown, basidia 9–12 wide .... *alutaceo-umbrinum*
Fruitbodies mid to dark brown, basidia 7–10 wide 1
1. Spores broadly ellipsoid or irregular, no clamps .... *badium*
Spores spherical or subspherical, a few clamps in the subhymenium .... *litschaueri*

Tomentellastrum alutaceo-umbrinum (Bres.) M.J. Larsen (Fig. 411)
Fruitbodies membranous, up to 1 mm thick, smooth, ochraceous or pale brown, with darker subiculum. Hyphae yellowish brown, always without clamps. Basidia 50–70 × 9–12. Spores spherical or subspherical, 9–14 diam., brown, finely echinulate. On leaf litter and soil.

Tomentellastrum badium (Link ex Steudel) M.J. Larsen
Fruitbodies arachnoid to membranous, 0.5 mm thick, smooth or finely granular, mid to dark brown. Hyphae yellowish to brown, without clamps. Basidia 40–70 × 7–9. Spores broadly ellipsoid or irregular in outline, 7–11 diam., brown, echinulate. On rotten wood.

Tomentellastrum fuscocinereum (Pers. ex Fr.) Svrček (Fig. 412)
Fruitbodies compact, up to 1 mm thick, smooth, greyish brown with dark brown subiculum. Hyphae all without clamps. Basidia clavate, hyaline, 50–70 × 10–15, sterigmata 12–15 long. Spores broadly ellipsoid or slightly depressed on one side, 8–12 × 7–9, brown, with rather short, conical, hyaline spines. On firm bare ground and on fallen sticks and other debris.

Tomentellastrum litschaueri (Svrček) M.J. Larsen
Fruitbodies membranous, soft and spongy, smooth, mid to dark brown or chestnut brown. Hyphae mostly without clamps but a few found occasionally in the subhymenium. Basidia 40–60 × 8–10, sterigmata 5–8 long. Spores spherical or subspherical, 8–12 diam., brown, finely echinulate. On debris and overgrowing mosses on the ground.

Tomentellina *fibrosa* (Berk. and Curt.) M.J. Larsen (Fig. 413)
Fruitbodies resupinate, effused, arachnoid or hypochnoid, loosely attached, smooth or almost so, rust brown to dark brown, with pale border and rhizomorphs. Hyphae hyaline to brown, without clamps or with very few and those only in the subiculum. Yellowish skeletal hyphae present. Cystidia solitary or in clusters, protruding above the surface and seen easily under a lens, rust brown, septate, thick-walled, smooth, 100–200 × 5–8. Basidia clavate, hyaline, 40–60 × 6–9, with sterigmata 5–7 long. Spores roughly spherical but irregular in outline, sometimes lobed or angular, 7–11 diam., brownish, verrucose, with warts sometimes dichotomously notched. On rotten wood and bark.

Tomentellopsis

Fruitbodies resupinate, effused, arachnoid to membranous or pellicular, smooth, mostly pale. Hyphae without clamps. No cystidia. Basidia hyaline, clavate or suburniform. Spores not more than 7 diam., hyaline or pale straw, pinkish or pale brown, thin-walled, echinulate, J–.

KEY

Spines on spores up to 1.5 long, hymenium cream to buff or pale brown .... *zygodesmoides*
Spines short, always less than 1, hymenium other colours 1
1. Hymenium becoming pale sulphur yellow or greenish yellow
.... *echinospora*

Hymenium cream becoming rose pink or salmon pink, with red flecks .... *submollis*

Tomentellopsis echinspora (Ellis) Hjortst (Fig. 414)
Fruitbodies arachnoid to membranous, thin, soft, loosely attached, smooth, hymenium at first cream, then pale sulphur yellow to greenish yellow, covering a white, cobwebby subiculum. Hyphae hyaline. Basidia 20–30 × 6–9, clavate to urniform, with sterigmata 3–5.5 long. Spores spherical or subspherical, 4–6.5 diam., hyaline to pale straw colour, densely echinulate. On rotten wood, debris, mosses etc.

Tomentellopsis submollis (Svrček) Hjortst
Fruitbodies arachnoid to membranous, thin, smooth, cream to rose pink or salmon pink with reddish flecks, subiculum and margin white or whitish. Basidia 30–35 × 5–7, with short sterigmata. Spores spherical to broadly ellipsoid, 5.5–7 × 4–6, hyaline or tinged pinkish, densely and shortly echinulate. On rotten wood, bark and dead leaves in damp places.

Tomentellopsis zygodesmoides (Ellis) Hjortst (Fig. 415)
Fruitbodies arachnoid to thinly membranous, smooth, cream to buff or pale brown, subiculum and margin brown. Hyphae yellowish to brownish. Basidia 30–40 × 5–9, with sterigmata 4–7 long. Spores angularly sub-spherical to ellipsoid, 4.5–7 × 4–5, hyaline to pale straw, brownish in mass, spines up to 1.5 long. On rotten wood and bark.

Trametes
Fruitbodies pileate or, rarely, effuso-reflexed, often imbricate, annual or perennial, flat and mostly thin, softly leathery brackets, frequently with a velvety, zonate upper surface and with the pores on the lower surface most commonly isodiametric, regularly radially elongated and slot-like only in *T.gibbosa*. Generative hyphae with clamps. Skeletal hyphae hyaline. No true cystidia. Basidia small. Spores cylindrical to ellipsoid, hyaline, thin-walled, J–. On wood and bark mostly of deciduous trees, rarely of conifers.

KEY

    Pores large, not more than 1–2 per mm, fruitbodies large, often more than 10 × 7 cm, and more than 2 cm thick 1
    Pores small, mostly 3–5 per mm, fruitbodies not more than 10 × 7 cm, less than 1.5 cm thick 2
1. Pores radially elongated, slot-like, spores 4–5 long .... *gibbosa*
    Pores isodiametric, spores 7–10 long .... *suaveolens*
2. Fruitbodies very thin, 1–3 mm, upper surface markedly zonate with variously coloured and often quite dark zones; black line separates flesh from tomentum .... *versicolor*

Fruitbodies 0.5–1.5 cm thick, mostly paler and often only faintly zonate 3

3. Lower surface becoming greyish, especially when old .... *hirsuta*
    Lower surface cream to straw colour or ochraceous 4

4. Upper surface shortly silky velvety and white when young becoming cream or straw colour, zonation very weak .... *pubescens*
    Upper surface velvety to tomentose, with bare brownish zones, other zones greyish white or greyish ochraceous .... *multicolor*

Trametes gibbosa (Pers. ex Fr.) Fr. (Fig. 416)

Fruitbodies solitary or gregarious, sometimes imbricate, broadly attached, semicircular, flat or somewhat humped, corky, tough brackets 6–20 × 8–15 cm, 1–4 cm thick, upper surface undulating to tuberculate, zonate, at first whitish and velvety, becoming smooth and ochraceous or greyish, often tinged green by algae, lower surface white to pale ochraceous, pores radially elongated, slot-like, especially near base, 1–5 × 0.5–2 mm, tubes up to 1.5 cm long, usually not stratified. Spores cylindrical or oblong ellipsoid, very slightly curved, 4–5 × 2–2.5. Common on dead wood of deciduous trees, especially on stumps of *Fagus* and *Ulmus*.

Trametes hirsuta (Wulf. ex Fr.) Pilát

Fruitbodies solitary or gregarious, often imbricate, broadly or narrowly attached, flat, semicircular or fan-shaped, sharp-edged, leathery brackets, 4–10 × 2–7 cm, 0.5–1 cm thick, upper surface densely hairy, usually somewhat zonate, whitish or pale grey at first, becoming darker grey or brownish when old and then often tinged green by algae, the silvery hairs in different zones are in some adpressed and in others erect and more pronounced, lower surface white to cream at first but always becoming greyish, pores round to angular, 2–4 per mm, tubes 1–4 mm long; flesh in section 2-zoned, the upper zone darker than the lower. Spores ellipsoid to cylindrical, often slightly curved, 5–7 × 1.5–2.5. Mostly on fallen dead trunks and branches of deciduous trees.

Trametes multicolor (Schaeff.) Jülich

Fruitbodies mostly imbricate, sometimes fused laterally, broadly attached, flat and semicircular or more narrowly attached, humped at base and fan-shaped, tough brackets, 3–7 × 2–4 cm, 1–1.5 cm thick and somewhat triquetrous in section with a thin brownish orange line just below the tomentum; upper surface velvety to tomentose, zonate and sometimes with bare zones which are often brownish, other zones greyish white, greyish ochraceous or reddish brown; lower surface white to straw colour or ochraceous, pores angular, 3–4 per mm, tubes up to 4 mm long. Spores cylindrical or oblong-ellipsoid, 5.5–7 × 2.5–3. On dead wood of deciduous trees.

Trametes pubescens (Schum. ex Fr.) Pilát
Fruitbodies mostly in rows or imbricate, often crowded, broadly attached
and semicircular or narrowly attached with short stalk-like base, soft, fleshy
brackets which dry very light and brittle, 4–10 × 2–7 cm, 0.5–1.5 cm thick,
with sharp margin, upper surface white and shortly silky velvety when
young, becoming cream or straw colour, sometimes with weak brownish
zonation, almost smooth when old, lower surface cream to ochraceous,
pores at first angular and 3–4 per mm, widening with age, the walls then
becoming lacerate. Spores cylindrical, 5–7 × 1.5–2.5. On dead wood of
deciduous trees, e.g. *Alnus* and *Betula*; very attractive to and often eaten by
insects.

Trametes suaveolens (Fr.) Fr.
Fruitbodies solitary or in small groups, broadly attached, semicircular, flat
or convex brackets, 3–15 × 2–10 cm, 1–5 cm thick, triquetrous in section;
upper surface undulating, at first velvety and white or cream, becoming
smooth and somewhat greyish or brownish; lower surface white to pale
cream or ochraceous, pores angular, 1–2 per mm or sometimes elongated
up to 3 mm. Spores 7–10 × 3–4, narrowly ellipsoid. On both living and dead
wood of deciduous trees, mainly on *Populus* and *Salix*. Smells of aniseed.

Trametes versicolor (L. ex Fr.) Pilát (Fig. 417)
Many-zoned Polypore
Fruitbodies pileate or sometimes effuso-reflexed, commonly in large
groups, imbricate or forming rosettes, sessile, tongue- or fan-shaped, flat,
leathery brackets 4–8 × 3–5 cm, 1–2 mm thick; upper surface velvety,
smooth when old, concentrically zonate, zones variously coloured,
ochraceous brown, bluish, rust brown, greyish or blackish, with a wavy,
whitish or pale margin; lower surface white, cream or pale straw colour,
with angular pores mostly 3–5 per mm; in section a black or blackish brown
line can be seen separating the tomentum from the flesh. Spores cylindrical,
slightly curved, 5–6 × 1.5–2. Very common on wood and bark of deciduous
trees.

Trechispora

Fruitbodies resupinate, cobwebby or membranous, soft, smooth, tuberculate, poroid or with teeth. Hyphae with clamps and often swollen near
septa. No cystidia. Basidia short, 4-spored. Spores mostly small, hyaline or
slightly yellowish, smooth, echinulate or verruculose.

KEY

Hymenium poroid .... *mollusca*
Hymenium covered with teeth up to 1.5 mm long .... *farinacea*

Hymenium smooth or tuberculate or, rarely, with a few short teeth . 1
1. Spore walls echinulate or verruculose 2
    Spore walls smooth 4
2. Spores 5–9 long .... *fastidiosa*
    Spores 4–6 long .... *vaga*
    Spores 2.5–3.5 long 3
3. Conidia present and rhizoids formed .... *alnicola*
    No conidia or rhizoids .... *microspora*
4. Spores 3–4 × 2 .... *byssinella*
    Spores more than 2 wide 5
5. Spores 4–7 × 2–3.5, fruitbodies with rhizoids at margin ...... *amianthina*
    Spores 3–5 × 2–4, fruitbodies without rhizoids .... *cohaerens*

Trechispora alnicola (Bourd. and Galz.) Liberta
Fruitbodies resupinate, membranous, up to 0.2 mm thick, tuberculate, ochraceous or pale yellowish, margin fringed, often with slender rhizomorphs. Basidia 10–25 × 4–5.5. Spores broadly ellipsoid, finely echinulate, 2.5–3.5 × 2–3. Conidia terminal on hypha-like conidiophores, hyaline, subspherical to ellipsoid, basally flattened, 4.5–7 × 3–4.5, smooth, with thin or thick walls. On dead wood of deciduous trees.

Trechispora amianthina (Bourd. and Galz.) Liberta
Fruitbodies resupinate, pellicular to membranous, up to 0.2 mm thick, smooth, pale yellowish, margin sometimes with thin rhizomorphs. Hyphae swollen in places to 10, thin-walled. Basidia 15–20 × 4–4.5. Spores narrowly ellipsoid, 4–7 × 2–3.5, smooth. On rotten wood and on leaf mould.

Trechispora byssinella (Bourd. and Galz.) Liberta
Fruitbodies arachnoid to pellicular, very thin, loosely attached to substrate, smooth, white, margin with thin rhizomorphs. Hyphae with some swellings to 8 wide. Basidia 10–15 × 3–4.5. Spores narrowly ellipsoid, 3–4 × 2, smooth. On wood and bark and overgrowing mosses.

Trechispora cohaerens (Schw.) Jülich and Stalpers (Fig. 418)
Fruitbodies resupinate, arachnoid to membranous, up to 0.2 mm thick, smooth, white to ochraceous, some hyphae swollen to 8 wide. Basidia 10–20 × 4–6. Spores broadly ellipsoid to tear shape, 3–5 × 2–4, smooth. On wood of deciduous trees and conifers.

Trechispora farinacea (Pers. ex Fr.) Liberta (Fig. 419)
Fruitbodies resupinate, up to 0.2 mm thick, arachnoid, pellicular or membranous, usually covered with soft teeth, white, cream, ochraceous or pale grey, teeth up to 1.5 mm long, cylindrical to subulate, margin sometimes with thin rhizomorphs. Hyphae swollen occasionally to 8 wide. Basidia 10–18 × 4–6. Spores ellipsoid, 3–4.5 × 2.5–3.5, echinulate. Common on rotten wood and bark.

Trechispora fastidiosa (Pers. ex Fr.) Liberta (Fig. 420)
Fruitbodies resupinate, up to 0.4 mm thick, membranous to cottony, loosely attached to substrate, smooth, tuberculate or with some short teeth, white, cream or pale yellow. Hyphae with swellings up to 8 wide, occasionally with sharp-pointed crystals. Basidia 15–30 × 5–7. Spores ellipsoid, apiculate, 5–9 × 3–5.5, hyaline to yellowish, echinulate or verruculose. On wood and debris on the ground and overgrowing mosses.

Trechispora microspora (P. Karst.) Liberta
Fruitbodies arachnoid or thinly membranous, smooth or granular, white, cream or honey colour. Some hyphae with swellings up to 7 wide. Basidia 8–12 × 4–5. Spores 3–3.5 × 2–3, verruculose. On rotten wood and overgrowing mosses.

Trechispora mollusca (Pers. ex Fr.) Liberta (Fig. 421)
Fruitbodies resupinate, poroid, softly membranous, up to 2 mm thick, white or cream, pores angular, 2–5, mostly 3–4 per mm, margin often thin, with rhizomorphs. Hyphae narrow but often with swellings up to 8 wide. Basidia 10–15 × 5–6. Spores broadly ellipsoid or subspherical, 3–5 × 2.5–4, hyaline, echinulate. On rotten wood and debris.

Trechispora vaga (Fr.) Liberta
Fruitbodies filamentous or membranous, up to 0.5 mm thick, smooth or tuberculate, sulphur yellow, ochraceous or honey colour to brown or greyish brown, turning wine red in KOH; margin with rhizomorphs. Hyphae yellowish with some parts swollen to 8–9. Basidia 10–20 × 5–7. Spores hyaline to pale yellowish, broadly ellipsoid, 4–6 × 3–4, echinulate. On rotten wood of deciduous trees.

Tremella

Fruitbodies of various shapes, often pulvinate, cerebriform or lobed, gelatinous, surface smooth or undulating. Hyphae mostly with clamps. No cystidia. Basidia hyaline, spherical or ellipsoid, 2–4-spored, longitudinally or obliquely septate, usually with clamp at base, sterigmata mostly cylindrical, often long. Spores spherical or broadly ellipsoid, hyaline, thin-walled, smooth, J–, often forming secondary spores and/or conidia.

KEY

```
      Not growing on other fungi ........................................................ 1
      Always growing on other fungi .................................................. 6
1.   Fruitbodies conspicuous, more than 3 cm across ........................... 2
      Fruitbodies inconspicuous, always less than 3 cm across, even when
            forming confluent masses ...................................................... 4
2.   Fruitbodies bright yellow, cerebriform ........................... mesenterica
```

Fruitbodies duller, brown or blackish, tufted 3
3. Fruitbodies brown or reddish brown, spores 9–12 long .... *foliacea*
    Fruitbodies blackish when fresh, spores 7–9 long .... *intumescens*
4. Fruitbodies aggregated, together looking like mulberries, red then black .... *moriformis*
    Fruitbodies not so aggregated, olivaceous, greenish or ochraceous brown 5
5. Spores 10–12 × 8–11 .... *exigua*
    Spores 6–10 diam. .... *virescens*
6. On *Cucurbitaria berberidis*, as well as on dead branches .... *exigua*
    Growing in the hymenium of *Dacrymyces* and *Ditiola* .... *obscura*
    On *Diaporthe*, especially species found on *Quercus* .... *globospora*
    On stromata of *Diatrype* .... *indecorata*
    On *Lophodermium* spp. on *Pinus* .... *translucens*
    On *Peniophora nuda* .... *versicolor*
    On *Postia caesia* and *P. lactea* .... *polyporina*
    On or, if not, very closely associated with *Stereum sanguinolentum* .... *encephala*

Tremella encephala Pers. (Fig. 422)
Fruitbodies solitary or gregarious, 1–3 cm diam., closely attached to the substrate; upper surface undulating and often cerebriform, white, ochraceous or pinkish brown, sometimes pearly opalescent, compact, with a hard white core surrounded by gelatinous flesh. Basidia spherical or subspherical, 12–20 × 10–18, with clamp, sterigmata up to 140 long. Spores spherical or subspherical, 8–12 × 6–9, secondary spores 6–8 × 5–6. Grows parasitically on or associated constantly with *Stereum sanguinolentum* on conifers.

Tremella exigua Desm.
Fruitbodies developing under bark and erumpent or growing on *Cucurbitaria berberidis*, 1–2 mm diam., gregarious and, when confluent, up to 2–3 cm across, gelatinous, olivaceous. Basidia 20–27 × 12–15, often with oblique septa. Spores subspherical or ellipsoid, 10–12 × 8–11. On deciduous trees and shrubs and on *Cucurbitaria berberidis*.

Tremella foliacea (Pers. ex S.F. Gray) Pers. (Fig. 423)
Fruitbodies tufted and made up of a number of broad, leafy lobes 3–10 cm long, gelatinous, brown or reddish brown. Basidia 12–20 × 10–15, sterigmata up to 50 long. Spores subspherical with protuberant hilum, 9–12 × 6–8. Conidia 3–4 × 2–3. Mostly on dead branches and stumps of deciduous trees, rarely on conifers.

Tremella globospora Reid (Fig. 424)
Fruitbodies pulvinate or drop-like, hyaline to whitish, 3–5 mm diam., sometimes confluent and then up to 1 cm, gelatinous, only obvious when

fresh and moist. Basidia $10-20 \times 10-14$, sessile or with stalk-like base with clamp, sterigmata up to 30 long. Spores $7-9 \times 5-8$. On *Diaporthe* species, especially on fallen branches of *Quercus* which have been lying for some time on the ground.

Tremella indecorata Sommerf.
Fruitbodies hemispherical or pulvinate, sometimes lobed, 0.5-2 cm diam., gelatinous, hyaline at first, finally becoming rather dark brown. Basidia almost spherical, 10-20 diam., sterigmata up to 70×3. Spores almost spherical, 10-15 diam., or $10-15 \times 8-12$. On stromata of *Diatrype*.

Tremella intumescens Sm. ex Hook.
Fruitbodies tufted with swollen, contorted lobes 3-6 cm long, blackish, gelatinous. Basidia $10-15 \times 10-12$, with sterigmata up to 20 long. Spores $7-9 \times 6-8$. On fallen branches of deciduous trees.

Tremella mesenterica Retz. (Fig. 425)
Golden Jelly Fungus
Fruitbodies mostly solitary, cerebriform or lobed, gelatinous, bright golden yellow and very conspicuous in wet weather, 3-10 cm wide and up to 4 cm high. Basidia $10-25 \times 10-20$, with clamps, sterigmata up to 150×3. Spores broadly ellipsoid to subspherical, $8-14 \times 5-10$. Conidia spherical and 3-5 diam., or ellipsoid $3-4 \times 2$. On dead fallen branches of deciduous trees, especially *Corylus, Fagus* and *Quercus*, autumn and winter, very common; sometimes only the conidial state is found.

Tremella moriformis Sm. ex Purt.
Fruitbodies spherical, 1-3 mm diam., clustered together in groups up to 1 cm across and looking like mulberries, ruby red at first but soon turning black, gelatinous. Basidia $10-20 \times 8-15$, sterigmata long and sometimes clavate. Spores subspherical, 6-10 diam. On fallen branches of deciduous trees, recorded on *Acer, Castanea* and *Ulmus*.

Tremella obscura (Olive) Christ.
Fruitbodies not developed separately, all growth taking place inside the hymenium of the fungal host. Basidia spherical or broadly ellipsoid, $8-15 \times 6-12$, sterigmata up to 50 long. Spores $5-10 \times 4-7$. In hymenium of *Dacrymyces* and *Ditiola*.

Tremella polyporina Reid
Fruitbodies gelatinous, scarcely visible to the naked eye. Basidia spherical or subspherical, $9-14 \times 8-11$, sterigmata up to 15 long. Spores $5-7 \times 5-6$, with protuberant hilum. Growing inside the tubes and on rotting fruitbodies of *Postia caesia* and *P. lactea*.

Tremella translucens Gordon (Fig. 426)
Pseudostypella translucens
Fruitbodies parasitic on *Lophodermium* species on pine needles, conspicuous only when wet and then 2–3 mm diam., pulvinate, white, gelatinous, with smooth or wrinkled surface. Hyphae with clamps at septa, markedly swollen on the side of the septum away from the clamp. Basidia subspherical, 10–14 × 9–12, each with a basal clamp, sterigmata mostly 12–20 long. Spores ellipsoid, slightly curved, 7–10 × 3–5, hyaline, smooth, forming secondary spores.

Tremella versicolor Berk. and Br.
Fruitbodies gregarious, sometimes confluent, round or lenticular, 2–3 mm diam., gelatinous, orange when young, later turning brown. Basidia almost spherical, 9–15 diam., sterigmata up to 50 × 3. Spores ellipsoid, 5–6.5 × 4. Conidia 3–7 × 2–5. On *Peniophora nuda*.

Tremella virescens (Schm. ex Fr.) Bref.
Fruitbodies pulvinate, 2–3 mm diam., sometimes with a few becoming agglutinated but not confluent in large masses, greenish or olivaceous, becoming ochraceous brown. Basidia 12–19 × 10–15, sterigmata up to 40 long. Spores almost spherical, 6–10 diam. On branches of deciduous trees.

Tremellodendropsis *tuberosa* (Grev.) Crawf. (Fig. 427)
Fruitbodies solitary, gregarious or tufted, erect, *Clavaria*-like, branched, tough gelatinous, grey or greyish brown, 2–7 × 2–4 cm, clearly stalked, stalk 2–4 mm thick, branches often long, narrow and compressed. Hyphae with clamps, some skeletal, all hyaline. No cystidia. Basidia hyaline clavate, up to 80 × 16, the upper part only longitudinally septate, 2–4 sterigmata up to 15 × 3. Spores fusiform, 10–22 × 5–8, hyaline, smooth, thin-walled, J–. On ground in woods and parks.

Tremiscus *helvelloides* DC (Fig. 428)
Fruitbodies erect, gelatinous, trumpet-shaped or elongated ear-shaped, tapered towards the base, split down one side, 5–14 cm tall, salmon pink, sometimes flushed orange, with the hymenium on the outside, sterile stalk-like base paler or whitish, about 1.5 cm wide. Hyphae with clamps. Basidia ellipsoid, 10–20 × 8–10, stalked, with longitudinal septa, 2–4 sterigmata 15–50 long, stalks up to 25 long. Spores cylindrical or narrowly ellipsoid, often slightly curved, 10–12 × 4–6, hyaline, thin-walled, smooth, J–. On the ground, mostly on buried wood or sawdust heaps along roadside verges, not common but seen occasionally in very large numbers.

Trichaptum
Hirschioporus
Fruitbodies effuso-reflexed to resupinate, tough, leathery. Brackets

imbricate, narrow, shelf-like, hymenium violet when growing actively, later cocoa-brownish, poroid, hydnoid or irpicoid. Hyphae with clamps, skeletal ones abundant. Cystidia present in hymenium. Spores hyaline, thin-walled, smooth, J−.

KEY

Hymenium mostly poroid .... *abietinum*
Hymenium irpicoid or hydnoid .... *fusco-violaceum*

Trichaptum abietinum (Pers. ex Fr.) Ryvarden (Fig. 429)
Fruitbodies tough, leathery, pileate to effuso-reflexed or resupinate; brackets numerous, imbricate and fused laterally, 1–4 × 0.5–2.5 cm, 1–3 mm thick, with sharp, wavy margin, upper surface velvety, somewhat zonate, whitish to grey, lower surface and decurrent hymenium violet when growing actively, later turning cocoa brownish, pores at first isodiametric, angular, 3–5 per mm, later becoming toothed or lacerate. Cystidia mostly narrowly clavate, 15–35 × 4–7, often coated with crystals. Basidia 15–20 × 4–6. Spores cylindrical to ellipsoid, slightly curved, 6–9 × 2–3.5. On conifer wood, very common.

Trichaptum fusco-violaceum (Ehrenb. ex Fr.) Ryvarden
Differs from *T. abietinum* in having slightly larger brackets, with the lower surface and decurrent hymenium hydnoid to irpicoid with teeth 4–5 mm long, radially arranged at the margin, but is similarly coloured and microscopically identical. Found mostly on *Pinus sylvestris*.

Tubulicrinis

Fruitbodies resupinate, effused, pale, thin, often waxy. Hyphae thin-walled, with clamps. Cystidia each with two or more roots, walls thick except at apex, dissolve readily in KOH. Basidia hyaline, clavate or suburniform. Spores hyaline, thin-walled, smooth, J−.

KEY

Cystidia capitate, spores broadly ellipsoid .... *accedens*
Cystidia not capitate, spores narrowly cylindrical to allantoid 1
1. Cystidia sharply pointed at apex .... *subulatus*
   Cystidia only slightly tapered to a rounded apex .... *glebulosus*

Tubulicrinis accedens (Bourd. and Galz.) Donk (Fig. 430)
Colonies effused, membranous to somewhat waxy, smooth, very thin, mostly greyish white. Cystidia 40–80 long, cylindrical to subulate, capitate, thick-walled and 3–5 thick except for the thin-walled head which is swollen to 6–8. Basidia 8–12 × 3–5. Spores broadly ellipsoid, 4–5 × 3–3.5. On rotten wood of conifers.

Tubulicrinis glebulosus (Bres.) Donk (Fig. 431)
Fruitbodies effused, thin, membranous to waxy, smooth, cream or ochraceous. Cystidia almost cylindrical but tapered slightly towards the rounded tip, 100–150 × 8–12, thick-walled except at apex, slightly amyloid. Basidia 15–20 × 4–5. Spores narrowly cylindrical to allantoid, 6–8 × 1.5–2. On rotten wood of conifers.

Tubulicrinis subulatus (Bourd. and Galz.) Donk (Fig. 432)
Fruitbodies effused, firmly attached to substrate, floccose to crustose, smooth or tuberculate, white to cream, waxy. Cystidia subulate, 70–150 × 8–15, hyaline, for the most part very thick-walled but thin-walled at the sharply pointed apex where there is occasionally some incrustation. Basidia 10–20 × 4–5, 4-spored. Spores allantoid 6–9 × 1.5–2.5, with a guttule at each end. On rotten wood of conifers.

Tulasnella

Fruitbodies resupinate, cobwebby, membranous, waxy or gelatinous, hymenium smooth. Cystidia present in some species. Basidia hyaline, stalked, clavate, pyriform or subspherical, bearing 2–4 swollen, often broadly fusiform sterigmata, each cut off eventually from the basidium by a septum. Spores hyaline, smooth, thin-walled, J−, often forming secondary spores.

KEY

Parasitic in hymenium of Corticiaceae .... *inclusa*
Not so, growing on dead wood and bark or on the surface of old
    polypores 1
1. Spores more than 20 long .... *calospora*
    Spores more than 10 but less than 20 long .... *violacea*
    Spores not more than 10 long 2
2. Spores 5–7 long .... *pruinosa*
    Spores 7–10 long 3
3. Spores allantoid .... *allantospora*
    Spores ellipsoid to subspherical 4
4. Irregular cystidia 40–70 × 6–9 present .... *cystidiophora*
    No cystidia 5
5. Fruitbodies purple or violaceous 6
    Fruitbodies white or grey 7
6. Fruitbodies pale violaceous, spores broadly ellipsoid to subspherical,
        6–7.5 wide .... *violea*
    Fruitbodies dark purple, drying blackish, spores 4–5.5 wide
        .... *tremelloides*

7.   Fruitbodies thin, white or whitish .... *albida*
     Fruitbodies rather thick, hyaline to dark grey .... *pinicola*

Tulasnella albida Bourd. and Galz.
Fruitbodies cobwebby, thin, white or whitish, often almost invisible.
Basidia clavate, 15–20 × 7–9, sterigmata 8–14 × 5–6.5. Spores 7–9 × 4–5. On
dead branches of deciduous trees.

Tulasnella allantospora Wakef. and Pearson (Fig. 433)
Fruitbodies effused, very thin, waxy or pruinose, greyish or pale lilac.
Basidia 7–10 × 4–6, sterigmata 6–7 × 4–5. Spores allantoid, 7–9 × 3–3.5. On
dead branches of deciduous trees and conifers.

Tulasnella calospora (Boud.) Juel
Fruitbodies widely effused, membranous to waxy, whitish to pale cream.
Hyphae without clamps. Basidia pyriform or clavate, 15–25 × 9–15,
4-spored, sterigmata broadly fusiform, 10–16 × 7–9. Spores narrowly fusiform, straight or S-shaped, 20–30 × 4.5–5.5. On dead branches of deciduous
trees and on old polypores.

Tulasnella cystidiophora Höhn and Lasch
Fruitbodies effused, thin, waxy, almost hyaline. Cystidia (g) present,
hyaline, often clavate or irregularly moniliform, 40–70 × 6–9. Basidia
broadly clavate, 12–20 × 8–10, sterigmata 10–16 × 6–6.5. Spores subspherical, 7–8 × 6–7. On fallen branches of deciduous trees.

Tulasnella inclusa (Christ.) Donk
Parasitic in the hymenium of Corticiaceae, e.g. *Athelia, Phlebia* and
Sistotrema, not forming separate fruitbodies. Basidia 10–15 × 7–10, sterig-
mata 10–20 × 4–5. Spores ellipsoid, 6.5–8 × 4.5–5.5.

Tulasnella pinicola Bres.
Fruitbodies effused, rather thick, gelatinous to waxy, surface often
undulating or tuberculate, in thickest parts dark grey, elsewhere almost
hyaline. Basidia clavate, 15–30 × 7–8, sterigmata 10–15 × 5–6. Spores ellipsoid, sometimes curved, 7–10 × 5–5.5, forming secondary spores. On dead
branches of conifers and deciduous trees and occasionally also on old polypores.

Tulasnella pruinosa Bourd. and Galz. (Fig. 434)
Fruitbodies effused, cobwebby, waxy, thin, pale grey with white pruina.
Basidia pyriform, 9–11 × 5–7. sterigmata 7–8 × 4–5.5. Spores ellipsoid-
oblong, 5–7 × 3–4. On fallen branches of deciduous trees.

Tulasnella tremelloides Wakef. and Pearson
Fruitbodies resupinate, gelatinous, surface undulating or wrinkled, dark
purple, drying blackish. Basidia clavate, 15–20 × 6–8, sterigmata 15–20 × 3–
4. Spores ellipsoid, 8–10 × 4–5.5. On wood and pine needles.

Tulasnella violacea (J. Olsen) Juel (Fig. 435)
Fruitbodies effused, membranous or pellicular, waxy or gelatinous, pale
pinkish lilac to violet, drying cream. Basidia clavate, 15–25 × 9–12. sterigmata 13–17 × 6–7.5. Spores limoniform or broadly fusiform, 10–15 × 5–7,
forming secondary spores. On fallen branches of deciduous trees and
conifers.

Tulasnella violea (Quél.) Bourd. and Galz. (Fig. 436)
Fruitbodies effused, membranous, waxy, smooth or slightly tuberculate,
rather pale violaceous, more pinkish or sometimes cream when dry. Basidia
clavate, 15–20 × 7–9, sterigmata 10–15 × 5–7. Spores broadly ellipsoid or
subspherical, 6–9 × 6–7.5. On wood and bark of fallen branches, especially
Betula, rarely found on conifers, occasionally on old polypores.

Tylopilus *felleus* (Bull. ex Fr.) P. Karst. (Fig. 437)
Bitter Bolete
Fruitbodies erect, pileate, composed of cap and central stalk; cap 6–15 cm
diam., fleshy, convex or somewhat flattened, upper surface honey colour to
greyish brown or snuff brown, dull, slightly velvety when young, pore
surface white becoming pink, bruising brownish, pores rather broad,
angular, stalk 6–10 × 2–5 cm, often swollen at base, cap colour or paler,
with distinct brown network, flesh white or cream with bitter taste. Cystidia
fusiform or lageniform. Spores 12–15 × 3–5, pink or pinkish in mass. Fairly
common, on the ground mostly under *Fagus* and *Quercus*, late summer–
autumn.

Tylospora
Fruitbodies resupinate, effused, arachnoid to membranous, loosely
attached to substrate, soft, white or cream. Hyphae hyaline or pale yellowish, with clamps. No cystidia. Basidia hyaline. Spores triangular, 3-lobed,
smooth or more irregularly 3-lobed and verrucose.

KEY

Spores triangular, 3-lobed, smooth .... *asterophora*
Spores irregularly 3-lobed, verrucose .... *fibrillosa*

Tylospora asterophora (Bonord.) Donk (Fig. 438)
Basidia 20–25 × 4–6. Spores 4–6 diam. On rotten wood and leaf litter.

Tylospora fibrillosa (Burt) Donk (Fig. 439)
Basidia 20–25 × 4–6. Spores 6–7 × 4–6, coarsely verrucose. On rotten wood
of conifers and on plant debris.

Typhula

(Key includes one species still in *Pistillaria*)
Fruitbodies erect, unbranched, often but not always arising from sclerotia
which may be superficial or embedded in the substrate, cylindrical to
narrowly or broadly clavate, mostly with a fertile head and sterile stalk.
Hairs on stalk often distinctive. Hyphae mostly with clamps, frequently
rather thick. Basidia long, clavate, mostly 4-spored. Spores smooth, hyaline,
thin-walled, J+ or J−.

KEY

Fruitbodies with white, clavate or cylindrical heads and clearly
    contrasted reddish brown stalks .... *erythropus*
Fruitbodies not so 1
1. Head and stalk yellowish brown, tall and slender .... *phacorrhiza*
   Head rose pink when fresh 2
   Head white or whitish when fresh 3
2. Heads covered with shining granules, on leaves and stems but not of
   grasses .... *micans*
   No shining granules, always on grasses .... *incarnata*
3. On *Pteridium aquilinum*, sclerotia yellowish, embedded in host
   .... *quisquiliaris*
   Not on *Pteridium*, sclerotia, when present, darker 4
4. Fruitbodies arising from sclerotia 5
   Fruitbodies not arising from sclerotia 10
5. Nearly always more than one and often many fruitbodies formed on
   each sclerotium .... *corallina*
   Always or almost always only one fruitbody formed on each
   sclerotium 6
6. On grasses 7
   Not on grasses 8
7. Hairs on stalk branched, thread-like, only 1 thick .... *capitata*
   Hairs on stalk subulate, swollen part 5 thick .... *graminum*
8. Fruitbodies 3–4 mm tall .... *setipes*
   Fruitbodies 1–2 cm tall 9
9. Spores 8–11 × 3.5–4, J− .... *sclerotioides*
   Spores 10–12 × 5–6, J+ .... *hollandii*
10. Fruitbodies never more than 1 mm tall 11
   Fruitbodies more than 1 mm tall 12
11. Spores 9–12 long, on Cyperaceae .... *hyalina*
   Spores 4.5–5.5 long .... *Pistillaria pusilla*
12. Hairs on stalk plentiful, subulate up to 100 × 6 .... *setipes*
   Hairs on stalk very few and depauperate or none 13
13. Spores 5–6.5 × 4.5–6 .... *culmigena*
   Spores 6–9 × 3–3.5 .... *uncialis*

Typhula capitata (Pat.) Berthier
Fruitbodies up to 1 cm tall with sclerotium, head clavate or pyriform, up to
1 × 0.7 mm, white, stalk up to 7 × 0.1–0.15 mm, yellowish brown, hairy.
Sclerotia each with one fruitbody, lenticular and 1–1.5 mm diam. when in
substrate, when free ventricose. Hairs on stalk plentiful, mostly branched,
thread-like, thin-walled, up to 70 × 1. Basidia 30–40 × 6–8. Spores narrowly
ellipsoid, 10–16 × 3.5–4.5, J+. On dead stems of grasses, e.g. *Calamagrostis,*
rarely on leaves.

Typhula corallina Quél. and Pat.
Fruitbodies 2–10 cm tall, head cylindrical or tapered towards the apex and
there sometimes divided into 2, white when fresh, drying grey, brownish or
olivaceous grey, stalk very short, hyaline or pale grey, smooth or with a very
few much reduced hairs. Sclerotia superficial, lenticular or bulbous, 1–
3 mm diam., ochraceous, brownish or blackish, each bearing up to 12 fruitbodies. Hyphae without clamps. Basidia 30–40 × 7–8, 2–4-spored. Spores
ellipsoid, 8–10 × 4–5.5. On leaves, dead herbaceous stems, fern fronds etc.
Host plants include: *Chaerophyllum, Cirsium, Epilobium, Eupatorium, Mentha,
Petasites, Alnus, Corylus* and *Fraxinus.*

Typhula culmigena (Mont. and Fr.) Schroet.
Fruitbodies in groups, without sclerotia, 1–3 mm tall, white or hyaline,
head clavate or pyriform, 0.5–2 × 0.3–1 mm, stalk shorter or longer than
head, 0.1–0.2 mm thick, with occasionally a few thin-walled, thread-like
hairs 40–80 × 2. Hyphae with clamps. Basidia 30–40 × 5–7. Spores ellipsoid,
5–6.5 × 4.5–6, J−. Saprophytic on grasses, e.g. *Calamagrostis, Molinia, Secale*
and *Triticum,* and occasionally on leaves of *Fraxinus* and other deciduous
trees.

Typhula erythropus (Pers.) Fr. (Fig. 440)
Fruitbodies 1–3 cm tall, with sclerotium, head cylindrical to clavate with
rounded apex, 3–8 × 0.5–1 mm, white, stalk longer than head, 0.1–0.2 mm
thick, cartilaginous, reddish brown, darker towards the base, finely hairy.
Hairs on stalk subulate, about 100 × 5, thick-walled, septate, mostly
unbranched, hyaline or pale brown, occasionally appendaged. Sclerotia
round or lenticular, 1–2.5 mm diam., reddish brown or blackish, each bearing a single fruitbody. Hyphae up to 15 wide, with clamps. Basidia 20–
40 × 5–6, 4-spored. Spores narrowly ellipsoid, 6–8 × 2.5–3.5, weakly J+. On
veins and petioles of previous year's fallen dead leaves of various trees
including *Acer, Alnus, Fraxinus* and *Populus,* very common.

Typhula graminum P. Karst.
Fruitbodies 2–8 mm tall, with sclerotium, head cylindrical to narrowly
ellipsoid, 0.5–2.5 × 0.1–0.4 mm, white, stalk twice as long as head, hyaline or
white, smooth to finely hairy. Hairs on stem subulate, thin-walled, up to
100 × 5. Sclerotia embedded or erumpent, 1 × 0.5 mm, dark reddish brown

or blackish, each with a single fruitbody. Hyphae without clamps. Basidia 25–30 × 6–8, 4-spored. Spores narrowly ellipsoid, 9–12 × 3.5-4, J+. On dead stems and leaves of grasses.

Typhula hollandii Reid
Fruitbodies up to 1.5 cm tall, thread-like, arising singly from sclerotia, head cylindrical, up to 3 mm long, white, drying yellowish, stalk hairy, about 4 times as long as head. Hairs on stalk cylindrical to subulate, thin-walled, 80–100 × 4-6.5. Sclerotia round, dark brown, 0.5 mm diam. Hyphae up to 20 wide, with clamps. Basidia 25–40 × 10–13, 2-4-spored. Spores narrowly ellipsoid, 10–12 × 5–6, J+. On plant debris in the fork of a tree.

Typhula hyalina (Quél.) Berthier
Fruitbodies 0.7–1 mm tall, without sclerotia, head hemispherical, hyaline to white, up to 0.45 × 0.3 mm, stalk up to 0.1 mm thick, hyaline with a few thread-like hairs 30–80 × 1, thin-walled, rarely branched. Basidia 40–60 × 7-8, 4-spored. Spores narrowly ellipsoid, 9–12 × 3.5-4, J–. Parasitic on leaves of Cyperaceae.

Typhula incarnata Lasch ex Fr.
Fruitbodies 0.5–3 cm tall, arising singly from sclerotia, head cylindrical, up to 2 × 0.3 mm, rose pink, stalk narrower and much longer than head, white or pale grey. Hairs on stalk subulate, thick-walled, 20–80 × 5–10. Sclerotia round or elongated, 0.5–4 mm across, reddish brown. Hyphae with clamps. Basidia 4-spored. Spores ellipsoid, 9–11 × 3.5–5, weakly J+. On grasses.

Typhula micans (Fr.) Berthier
Fruitbodies gregarious, 1–3 mm tall, without sclerotia, head occasionally forked, 0.8–2 × 0.2–0.4 mm, rose pink when fresh, drying paler or whitish, pruinose, covered with shining granules, stalk usually short, 0.1–0.2 mm thick, smooth. Hyphae up to 15 thick, with clamps. Basidia 30–40 × 8-15, 2-4-spored. Spores broadly ellipsoid, 9–12 × 4.5–6, J+. Saprophytic on dead leaves and stems.

Typhula phacorrhiza Fr. (Fig. 441)
Fruitbodies elastic, tough, mostly 5–10 cm tall, 0.5–1 mm thick, thread-like, yellowish brown or honey colour, head narrowly cylindrical, 2–4 cm long, tip sterile, stalk up to 8 cm long, darker than head and slightly narrower. Hairs on stalk up to 2 mm long, thin-walled. Sclerotia irregular in shape, 3–6 × 1.5–3.5 mm, greyish brown, each with one or, rarely, 2 or 3 fruitbodies. Hyphae up to 15 wide, with clamps. Basidia 25–35 × 9-10, 4-spored. Spores narrowly ellipsoid, 14–16 × 4.5-5.5, J–. On fallen dead leaves of deciduous trees, e.g. *Alnus, Fraxinus* and *Populus.*

Typhula quisquiliaris (Fr.) P. Henn (Fig. 442)
Fruitbodies often in rows but arising singly from embedded sclerotia, up to 7 mm tall, white, head clavate, 1.5–4 × 1-2.5 mm, stalk 0.3–0.4 mm thick,

finely hairy. Hairs on stalk thick-walled, often swollen towards base, 15–60 × 3–7. Sclerotia 1.5–3 × 0.5 mm, pale yellowish. Hyphae with clamps. Basidia 50–70 × 7–8, 4-spored. Spores narrowly ellipsoid, 9–14 × 4–5.5, strongly J+. On dead stalks of bracken, *Pteridium aquilinum.*

Typhula sclerotioides (Pers.) Fr.
Fruitbodies often in rows, arising singly from superficial, rather large sclerotia, up to 1 cm or, rarely, 2 cm tall, white, head cylindrical to narrowly clavate, 0.5–1 mm thick, stalk shorter or longer than head, up to 0.5 mm thick. Hairs on stalk thin-walled, with swollen lower part 20–30 × 6–7, terminal cylindrical part about 50–70 × 2. Sclerotia 1–2.5 mm diam., round, often flattened, blackish brown to black. Hyphae with clamps. Basidia 25–40 × 6–7, 4-spored. Spores narrowly ellipsoid, 8–11 × 3.5–4, J–. On dead petioles and stems of herbaceous plants, including *Chaerophyllum* and *Petasites.*

Typhula setipes (Grev.) Berthier
Fruitbodies gregarious, seldom more than 3–4 mm tall, with anchoring hyphae or with sclerotia, head broadly clavate or pyriform, white, 1–2 × 0.3–0.7 mm, stalk 0.1–0.2 mm thick, hyaline with brownish base, finely hairy. Hairs on stalk narrowly subulate, simple or branched, thin-walled, up to 100 × 6, upper part thread-like and only 1 thick. Sclerotia, when present, lenticular, brownish, 0.5–1 mm diam. Hyphae up to 22 wide, without clamps. Basidia 20–30 × 7–8. Spores ellipsoid, 7–10.5 × 3–5, J+. On dead leaves of deciduous trees.

Typhula uncialis (Grev.) Berthier
Fruitbodies tough, elastic, gregarious, mostly 3–5 mm tall, but occasionally up to 1 cm, without sclerotia, head clavate or spatulate, milky white, 0.5–1.5 mm wide, smooth, stalk shorter or longer than head, 0.2–0.4 mm thick, hyaline. Few or no hairs on stalk. Hyphae with clamps. Basidia 30–35 × 5–6. Spores 6–9 × 3–3.5, J–. On rotting stems of herbaceous plants, e.g. *Chaerophyllum*, *Epilobium* and *Petasites*, also on *Juncus* and ferns.

Tyromyces
Many species formerly in this genus have been transferred to the genus *Postia.*
Fruitbodies pileate or resupinate, soft, fleshy or leathery when fresh, drying hard, pores 2–4 per mm, pore surface white, cream, pinkish cream or salmon pink. Hyphae hyaline, with clamps, not metachromatic with cresyl blue; skeletal hyphae present. No true cystidia. Spores hyaline, thin-walled, smooth, J–.

KEY

Fruitbodies pileate, pore surface white or cream .... *chioneus*
Fruitbodies resupinate, pinkish cream .... *placenta*

Tyromyces chioneus (Fr.) P. Karst. (Fig. 443)
Fruitbodies solitary or in small groups, pileate, sessile, flat, dimidiate or semicircular, sharp-edged brackets, 2–8 × 4–10 cm, 1–4 cm thick, soft fleshy when fresh, drying hard and brittle, upper surface at first white and tomentose, becoming smooth to warted or wrinkled and cream, yellow-ochre or greyish, lower surface white or cream, pores round or angular, 3–4 per mm, flesh thicker than tube layer. Basidia 10–15 × 4–5. Spores cylindrical to ellipsoid, slightly curved, 4–5 × 1.5–2. On wood of deciduous trees, seldom on conifers.

Tyromyces placenta (Fr.) Ryvarden
Fruitbodies resupinate, sometimes with sterile nodules, firmly attached to substrate, fleshy or leathery fresh but drying hard, up to 1–2 cm thick, flat-cushiony, variable in colour, whitish when quite young, soon becoming pinkish cream to salmon pink, finally dirty ochraceous or greyish, pores angular, 2–4 per mm. Basidia 15–25 × 5–6, often 2-spored. Spores 5–7 × 2–3, cylindrical, often curved slightly. On wood of conifers, e.g. *Larix* and *Picea*.

Uloporus lividus (Bull. ex Fr.) P. Karst. (Fig. 444)
Fruitbodies pileate, composed of cap and stalk, often asymmetrical, clustered, with stalks joined together; cap 4–10 cm diam., convex or flattened, upper surface muddy straw colour, pale ochraceous or cinnamon, viscid when wet, pore surface sulphur to olivaceous yellow, bruising bluish or greyish green, pores large, irregular, tubes short, decurrent; stalk 3–7 × 1–2 cm, more or less cap colour but may be darker or lighter, flesh pale yellow in cap, somewhat brownish or rusty brown in base of stalk, becoming blue in parts when cut. Cystidia cylindrical to fusiform. Spores ellipsoid or ovoid, 4.5–6 × 3–4, in mass olivaceous brown. Always under *Alnus* in carrs, often hidden amongst grass and difficult to find.

Uthatobasidium fusisporum (J. Schroet.) Donk
(Keyed out with *Thanatephorus*)
Fruitbodies resupinate, cobwebby or membranous, whitish, yellowish or with olivaceous tints. Basidia 15–30 × 9–12, 2–4-spored, sterigmata up to 15 long. Spores fusiform, 10–20 × 5–8. On deciduous trees and on humus.

Uthatobasidium ochraceum (Massee) Donk (Fig. 445)
Fruitbodies resupinate, soft, membranous or floccose to fibrous, ochraceous. Basidia cylindrical or slightly clavate, 15–25 × 8–10, 4-spored, sterigmata up to 15 long. Spores broadly ellipsoid or subspherical, 8–10 × 5–7. On wood and plant debris.

Vararia *ochroleuca* (Bourd. and Galz.) Donk (Fig. 446)
Fruitbodies resupinate, effused, thin, membranous, smooth, cream to pale
ochraceous. Hyphae without clamps. Narrow, hyaline to yellowish, thick-
walled, repeatedly dichotomously branched hyphae abundant in subiculum
and present also in hymenium, with tips of end branches sharply pointed.
Cystidia obclavate, rostrate, 25–70 × 8–20, towards the apex 3–4 wide,
smooth, with oily contents. Basidia cylindrical to utriform, 15–30 × 3–4,
4-spored. Spores broadly oblong-ellipsoid, 3–4 × 2.5–3, smooth, thin-
walled, J−. On fallen branches and litter in woods.

Vesiculomyces *citrinus* (Pers.) Hagström (Fig. 447)
Fruitbodies resupinate, firmly attached to substrate, up to 0.5 mm thick,
waxy to membranous, at first forming small patches which become confluent and are then widely effused, surface smooth or tuberculate, whitish
to lemon yellow, when old ochraceous, margin white, fringed. Hyphae
without clamps. Cystidia (g) vesicular, tapering towards apex, 30–80 × 7–20,
thin-walled, smooth, some with oil drops, completely embedded ones very
broad, those in the hymenium much narrower. Basidia narrowly clavate,
40–50 × 5–7, 4-spored. Spores spherical or subspherical, apiculate, smooth,
thin-walled, hyaline, mostly 4–6 diam., J+, with oil drops. On wood of
conifers and deciduous trees.

Vuilleminia *comedens* (Nees ex Fr.) Maire (Fig. 448)
Fruitbodies resupinate, widely effused, developing below the bark which
rolls back, rather gelatinous or waxy, smooth, whitish or cream to greyish
ochraceous or flesh colour, often with a violet sheen, up to 0.5 mm thick.
Hyphae with clamps. Basidia narrowly clavate, up to 150 × 8–14, with 2–4
stout sterigmata. The basidia are separated by unbranched, cylindrical, thin-
walled hyphae. Spores cylindrical, curved, 15–23 × 5.5–7, hyaline, thin-
walled, smooth, J− or weakly J+. On dead, often attached branches of
deciduous trees, especially *Quercus*.

Xenasma
Fruitbodies resupinate, effused, thin. Hyphae hyaline, with clamps. Basidia
(pleuro-) 4–6-spored. Spores hyaline or yellowish, with thin or slightly
thickened walls, verrucose or verruculose, with warts disappearing in KOH,
J−.

KEY

Spores 5–7.5 × 3–4 .... *pruinosum*
Spores 9–12 × 5–6 .... *pulverulentum*

Xenasma pruinosum (Pers.) Donk (Fig. 449)
Fruitbodies effused, thin, smooth, somewhat gelatinous to waxy, greyish

white or with bluish tinge, pruinose. Cystidia cylindrical to subulate, hyaline, $50-100 \times 5-10$. Cystidioles $30-50 \times 2-5$, capitate and some with several short outgrowths from the head. Basidia $15-20 \times 6-8$, often 6-spored. Spores ellipsoid, $5-7.5 \times 3-4$, verruculose. On rotten wood of deciduous trees.

Xenasma pulverulentum (Litsch.) Donk (Fig. 450)
Fruitbodies effused, smooth or slightly granular, gelatinous to waxy, whitish to bluish grey. Cystidioles cylindrical to capitate, $15-40 \times 3-4$. Basidia $20-40 \times 5-8$, 4-spored. Spores ellipsoid, $9-12 \times 5-6$, verrucose with warts arranged spirally in rows. On rotten wood of deciduous trees.

Xenasmatella *tulasnelloidea* (Höhn. and Litsch.) Oberw. ex Jülich (Fig. 451)
Fruitbodies resupinate, effused, firmly attached, membranous, waxy or gelatinous, surface smooth or tuberculate, greyish white, bluish grey or ochraceous. Hyphae with clamps. No cystidia. Basidia (pleuro-) $10-20 \times 5-8$, 4-spored. Spores broadly ellipsoid, $5-6 \times 3.5-4$, hyaline, verruculose, warts not disappearing in KOH, J–. On rotten wood.

Xenolachne *longicornis* Hauerslev (Fig. 452)
Fruitbodies gelatinous, white to pale buff, glistening and somewhat pruinose, replacing the hymenium of the discomycete host. Hyphae with clamps. Basidia ellipsoid, $7-10 \times 3-4$, each with a single longitudinal septum and with a clamp at its base, two sterigmata tapered, $30-50$ long. Spores obclavate-fusoid, mostly $8-10 \times 3-4$, hyaline, smooth, biguttulate. On apothecia of *Cudoniella clavus*, *Discinella margarita* and *Hymenoscyphus vernus*.

GASTEROMYCETES

In this group the basidia and basidiospores mature within fleshy, slimy or gelatinous tissue referred to as the gleba, in which there are often cavities lined with hymenium and which is sometimes traversed by a central columella and/or by sterile tramal plates which give it a marbled appearance. The gleba is most commonly enclosed by two walls, the inner one generally forming a spore sac, but it may be on the outside of a honeycomb-like structure as in *Phallus*, line a latticed network as in *Clathrus* or be contained in small, often projectile bodies called peridioles as in bird's-nest fungi. The hymenium with spores attached to their basidia directly or by sterigmata is seen only in young stages of fruitbodies. The basidia disintegrate and the spores in most genera form a powdery mass often mixed with capillitium threads and other sterile material. In a few genera such as *Phallus*, *Mutinus* and *Clathrus*, the spores accumulate in slimy or sticky mucus. Sterile tissue below the gleba is called the subgleba and the two are separated occasionally by a parchment-like diaphragm. Mature fruitbodies often differ markedly from immature ones and if the latter are collected they should be kept until they ripen.

KEY TO GENERA

Mature fruitbodies growing on the surface of soil, dung, dead wood or plant debris 1

Mature fruitbodies growing in the soil, only occasionally just breaking through to and becoming visible on the surface 27

1. Fruitbodies erect and clearly much taller than broad, mostly 3–15 times 2

    Fruitbodies squat, rarely more than twice as tall as broad 10

2. Apex of fruitbody with a number of arms or lobes 3

    Apex of fruitbody without arms or lobes 4

3. Arms bright red, spreading horizontally .... *Aseroe*

    Arms dark reddish brown, incurved .... *Lysurus*

4. Tapered apex of fruitbody red or orange–red .... *Mutinus*

    Apex not red or orange–red 5

5. Head covered with olivaceous slime, stalk spongy, pale, smell strong, often very unpleasant 6

    Not with this combination of characters 7

6. Distinct net-like veil hanging down below head .... *Dictyophora*

    No such veil present .... *Phallus*

7. Fruitbody always small, never more than 8–9 cm high, with almost
    spherical head and slender stalk .... *Tulostoma*
    Fruitbodies much larger, mostly 15–30 cm high 8

8. Stem not scaly, head at first full of pea-shaped peridioles which
    eventually disintegrate to form a powdery mass .... *Pisolithus*
    Stem distinctly scaly, no peridioles 9

9. Volva at base of stem, whole fungus rust-coloured .... *Battarraea*
    No volva, fruitbody whitish or brownish, gleba orange–brown
    .... *Queletia*

10. Mature fruitbody bright red, shaped rather like a squid with a short
    body and 4–6 arms .... *Anthurus*
    Mature fruitbody in the form of a hollow latticed sphere 11
    Mature fruitbody other shapes 12

11. Latticed sphere white .... *Ileodictyon*
    Latticed network bright red .... *Clathrus*

12. Gleba inside little lenticular, ellipsoid or spherical peridioles;
    fruitbodies small or very small (less than 2 cm), often cup-shaped or
    like birds' nests 13
    Gleba never inside peridioles, fruitbodies mostly larger and other
    shapes 17

13. Peridiole one only, ellipsoid or spherical, fruitbody opening in a
    stellate manner with 5–9 acutely pointed lobes .... *Sphaerobolus*
    Peridioles usually several to many, if only one distinctly lenticular, and
    fruitbody not opening in a stellate manner 14

14. Peridioles not attached to inside wall of fruitbody by slender cords
    called funicles 15
    Peridioles attached to inside wall of fruitbody each by its own little
    funicle 16

15. Fruitbodies minute, not more than 2 mm diam., walls evanescent
    .... *Mycocalia*
    Fruitbody up to 2 × 1.5 cm; when mature the whole upper portion of
    wall breaking away to form a large opening .... *Nidularia*

16. Fruitbodies cup- or crucible-shaped; walls containing branched,
    brown, spiny hyphae .... *Crucibulum*
    Fruitbodies mostly like little flower-pots or funnel-shaped, occasionally
    barrel-shaped; walls not containing spiny hyphae .... *Cyathus*

17. Fruitbodies with outer wall at maturity split in a stellate manner into
    rather pointed, petal-like lobes which splay out or become reflexed,
    inner wall forming a spore-sac open at its apex 18
    Fruitbodies not so 20

18. Spore-sac with numerous mouths .... *Myriostoma*
    Spore-sac with one mouth 19

19. Lobes always strongly hygroscopic; spores 8–13 diam. .... *Astraeus*
    Lobes only in a few species hygroscopic; spores 3–7 diam. ... *Geastrum*

20. Spores 8–15 diam., walls reticulate or with large spines 21
    Spores 3–6.5 diam., smooth or minutely warted or spiny 22
21. Capillitium threads with numerous sharply pointed spines
    .... *Mycenastrum*
    Capillitium threads few or none, never with spines .... *Scleroderma*
22. Fruitbodies large, often 10 cm or more diam., never with a true mouth,
    the whole of the upper part breaking away or the wall splitting
    and disintegrating to release the spores 23
    Fruitbodies smaller, each with a mouth when mature 24
23. Fruitbodies subspherical, very large, commonly 20–50 cm diam. or
    even larger, with little or no subgleba .... *Langermannia*
    Fruitbodies often pear- or pestle-shaped, up to 15 cm diam., with
    cellular, spongy subgleba .... *Calvatia*
24. Subgleba none or dense and without cavities. Outer rather thick wall
    breaking up and disappearing, leaving behind the papery or
    parchment-like inner wall forming a spore-sac with apical mouth
    .... *Bovista*
    Subgleba spongy or pumice-like with large cavities 25
25. Parchment-like diaphragm separating gleba from subgleba, mature
    fruitbody at first with a small mouth but later with most of the
    upper wall coming away to form a large opening .... *Vascellum*
    No true diaphragm, mouth remaining relatively small 26
26. Spores smooth-walled, each with a large guttule and long pedicel
    .... *Bovistella*
    Spores mostly rough-walled, rarely with pedicels .... *Lycoperdon*
27. Spores mostly spherical or subspherical 28
    Spores always longer than broad, ellipsoid, lemon- or spindle-shaped
    33
28. Spores less than 8 diam., smooth or with small warts or spines 29
    Spores 10–20 diam., walls reticulate or with large warts or spines ... 30
29. Fruitbody wall flocculent, containing calcium oxalate crystals
    .... *Gastrosporium*
    Fruitbody wall soft, woolly or cottony, without calcium oxalate
    crystals .... *Sclerogaster*
30. Spore wall reticulate .... *Leucogaster*
    Spore walls bearing truncate warts 2–3 broad, 1–1.5 high ... *Wakefieldia*
    Spore walls spiny 31
31. Hyphae with clamps, gleba pale rose or flesh-coloured .... *Hydnangium*
    Hyphae without clamps, gleba other colours 32
32. Spines amyloid .... *Gymnomyces*
    Spines not amyloid .... *Octaviania*
33. Spores with very broad spines and a distinctive flared collar at the base
    .... *Stephanospora*
    Spores not so 34

34. Mature gleba black, spores always dark, flattened at base, with pore, smooth .... *Melanogaster*
    Never with this combination of characters, gleba other colours, very rarely black when old 35
35. Spores never more than 8 wide 36
    Spores always 8 or more wide 37
36. Surface of fruitbody usually covered with mycelial strands or rhizomorphs which sometimes form a network .... *Rhizopogon*
    Surface of fruitbody bald, silky and smooth or woolly, without rhizomorphs .... *Hysterangium*
37. Spores conspicuously ribbed with 8–10 ribs, fruitbody losing single wall at maturity, the surface then appearing strongly pitted like *Morchella* .... *Gautieria*
    Not so, if spore surface rugose then fruitbody surface not pitted 38
38. Fruitbodies not yielding white latex then cut, spores smooth, warted or rugose .... *Hymenogaster*
    Fruitbodies yielding white latex (from trama plates) when cut, spore surface ornamented with isolated amyloid rods or spines up to 1 high .... *Zelleromyces*

DESCRIPTIONS OF GENERA AND SPECIES

Anthurus *archeri* (Berk.) Fischer (Fig. 453)
Fruitbody when young 3–4 cm high, dirty white, when mature shaped rather like a squid with a red body 3–5 × 1–2.5 cm and 4–6 splayed out red arms or tentacles 4–7 cm long. Gleba slimy, dark olivaceous, on inner side of arms. Spores narrowly ellipsoid, hyaline, greenish in mass, smooth, 5–7 × 2–2.5. On rather acid soil in gardens, meadows and occasionally woods. A very rare species introduced from Australia about 70 years ago.

Aseroe *rubra* Labillard (Fig. 454)
Fruitbody when young spherical, whitish, about 3 cm diam., when mature composed of a more or less cylindrical, white to salmon pink stalk about 6 × 2 cm, sheathed by a volva at the base, expanded at the top into a bright red disc bearing at the margin 5–8 bifurcate, or up to 20 unbranched arms 3–7 cm long. Gleba slimy, olivaceous, on the disc and inner side of arms. Spores hyaline, 4–6 × 1.5–2. Foetid. A very rare species, probably introduced in soil from Australia.

Astraeus *hygrometricus* (Pers.) Morgan (Fig. 455)
Fruitbody 3–10 cm diam., spherical when young; outer wall many-layered, brown, dividing into 4–20 relatively hard, strongly hygroscopic lobes which open out when moist but close together and become leathery when dry; wall of spore-sac membranous, white, grey or black, opening by an

irregular tear at the apex. Spores spherical, 8–13 diam., cinnamon brown, cocoa in mass, warted, warts about 1 long. On sandy, stony or loamy soil in and along the borders of woods, autumn, rare.

Battarraea *phalloides* (Dicks.) Pers. (Fig. 456)
Fruitbody when young spherical and completely encased in outer wall which later splits, the lower part forming a volva and the upper part scales covering the inner wall, which itself eventually falls away in one piece, leaving a mass of spores mixed with elaters. Mature fruitbody rust-coloured, head hemispherical to somewhat conical, 1–3 cm diam., stalk very scaly, up to 22 × 1–2 cm. Spores subspherical, 5–6.5 diam., rusty brown, finely and densely warted. Elaters 50–80 × 4–6, with ring-like or spiral thickenings. On dry, sandy hedgebanks, sometimes growing amongst elm suckers, May–June, September–October, a rare species but found year after year in a few places.

Bovista
Fruitbodies mostly almost spherical but sometimes pear-shaped or capitate; outer wall 2-layered, without sphaerocysts, smooth, branny, granular or finely echinulate, eventually often breaking up and disappearing completely; inner wall forming spore-sac thin, papery or parchment-like, tough, with apical mouth. Subgleba compact or none, without cavities, never spongy or pumice-like. Spores spherical or broadly ellipsoid, walls smooth or rough, sometimes with long pedicels. Capillitium threads free or attached, with a thick main hypha bearing numerous forked, tapering branches, not spiny, with or without pores and septa.

KEY

Spores with curved, sometimes U-shaped pedicels .... *graveolens*
Spores without pedicels 1
Spores with straight or almost straight pedicels 2

1. Subgleba absent or poorly developed .... *pusilla*
    Subgleba up to 1.5 cm high, common on sand-dunes .... *aestivalis*
2. Subgleba well developed, grows on or amongst mosses on alkaline moors .... *paludosa*
    Subgleba absent 3
3. Fruitbodies about one third submerged in soil, remaining *in situ*, capillitium threads with a few large pores .... *tomentosa*
    Fruitbodies on soil surface, becoming free and often blown about by the wind, capillitium threads without pores 4
4. Fruitbodies always small, 0.5–1.5 cm, mouth raised and somewhat conical and sulcate .... *limosa*
    Fruitbodies larger, mouth not raised, conical or sulcate 5

5.   Fruitbody with single mycelial cord at base, inner wall dark purplish
     brown to black .... *nigrescens*
     Fruitbody with several mycelial cords at base, inner wall lead-coloured
     .... *plumbea*

Bovista aestivalis (Bonord.) Demoulin (Fig. 457)
Fruitbody 1–4.5 cm diam., spherical to pear-shaped or capitate, with an
unbranched, whitish mycelial cord like a tap root, often coated with sand
grains; outer wall smooth at first, becoming branny or tesselated, white to
ochraceous brown; inner wall papery, greyish brown, often orange or
coppery red at base, mouth round, lobed, 4–5 mm wide. Gleba olivaceous
brown to umber. Subgleba well-developed, compact, cottony, up to 1.5 cm
high, gradually passing over into the gleba. Spores 3.5–5 diam., finely
warted, without pedicels. Capillitium threads dichotomously branched, up
to 12 diam., with minute pores and few septa. Common on coastal dunes.

Bovista graveolens K. Schwalb
Fruitbody 3–6 cm diam., almost spherical, without basal mycelial attachment cord; outer wall white, smooth, gradually disappearing; inner wall
shining, greyish brown, occasionally bronze, parchment-like, often with
pale or dark flecks. Subgleba absent. Spores spherical or subspherical, 4–5.5
diam., smooth or punctate, often encrusted with colourless amorphous
matter, pedicel curved, almost U-shaped or encircling spore, 8–13 long,
expanded at free end, carrying away a part of the basidium wall apparently.
Capillitium threads without pores or septa, 10–25 thick, dichotomously
branched, reddish brown or blackish brown. In fields, especially those of
cereals and occasionally in gardens.

Bovista limosa Rostrup
Fruitbody spherical, always small, 0.5–1.5 cm diam., without basal mycelial
cord; outer wall whitish, at first smooth then branny; inner wall matt or
glossy, papery, reddish brown to umber, mouth up to 3 mm wide, raised,
somewhat conical and sulcate. Subgleba absent. Spores spherical, 4–6 diam.,
finely warted, pedicel more or less straight, 5–10 long. Capillitium threads
yellowish or reddish brown, without pores, scarcely branched. Usually on
dry, calcareous sandy soil; reported in Britain only from Lancashire dunes.

Bovista nigrescens Pers. (Fig. 458)
Fruitbody free when ripe, almost spherical, 3–8 cm diam., slightly pointed
below, with single, inconspicuous mycelial attachment cord; outer wall
white, almost smooth, disappearing at maturity to expose the shining, dark
purplish brown to black, parchment-like inner wall which breaks open at
the apex, mouth irregular, 1–3 mm wide. Gleba dark purplish brown.
Subgleba absent. Spores 4–6.5 diam., spherical, brown, finely warted, with
straight, hyaline pedicel 4–13 long, not thickened at free end. Capillitium
threads up to 30 thick, reddish or purplish brown, thick-walled, dichot-

omously branched, without pores or septa. Mostly in fields, on golf courses and grassy dunes, solitary or gregarious.

Bovista paludosa Lév.
Fruitbody sometimes spherical but mostly pear-shaped or capitate, 1.5–6 × 1–3 cm, false stalk up to 3.5 cm long, often wrinkled at base, mycelial attachment cord white, inconspicuous; outer wall white, smooth, thickish, sometimes retained as a thin crust; inner wall yellowish or reddish brown, becoming blackish brown, glossy or matt, mouth irregular. Subgleba well-developed, compact. Spores spherical or subspherical, 3.5–5.5 diam., punctate, with straight, cuspidate pedicels 7–15 long. Capillitium threads olivaceous brown to brown, up to 12 thick, without pores. Solitary, growing on or amongst mosses e.g. *Sphagnum*, on alkaline moors.

Bovista plumbea Pers. (Fig. 459)
Fruitbody free when ripe, spherical or subspherical, 1–5 cm diam., with several attachment cords; outer wall white or whitish, smooth, disappearing at maturity to expose the lead-coloured, matt inner wall, mouth 3–4 mm or more wide. Gleba clay-coloured or olivaceous brown. Subgleba absent. Spores ovoid or subspherical, smooth or almost so, mostly 4.5–6.5 × 4–5.5, pedicel straight, 5–18 long. Capillitium threads up to 25 thick, reddish brown, thick-walled, dichotomously branched, without pores or septa. Mostly gregarious on lawns and golf courses.

Bovista pusilla (Batsch) Pers.
Fruitbody almost spherical, flattened above, 1–3 cm diam., wrinkled at the base, with conspicuous, unbranched, greyish brown mycelial attachment cords; outer wall creamy white, drying brownish, at first smooth then branny or flaky to tessellated; inner wall papery, greyish brown, matt or glossy, mouth round or irregular. Subgleba absent or very little developed. Spores spherical, 3.5–5.5 diam., distinctly warted, without pedicels. Capillitium threads rather thin walled, fragile, 4–6 thick, olivaceous brown to brown, with numerous small pores. On dry, open, acid grassland, *Calluna* heaths, sand dunes etc. but not in woods.

Bovista tomentosa (Vitt.) Quél.
Fruitbodies sometimes clustered, about one third submerged in the ground and remaining in situ, spherical, 1–3 cm diam., with whitish mycelial cord; outer wall white, smooth or slightly tomentose, becoming crusty and disappearing; inner wall papery, shining, chestnut brown, blackening from the base, mouth round, lobed, 1–3 mm diam. Subgleba absent. Spores mostly ellipsoid, 4–5.5 × 3.5–5, smooth to punctate, with straight, truncate pedicel 5–15 long. Capillitium threads 10–25 thick, dichotomously branched, septate, with a few large pores. In dry, open, calcareous areas.

Bovistella *radicata* (Dur. and Mont.) Pat. (Fig. 460)
Fruitbodies top- to pear-shaped or almost spherical, strongly tapered and puckered at the base, 3–7 × 2–5 cm, not becoming detached from the mycelium; outer wall with sphaerocysts, at first white then brownish, eventually decaying, sometimes with a few warts or spines which not uncommonly are cruciately or stellately grouped; inner wall smooth, papery or like parchment, pale greyish brown, reddish brown at base, old blue–grey, mouth irregularly lobed, up to 2.5 cm wide. Gleba at first white, then olive and finally brownish. Subgleba strongly developed, spongy, with large cavities, deeply cup-shaped, soft, brown, divided from the gleba by a felt-like pseudodiaphragm. Spores spherical or broadly ellipsoid, 3.5–5 diam., smooth, with large oil drop or guttule, pedicel 4–11 long. Capillitium threads fragile, dichotomously branched, with a few large round pores, main hypha 6–14 thick. On acid, sandy soils, e.g. old heaths.

Calvatia

Fruitbodies mostly rather large, without mouth, the whole of the upper wall breaking away to release the spores; outer wall containing sphaerocysts. Gleba without columella. Subgleba cellular, sometimes with large cavities, not separated from the gleba by a true diaphragm but there may be a thin pseudodiaphragm. Spores without or with a short pedicel. Capillitium threads very fragile, not made up of a main stem and thin branches, with pores and sometimes with septa.

KEY

Fruitbody 2–5 cm diam., subspherical .... *candida*
Fruitbody 5–15 cm diam., other shapes 1
1.   Fruitbody as broad as tall, spores smooth .... *utriformis*
Fruitbody taller than broad, more or less pestle-shaped, spores
     distinctly verrucose .... *excipuliformis*

Calvatia candida (Rostk.) Hollós
Fruitbody spherical to somewhat flattened, 2–5 cm diam., when mature white with reddish flecks, with conspicuous, up to 5 mm thick basal, sand-grain encrusted mycelial cord; outer wall almost smooth to finely tesselated; inner wall papery, yellowish to greyish brown, upper part soon disintegrating leaving a rather flat cup with crenate edge. Gleba pale olivaceous or greyish brown. Subgleba little developed, compact, passing gradually into the gleba. Spores spherical, 4–5.5 diam., finely warted, with large guttules, pedicel up to 2 long. Capillitium threads with few pores, brittle, dichotomously branched, up to 7 thick, swollen at septa where they break. In sandy, grassy places and stubble fields.

Calvatia excipuliformis (Schaeff. ex Pers.) Perdeck (Fig. 461)
Fruitbodies mostly pestle-shaped or capitate, rarely pear-shaped, 5–20 cm tall, 5–12 cm wide, stalk cylindrical, narrowed and grooved or wrinkled towards the base, with branched, white mycelial cords; outer wall at first white then olivaceous buff, or pale brown, with deciduous spines 1–2 mm long; inner wall soon yellowish to greyish brown, the upper part breaking away to expose the spores, leaving eventually just the stalk. Gleba and subgleba when ripe dark olivaceous brown to blackish brown, scarcely any trace of pseudodiaphragm. Spores spherical, 4–6.5 diam., with guttules, distinctly verrucose, pedicel 1–3 long. Capillitium threads up to 10 diam., dichotomously branched, without swelling at forks, fragile, with small pores, without septa. Mainly in and along the borders of woods.

Calvatia utriformis (Bull. ex Pers.) Jaap (Fig. 462)
Fruitbody roughly pear-shaped, as broad as tall, 5–15 cm each way, with very short false stalk, wrinkled at base; outer wall at first white, then greyish brown, scurfy, surface breaking up into polygonal areas; inner wall papery, greyish brown, the upper part breaking away to expose the spore mass. Gleba becoming olivaceous or greyish brown. Subgleba with large cavities, divided from the gleba by a reddish brown pseudodiaphragm. Spores more or less spherical, rarely lemon-shaped, 4–5.5 diam., smooth, thick-walled, pedicel 2 long. Capillitium threads brittle, thin-walled, with pores, without septa, up to 8 thick, swollen to 18 at forks. On dry fields.

Clathrus *ruber* Micheli (Fig. 463)
Fruitbody when young spherical to ovoid, 2–3 cm diam., white, when mature a hollow, red latticed sphere up to 10 cm diam. or ovoid and up to 10 × 6 cm. Lattice spongy with gleba on inner side. Gleba slimy, green to dark olivaceous. Spores cylindrical to ellipsoid, yellowish green, 5–6 × 1.5–2. A rare species which is seen occasionally in gardens and parks.

Crucibulum *laeve* (Huds. ex Relh.) Kambly et al. (Fig. 464)
Common Bird's-nest Fungus
Fruitbodies at first subspherical, soon becoming cup- or crucible-shaped, 0.5–1 cm tall, 5–8 mm wide, pale cinnamon to yellowish brown and hairy on the outside, greyish and smooth when old, at first closed by a lid which sloughs off at maturity revealing inside the cup or nest up to 15 lenticular, pale ochraceous or white peridioles, looking like little eggs and each attached to the smooth inner surface by a slender cord or funicle. Wall of fruitbody one-layered, containing some brown, branched, spiny hyphae. Spores ellipsoid, 7–11 × 3–5, hyaline, smooth. Peridioles with their basidia and spores are dispersed by falling rain drops. Gregarious on dead wood, stems and other debris lying on the ground, fairly common but inconspicuous, autumn to spring.

Cyathus

Fruitbodies mostly funnel-shaped or like little flower-pots, occasionally barrel-shaped; wall three-layered, without any spiny hyphae, lids white or whitish. Peridioles lenticular, attached by complex funicles or cords. Spores large, mostly more than 10 long and always more than 5 wide, smooth, thin-walled.

KEY

	Inner wall of fruitbody striate	*striatus*
	Inner wall smooth, not striate	1
1.	Spores not more than 14 × 8	*olla*
	Spores often 25 × 20 or more	*stercoreus*

Cyathus olla (Batsch) Pers. (Fig. 465)
Fruitbody broadly funnel-shaped, 10–18 mm tall and 8–12 mm wide, grey-ish yellow or greyish brown and felted on the outside, silver grey and smooth on the inside. Peridioles 8–10, white to grey, 2–4 mm diam. Spores 8–14 × 5–8. On soil and twigs and other debris, fairly common.

Cyathus stercoreus (Schw.) de Toni (Fig. 466)
Fruitbody funnel- or barrel-shaped, 6–15 mm tall, 4–8 mm wide, some-times short stalked, golden brown to blackish brown, outside hairy, inside smooth. Peridioles blackish, 1–2 mm diam. Spores 20–35 × 20–25. On dung, dungy soil and bonfire sites; has been recorded also on sand-dunes.

Cyathus striatus (Huds.) Pers. (Fig. 467)
Fruitbody flower-pot shaped, 8–15 mm tall, 6–8 mm wide, brown and shaggy on the outside, greyish and vertically striate or fluted on the inside. Peridioles 12–16, grey, about 2 mm diam. Spores 15–21 × 7–10. On fallen branches, twigs and other debris.

Dictyophora *duplicata* (Bose) E. Fischer (Fig. 468)
The Veiled Lady
Differs from *Phallus* in having a clearly well-developed, net-like pendent veil or indusium. Fruitbody when young spherical or ovoid, up to 7 × 5 cm, white at first then pale brownish or reddish, with white mycelial cord; when mature with a cylindrical, hollow, spongy stalk 15–20 × 2.5–4.5 cm and a netted, ribbed cap or head about 5 × 5 cm with apical disc perforated in the middle. Gleba covering head olive, slimy, stinking. Veil arising below the cap, net-like, perforated, hanging down as far as the middle of the stalk. Spores ellipsoid, 3.5–4.5 × 1.5–2. In conifer and deciduous woods, rare.

Gastrosporium *simplex* Mattirolo (Fig. 469)
Fruitbody hypogeous, 0.5–2.5 cm diam., spherical, ellipsoid or pear-shaped, with conspicuous mycelial cord 5–20 cm long and 1.5 mm thick; outer wall chalky white, dry, flocculent, composed of narrow, collapsed

hyphae mixed with calcium oxalate crystals; inner wall membranous, ochraceous brown, somewhat gelatinous, cracking apically at maturity to form an irregular opening. Gleba homogeneous, at first white, then ochraceous brown or olivaceous, soft and spongy, then flocculent and finally powdery. Subgleba absent. Spores spherical or subspherical, pale ochraceous, 3.5–5.5 × 3–5, with a few small, irregular warts, mostly without pedicels. No capillitium threads. In dry, grassy places, rare.

Gautieria *morchellaeformis* Vitt. (Fig. 470)
Fruitbody hypogeous to erumpent, with strong, fungussy and not unpleasant smell, spherical or bulbous, 1–4 cm diam., brownish, losing the single wall at maturity and surface then appearing deeply pitted, like *Morchella*. Gleba ochraceous brown, traversed by a branched white columella which eventually disappears, cavities 1–8 × 1–4 mm, white trama plates about 70 thick. Spores 15–30 × 8–18, broadly ellipsoid, spindle- or lemon-shaped, pale to dark brown, conspicuously ribbed, with 8–10 ribs. Mostly found in chalky, loamy soil or humus in deciduous woods, mainly *Quercus* and *Fagus*, possibly mycorrhizal.

Geastrum
Earth-stars
Fruitbodies at first subspherical or onion-shaped with two closely approximated but distinct walls; outer wall thick, fleshy to leathery, one layer of it fibrous, another fleshy, splitting at maturity in a stellate manner into a number of petal-like lobes which often become reflexed and are in some species hygroscopic; inner wall membranous or papery, forming a spore-sac which may be sessile or stalked and with or without an apophysis. The mouth is apical and that part of the wall which surrounds it is called the peristome. The spore-sac contains a pseudocolumella, unbranched capillitium threads, basidia and spores. Spores spherical, in the range 3–7, coloured and, in most species, warted.

KEY

Spore-sac supported by 4 almost perpendicular lobes, like legs, which are attached to a partially buried shallow cup 1
Lobes mostly more than 4, sometimes arched but not perpendicular, and not attached to a partially burred cup 2
1. Fruitbody large, spore-sac broader than tall, peristome without halo around base, spores 3–4 diam. .... *fornicatum*
Fruitbody small, 0.5–3 cm diam., spore-sac taller than broad, with distinct pale halo around base of peristome, spores 4–6 diam. .... *quadrifidum*
2. Peristome sulcate with distinct grooves 3

Peristome not sulcate or grooved 7
3. Spore-sac sessile .... *badium*
   Spore-sac stalked, sometimes obscurely so except when dry when it
   can be seen clearly **4**
4. Apophysis ribbed or striate and with a ring-like thickening a little
   way down the stalk .... *pectinatum*
   Apophysis collar-like, hanging down and with a sharp edge ... *striatum*
   Apophysis, when present, neither striate nor collar-like **5**
5. Fruitbody large, 6–10 cm diam. .... *berkeleyi*
   Fruitbody small, 2–4 cm diam. **6**
6. Wall of spore-sac smooth, lobes not hygroscopic .... *nanum*
   Wall of spore-sac verrucose, lobes somewhat hygroscopic .... *campestre*
7. Spore-sac stalked **8**
   Spore-sac sessile **10**
8. Fruitbody 1–3 cm diam., not flesh-coloured, stalk always clearly seen
   .... *minimum*
   Fruitbody 4–10 cm diam., stalk best seen in dry specimens **9**
9. Lobes at first pale flesh-coloured, later pinky brown .... *vulgatum*
   Lobes cream-coloured to ochraceous, surface cracked when old
   .... *coronatum*
10. Lobes markedly hygroscopic, closing up tightly when dry, opening
    again when wetted **11**
    Lobes not hygroscopic **12**
11. Distinct halo around peristome .... *recolligens*
    No halo present .... *floriforme*
12. Fruitbody with flesh-pink tint, spores 2.5–3.5, smooth, with guttule
    .... *sessile*
    Fruitbody without flesh-pink tint, spores larger, rough-walled and
    without guttule **13**
13. Spore-sac subtended by a thick collar .... *simplex*
    Spore-sac without collar **14**
14. Flesh of outer wall felted, spongy, made up of thick-walled hyphae
    without septa .... *saccatum*
    Flesh not so .... *lageniforme*

Geastrum badium Pers. (Fig. 471)
Fruitbody when young spherical, outer wall split at maturity into 5–7 lobes
and then 1.5–4 cm diam., with typical basal scar, lobes thin, leathery,
slightly hygroscopic, incurved when dry, slightly recurved when moist,
yellowish brown or brown. Spore-sac 6–13 mm diam., sessile, smooth or
finely floury, broadly attached, peristome conical, sulcate, up to 4.5 mm
tall. Spores shortly spiny, mostly 4.5–5.5 diam. In dry sandy or chalky
places, typically a dune species but found also on dry hedge-banks and on
heaths.

Geastrum berkeleyi Massee (Fig. 472)
Fruitbody when young spherical, at maturity split into 7–9 lobes and then 6–10 cm diam., ochraceous brown, lobes not hygroscopic, firm, unequal, divided to middle, brown on inner side. Spore-sac 2–3 cm diam., stalked, apophysis indistinct, wall clearly warted, peristome long, conical, sulcate-striate, surrounded at base by a smooth, depressed silky zone. Spores 4.5–6.5 diam., dark brown, with narrow warts. Under trees and amongst fir needles, rare.

Geastrum campestre Morgan
Fruitbody when young spherical, when mature 2–4 cm diam., whitish or brownish, encrusted, with 5–12 weakly hygroscopic lobes. Spore-sac 1–2 cm diam., short-stalked, spherical or flattened, surface, including that of the apophysis, granular or verrucose, peristome sulcate-striate, with 12–15 furrows, surrounded at base by distinct halo. Spores dark brown, 5–7 diam., spiny. In dry, sandy or grassy places, open woods and sand-dunes.

Geastrum coronatum Pers. (Fig. 473)
Fruitbody when young spherical, outer wall split at maturity into 5–14 rather thick, pointed lobes and then 5–10 cm diam., lobes not hygroscopic, slightly curved downwards, cream-coloured to ochraceous and cracked on the inside when old. Spore-sac 1–2.5 cm., lead grey or brownish grey, rather dark, smooth, apophysis and short stalk concolorous, peristome fimbriate, often flattened. Spores brown, with rather large warts, 5–6 (7) diam. In woods and parks and on hedge-banks.

Geastrum floriforme Vitt. (Fig. 474)
Fruitbody spherical when young, at maturity outer wall split into 5–7 lobes which frequently split again and then 2–5 cm diam., lobes strongly hygroscopic, when fresh ochraceous brown. Spore-sac sessile, depressed spherical, 3–4 × 7–9 mm, greyish sepia, peristome indeterminate, neither sulcate nor fimbriate. Spores brown, warted, 4.5–7 diam. In sandy places, along edges of woods or in open pine woods, rare in Britain. Two Suffolk records, both under *Cupressus.*

Geastrum fornicatum (Huds. ex Pers.) Hooker (Fig. 475)
Fruitbody brown or dark brown, 4–8 cm tall, 3–4 cm wide, when young spherical, at maturity outer wall split almost always into 4 lobes, the inner and outer layers of which separate. The outer layer forms a partially buried shallow cup with broad, pointed, upwardly directed lobes, the inner layer forms corresponding but longer downwardly directed and almost perpendicular arched lobes which support the spore-sac. Spore-sac broader than tall, 15–25 × 10–15 mm, short-stalked, with smooth apophysis, peristome short, fibrous, not surrounded by a pale halo. Spores brown, finely warted, 3–4 diam. Under bushes and in deciduous woods.

Geastrum lageniforme Vitt.
Fruitbody when young like a tapered onion bulb, smooth, pale, at maturity
outer wall split into 5–9 long, pointed, almost equal, sometimes recurved,
ochraceous to brownish lobes, and then 4–8 cm diam.; the soft fleshy inner
layer disappears and does not form a thick collar below the spore-sac as it
does in *G. triplex*. Spore-sac subglobose, sessile, ochraceous or pale brown,
1–2.5 cm diam., peristome well-defined, silky striate, forming a flattened
cone surrounded by a silky zone or halo. Spores brown, spiny or warted, 3–
6 diam. In hedges and woods, rare.

Geastrum minimum Schw. (Fig. 476)
Fruitbody when young spherical, at maturity outer wall dividing into 6–11
non-hygroscopic lobes and then 1–3 cm diam., pale brown. Spore-sac
round or ovoid, 0.5–1 cm diam., often covered with small white crystals of
calcium oxalate, with pale stalk and smooth apophysis, peristome fimbriate,
delimited by a furrow. Spores subspherical, brown, warted, 5–5.5 × 4–4.5.
In dry places, including sand-dunes, rare.

Geastrum nanum Pers. (Fig. 477)
Fruitbody when young spherical, at maturity outer wall splitting into 5–10
narrow, non-hygroscopic lobes and then mostly 2–3 cm diam., whitish,
pale brown or greyish. Spore-sac short-stalked, with small apophysis,
smooth, often greyish, peristome conical, 2–4 mm tall, sulcate, with 15–20
deep grooves, base surrounded by halo. Spores brown, finely warted, 4.5–
6.5 diam. On coastal dunes.

Geastrum pectinatum Pers. (Fig. 478)
Fruitbody spherical when young, at maturity outer wall splitting to form 5–
10 non-hygroscopic, subequal, acute, brown lobes, and then 3–7 cm diam.,
with flesh often cracking and leaving pale areas especially along the margins.
Spore-sac 5–15 mm tall, 10–25 mm wide, grey or greyish brown, smooth
or pruinose, sometimes marbled, rather long-stalked (5–10 mm), with
ribbed or striate apophysis, often with ring-like thickening a little way down
the stalk, peristome conical, 2–5 mm tall, sulcate with about 20 furrows,
without halo at base. Spores spherical, brown, 5–7.5 diam., with long,
distinctly truncate warts. On soil mainly in pine and spruce woods.

Geastrum quadrifidum Pers. (Fig. 479)
Fruitbody spherical when young, at maturity outer wall split to form usually
4 cream to yellowish brown, non-hygroscopic lobes, and then 0.5–3 cm
diam.; the lobes are downwardly directed, arched and almost perpendicular
and connect with a shallow cup half buried in the soil. Spore-sac taller than
broad, 5–10 × 4–7 mm, grey to greyish brown, short stalked and with
collar-like apophysis, peristome 0.5–1.5 mm tall, conical, fimbriate, with
distinct pale halo around base. Spores spherical, brown, 4–6 diam., with

well-separated, fairly large warts. Mostly on needle litter in pine and spruce woods.

Geastrum recolligens (Woodw. ex Sow.) Desvaux (Fig. 480)
Fruitbody with 6–11 narrow, very strongly hygroscopic lobes, 3–6 cm diam., when fresh and expanded, closing up into a tight ball when dry, opening out again when moistened. Lobes brown or dark brown on inside, silvery outside and becoming thin and woody. Spore-sac sessile, without apophysis, yellowish or pale brown, smooth, subspherical, flattened, 1–1.5 × 1–2 cm, peristome conical, 2–3 mm tall, silky fimbriate, surrounded at base by narrow, pale, silky halo. Spores spherical, 3–6 diam., brown, with well-spaced warts. In sandy woods and fields.

Geastrum saccatum Fr. (Fig. 481)
Fruitbody when young onion-shaped, outer wall yellowish or ochraceous, at maturity split into 6–9 equal, narrow, pointed lobes, and then 3–7 cm diam. The lobes are deeply saccate at the base, soft, the flesh felted, spongy, made up of thick-walled non-septate hyphae. Spore-sac not subtended by a thick collar as it is in *G. triplex*, sessile, 1–2 cm diam., yellowish, smooth to somewhat grooved, peristome conical, projecting, silky, surrounded at base by a wide, smooth, depressed zone. Spores spherical, 4–5 diam., dark brown, warted or spiny. On sandy ground, in hedgerows.

Geastrum sessile (Sacc.) Pouzar (Fig. 482)
(syn. *G. rufescens*)
Fruitbody when young spherical, at maturity outer wall split into 6–9 broad, acutely pointed lobes, and then 2–6 cm diam. When first opened out the lobes are often flesh pink on the inside but may be creamy pink; they often bend downwards and become cracked. Spore-sac sessile, 8–25 mm diam., subspherical, often with cone-like apex, peristome softly fimbriate, without halo at base. Spores spherical, 2.5–3.5 diam., pale brown, smooth or almost so, usually with a large guttule. In deep litter in woods, fairly common.

Geastrum striatum DC. (Fig. 483)
Fruitbody when young spherical, when mature outer wall split into 6–8 lobes, and then 2–6 cm diam., dirty yellowish or brown, with adhering soil particles on lower surface. Spore-sac 0.5–1 cm tall, 1–3 cm wide, at first whitish and finely mealy on surface, later turning grey, stalk 3–5 mm long, apophysis collar-like, hanging down, smooth and with a sharp edge, peristome 1–3 mm tall, conical, sulcate, with about 20 furrows, surrounded at base by a brownish halo. Spores spherical, brown, 5–6 diam., clearly warted. On rich, loamy soil in deciduous woods.

Geastrum triplex Jungh. (Fig. 484)
Fruitbody when young onion-shaped, at maturity outer wall split into 5–8 broad, non-hygroscopic, pointed lobes, and then 5–12 cm diam., lobes

creamy at first, soon ochraceous or olivaceous brown, fleshy inner layer peeling off with the exception of a thick, collar-like portion which remains around the base of the spore-sac. Spore-sac sessile, subspherical, somewhat flattened, 2.5–5 cm diam., smooth, concolorous with lobes, peristome at first silky fibrillose, becoming puckered and up to 14 mm diam., sometimes surrounded by a pale halo. Spores spherical, 4–5 diam., brown, with rather large warts. Usually in deep leaf litter in woods but we have found it also at the base of sandy cliffs. Probably the most frequently found species in Britain.

Geastrum vulgatum Vitt. (Fig. 485)
Fruitbody when young spherical, at maturity outer wall split into 5–9 rather broad, pointed, thick lobes, and then 4–10 cm diam., lobes at first pale flesh-coloured, later pinky brown. Spore-sac 1.5–2.5 cm diam., spherical or with a somewhat conical apex, pale buff or greyish brown, without crystals on surface, stalk very short and usually visible only in dry specimens, peristome short, silky fimbriate at first, becoming puckered or toothed, without halo around base. Spores spherical, pale brown, 4–6 diam., finely but distinctly warted. In woods and under hedges.

Gymnomyces *xanthosporus* (Hawker) A.H. Smith (Fig. 486)
Fruitbody hypogeous, remaining closed, spherical or ellipsoid, somewhat indented at base, without a stalk, 1–1.8 cm diam., wall thin, soft, not torn, eventually decaying, at first cream, later or on bruising flesh-coloured or darker red and finally dark brown. Hyphae without clamps. No latex. Gleba without columella, not cartilaginous, at first white, then brown, with large, irregular cavities lined by hymenium, and thin trama plates containing conspicuous sphaerocysts, especially at their junctions. Spores at first hyaline then yellowish brown, more or less spherical, 10–15 diam., without ornamentation, spines on surface up to 1.5 long, amyloid, no pedicel. In soil under *Picea* needles.

Hydnangium *carneum* Wallr. (Fig. 487)
Fruitbody hypogeous, round or irregular, 1–3.5 cm diam., without stalk but with a short, cushion-shaped sterile base and basal mycelial cord, wall dirty reddish to flesh colour, becoming brownish. Hyphae with clamps. Gleba fleshy, pale rose or flesh-coloured, with fairly large, irregular cavities, without columella. Spores 10–15 diam., without ornamentation, scattered conical warts on surface 0.5–2 long, hyaline to pale yellowish. On heaths and along sunny edges of woods, has been recorded in soil under *Eucalyptus*.

Hymenogaster
Fruitbodies hypogeous, variable in shape and colour, without rhizoids. Gleba not cartilaginous, fleshy and dry, cavities lined by hymenium and

forming a labyrinth, sometimes filled with spores, with thin, brittle trama plates. No columella, spores yellow to brown, ellipsoid to spindle-shaped, often with a pedicel, smooth or warted wall with an epispore and often also a perispore.

KEY

Spores smooth .... *luteus*
Spores warted or rugose 1
1. Spores up to 50 long 2
   Spores never more than 35 long 3
2. Fruitbody white to dark brown, smooth .... *olivaceus*
   Fruitbody yellowish brown, covered with fine silky hairs .... *sulcatus*
3. Spores with distinct apical papilla 4
   Spores without apical papilla 5
4. Fruitbody lemon to olivaceous .... *citrinus*
   Fruitbody white to grey .... *hessei*
5. Fruitbody hairy or tomentose 6
   Fruitbody smooth, bald 8
6. Spores 13–18 long .... *albus*
   Spores 19–25 long 7
7. Smell pleasant, like lily-of-the-valley .... *griseus*
   Smell stinking, sometimes oniony .... *lycoperdineus*
8. Spores not more than 18 long .... *arenarius*
   Spores up to 20 or more long 9
9. Spores not more than 11 broad .... *tener*
   Spores mostly more than 11 broad 10
10. Fruitbody becoming ochraceous .... *vulgaris*
    Fruitbody not becoming ochraceous 11
11. Fruitbody dirty white with brown flecks or spots .... *thwaitesii*
    Fruitbody evenly mid to dark brown .... *muticus*

Hymenogaster albus (Klotzsch) Berk. and Br. (Fig. 488)
Fruitbody 0.5–2.5 cm diam., white at first, becoming pale golden yellow, surface white-woolly or with silky hairs. Gleba white, then cinnamon brown to dirty ochraceous, with large cavities. Spores yellowish brown, 13–18 × 8.5–11, broadly ovoid, ellipsoid or lemon-shaped, always without a papilla, with finely rough epispore, without perispore. In gardens and woods.

Hymenogaster arenarius Tul. (Fig. 489)
Fruitbody 3–4 mm diam., pale grey or greyish brown, wall thin, smooth. Gleba white when young, then grey, never reddish. Spores broadly lemon-shaped, 11–18 × 9–11, dark yellow to golden brown, with warted epispore and thin, hyaline perispore. Has strong, unpleasant smell. In sandy soil in parks and woods.

Hymenogaster citrinus Vitt. (Fig. 490)
Fruitbody 0.5-4 cm diam., when fresh lemon to golden yellow, dry pale clay to olivaceous yellow. Gleba at first yellow, then brown to blackish brown. Spores 20-35 × 15-17, broadly spindle- to lemon-shaped, with elongated apical papilla and short pedicel, warted epispore and irregularly rough perispore. In woods.

Hymenogaster griseus Vitt. (Fig. 491)
Fruitbody 6-12 mm diam., white to cinnamon brown, surface floccose or white-hairy. Gleba at first grey then brown. Spores 20-25 × 11-13, ellipsoid, with short pedicel, without distinct apical papilla, perispore rugose or undulating. Smell like lily-of-the-valley. In soil of deciduous woods.

Hymenogaster hessei Soehner (Fig. 492)
Fruitbody white to dirty grey, never reddening, wall very thin. Gleba white, then brown and finally black. Spores 20-25 × 15-17, broadly ovoid to lemon-shaped, with distinct apical papilla, pedicel up to 5 long, with thick, markedly rugose perispore, at first golden but darkening later. In soil in woods.

Hymenogaster luteus Vitt. (Fig. 493)
Fruitbody 0.5-2 cm diam., at first white, then with yellowish to brown flecks, becoming brown on exposure to air, wall thin, brittle. Gleba when young white, later turning lemon yellow. Spores 18-25 × 8-12, ellipsoid, smooth-walled, without papilla. Mostly under *Fagus* and *Quercus* on chalky soil.

Hymenogaster lycoperdineus Vitt. (Fig. 494)
Fruitbody 2-5 cm diam., at first white, soon turning brown, wall finely tomentose. Gleba white when young, becoming brown, with rather large cavities. Spores 19-23 × 10-12, elongated ellipsoid to spindle-shaped, without distinct papilla, dark brown, with finely undulating or rough perispore. Stinking, sometimes oniony. In loamy soil, especially under *Quercus*.

Hymenogaster muticus Berk. and Br. (Fig. 495)
Fruitbody about 1 cm diam., bald, white when young, then mid to dark brown. Gleba reddish brown to dark brown. Spores 18-23 × 10-15, ovoid to broadly ellipsoid, without apical papilla, with thick, reddish brown, rugose perispore. In soil of deciduous woods.

Hymenogaster olivaceus Vitt. (Fig. 496)
Fruitbody 1-4 cm diam., at first white, then mid to dark brown, smooth. Gleba white when young, then pale yellow to yellowish brown or dark olivaceous brown. Spores 20-25 × 11-20, at first longish or ovoid-lanceolate, with cylindrical to clavate apical papilla, then pear-shaped without papilla, with short pedicel, dark yellowish brown, with thick, rugose perispore.

Hymenogaster sulcatus Hesse (Fig. 497)
Fruitbody 0.5–1.5 cm diam., tuberous, yellowish brown, never con-spicuously lemon-coloured, finely silky-haired. Gleba yellowish brown to brown. Spores 24–53 × 10–23, with papilla 9–14 long, with fairly long pedicel, at first yellowish brown and smooth, then reddish brown with rough perispore. In woods.

Hymenogaster tener Berk. and Br. (Fig. 498)
Fruitbody about 1 cm diam., white or dirty white, shining, smooth, wall 2-layered. Gleba at first white, then pale rose, finally brown, with narrow cavities. Spores 15–20 × 8–11, broadly lemon-shaped or ellipsoid, without papilla, yellowish brown to dark brown, perispore thin or missing, epispore clearly warty or rarely somewhat ribbed-reticulate. In woods.

Hymenogaster thwaitesii Berk. and Br. (Fig. 499)
Fruitbody pea-like, 1–1.5 cm diam., dirty white with brown spots or flecks, smooth. Gleba white at first, then brown. Spores 20–28 × 14–17, ovoid, broadly ellipsoid or subspherical, without an apical papilla, dark reddish brown, with thick, rough wavy or warted perispore. In deciduous woods.

Hymenogaster vulgaris Tul. (Fig. 500)
Fruitbody 1–1.5 cm diam., at first white, then dirty ochraceous, smooth. Gleba white when young, then lilac to brown and finally black. Spores 20–33 × 9–10, long spindle-shaped, narrowed or pointed at apex, without papilla, tapered towards base, with short pedicel, pale yellowish brown with thin, finely rough, rugose perispore. In humus under mosses in woods.

Hysterangium
Fruitbodies hypogeous, spherical or kidney-shaped, mostly whitish at first, tough wall with one or two layers. Gleba with cavities lined by hymenium; columella cartilaginous, simple or branched, bluish white. Spores hyaline to pale brownish or olive, cylindrical, ellipsoid or spindle-shaped, mostly smooth, often with perispore.

KEY

Spores 10–18 × 4–6 .... *nephriticum*
Spores 17–22 × 6–8 .... *thwaitesii*

Hysterangium nephriticum Berk. (Fig. 501)
Fruitbody kidney-shaped, 1–2.5 cm diam., white, woolly; wall rather thick, 2-layered, inner wall pseudoparenchymatous, outer one hyphal. Gleba at first rose-coloured then greyish brown. Spores pale yellowish brown, narrowly ellipsoid, 10–18 × 4–6. Gregarious and sometimes confluent, in woods, May–February, rare.

Hysterangium thwaitesii Berk. and Br. (Fig. 502)
Fruitbody spherical, 2–4 cm diam., bald, silky smooth, white becoming reddish with age or when bruised, drying brown; wall 1-layered. Gleba olivaceous brown. Spores yellowish brown, spindle-shaped, 17–22 × 6–8. In deciduous woods, August–October, rare.

Ileodictyon *cibarius* Tul. (Fig. 503)
Fruitbody a white, latticed sphere up to about 10 cm diam., which becomes free from the dirty white, reticulate volva and blown about by the wind. The meshes of the lattice are about 5 mm thick, tubular, hollow, not chambered. Gleba on the inside of the whitish mesh, olivaceous brown, viscous, smelling when fresh of Camembert cheese. Spores 4–6 × 2–2.5, greenish. Introduced from Australia and New Zealand and now established in England in a few places.

Langermannia *gigantea* (Batsch ex Pers.) Rostk. (Fig. 504)
Giant Puff-ball
Fruitbody very large, subspherical, mostly 10–50 cm diam., but specimens up to 150 cm across have been recorded, with basal mycelial cord; outer wall when young white or creamy, smooth with the feel of soft leather, partly breaking down and becoming free from the paper-thin inner wall, the whole becoming eventually greenish or olivaceous brown, splitting and disintegrating to release the spores. Gleba at first white, then sulphur yellow to olivaceous brown. Subgleba hardly any. Spores spherical or almost so, 3.5–6 diam., smooth or finely warted, olivaceous or brownish, with small guttule, pedicel 0.5–3 long. Capillitium threads fragile, up to 9 thick, rather thin walled, with many pores and septa. In meadows, pastures and gardens, usually on nutrient-rich soil, summer–autumn.

Leucogaster *floccosus* Hesse (Fig. 505)
Fruitbody hypogeous or occasionally erumpent, spherical or tuberous, lumpy, 1.5–3 cm diam., with branched mycelial cord, hairy, at first white, then flecked with yellow and finally ochraceous; wall unequally thickened. Gleba cartilaginous to waxy, brittle when dry, with somewhat angular cavities 1–1.5 mm diam.; trama plates white to ochre, thin. Spores 11–21 × 11–18, spherical to broadly ellipsoid, hyaline, surface clearly reticulate, with thick perispore. In woods, mostly under *Quercus* and *Fagus.*

Lycoperdon
Fruitbodies with two walls, outer wall containing sphaerocysts, inner one with apical mouth. Gleba with pseudocolumella in most species. Subgleba usually spongy-cellular or pumice-like, cavities seen easily. Diaphragm absent or little developed. Spores spherical or subspherical, not more than 6 diam., usually 3–5, mostly warted, pedicel present in some species. Capillitium threads well developed, elastic or brittle, with or without pores.

KEY

On wood, spores smooth .... *pyriforme*

On soil, spores not smooth 1

1. Outer wall of fruitbody with few or no spines, common on coastal dunes .... *lividum*

    Outer wall with numerous spines, not common on coastal dunes 2

2. Spines 3–6 mm long .... *echinatum*

    Spines never more than 3 mm long and usually much less 3

3. Spores with attached pedicels 4

    Spores without attached pedicels 6

4. Pedicels 10–35 long .... *pedicellatum*

    Pedicels much shorter 5

5. Pedicels 2–3 long, spines on fruitbody dropping off readily leaving a layer of brown sphaerocysts .... *marginatum*

    Pedicels 0.5–1.5 long, spines slender, persistent .... *umbrinum*

6. Fruitbody when young covered with large, white or creamy veil remnants which persist and form a ring-like zone around base of head .... *mammaeforme*

    No such veil remnants 7

7. Spines dark, often blackish 8

    Spines paler, never blackish 9

8. Spines dropping off, leaving areolate patches .... *foetidum*

    Spines persistent, no areolate patches .... *atropurpureum*

9. Spores coarsely warted, warts often scattered 10

    Spores minutely and closely warted 11

10. Fruitbodies whitish, growing on dry, calcareous grassland ..... *decipiens*

    Fruitbodies mostly creamy-coffee coloured, growing in woods, quite common .... *molle*

11. Subgleba purplish brown, a rare species found on damp, acid soil, e.g. moors with *Sphagnum* .... *ericaeum*

    Subgleba white to brown, very common in conifer and mixed woods and often also elsewhere .... *perlatum*

Lycoperdon atropurpureum Vitt.

Fruitbody top-shaped to almost cylindrical, 2–5.5 cm diam., firmly fixed to soil by large, agglutinated rhizomorphs; outer wall with dark yellowish brown to blackish spines well developed, persistent, brittle, erect, made up of sphaerocysts which can be thick-walled at the base; inner wall inconspicuous, yellowish, shining, visible here and there in old fruitbodies but not areolate. Subgleba with large cavities (up to 0.8 mm wide), remaining yellow for a long time but eventually brown. Spores 4–6 diam., with large, scattered warts, mixed with broken-off pedicels. Capillitium threads elastic, thick-walled, reddish brown, septate, pores few and small. Mostly in oak woods.

Lycoperdon decipiens Dur. and Mont.
Fruitbody usually spherical or broader than tall, whitish, attached to the soil
by rhizomorphs; outer wall with whitish, brittle spines not well formed and
looking more like stellate scales, made up of thin-walled, often collapsed
sphaerocysts; inner wall becoming brownish, shining, visible here and there
but not areolate. Subgleba little developed, brown. Spores spherical, 4–6
diam., with large, scattered warts. Capillitium threads brown, 4–7 thick,
often free, with numerous large, irregular pores. Mostly on dry, calcareous
grassland.

Lycoperdon echinatum Pers. (Fig. 506)
Fruitbody spherical to pear-shaped, hedgehog-like, 2–6 cm diam., at first
white then brown; outer wall with brown deciduous spines 3–6 mm long
which converge at their tips in groups of three or four; inner wall cream-
coloured, becoming distinctly areolate when the spines have dropped off,
and this is especially noticeable on the upper surface, mouth small. Gleba
purplish brown, never olivaceous. Subgleba little developed, rather
compact. Spores mixed with broken-off pedicels, spherical, brown, 4–5
diam., warted or somewhat spiny. Capillitium threads brown, elastic, 5–8
thick, with small pores. On rich soil, mainly in beech woods.

Lycoperdon ericaeum Bonord.
Fruitbody pear-shaped to somewhat stalked or top-shaped, rarely subspherical, 1.5–4.5 cm diam., pale brown, without rhizomorphs; outer wall
mealy and with cream-coloured, very brittle spines; inner wall cream-
coloured, normally not visible, not becoming areolate. Subgleba purplish
brown, with fairly large cavities. Spores 4–5.5 diam., closely and shortly
warted, mixed with few or no broken-off pedicels. Capillitium threads
fragile, yellowish-brown, 4.5–8 thick, septate, branched, with many large,
regular pores. On damp, acid soil, on moors with *Sphagnum*, rare in Britain.

Lycoperdon foetidum Bonord. (Fig. 507)
Fruitbody often foetid when young, 2–6 cm diam., top-shaped, pear-
shaped or subspherical tapered to a stalk-like base, ochraceous to very dark
brown; outer wall with fine, mostly blackish, in part deciduous spines
which converge at their tips in groups forming little pyramids; inner wall
cream-coloured, becoming areolate when spines have fallen. Subgleba well-
developed, occupying up to half the fruitbody, brown with violet or
olivaceous tints, cavities relatively large. Spores mostly 4–5 diam., punctate
to minutely warted, mixed with pedicel fragments, in mass yellowish or
olivaceous brown. Capillitium threads yellowish brown, 4.5–7 thick,
seldom septate, pores fairly abundant and irregular. Common in woods on
acid soils and on heaths.

Lycoperdon lividum Pers. (Fig. 508)
Fruitbody rather small, 1–3 cm diam., or up to 4.5. cm tall, pear-shaped or

capitate, with basal mycelial cord, hyphae mixed with sand grains or soil forming a sort of ball; outer wall whitish then brownish, mealy or warty-granular, short spines if present at all only few and just where the stalk joins the head; inner wall pale greyish brown, shining, papery, with small mouth. Subgleba white, then brownish with small cavities, passing over gradually into the gleba. Spores 3.5–5 diam., finely warted, not mixed with pedicel fragments. Capillitium threads yellowish brown, brittle, branched, 4–10 thick, with many large pores and few septa, commonly with humped or unciform outgrowths. Common on coastal dunes and found sometimes also on other open ground.

Lycoperdon mammaeforme Pers. (Fig. 509)
Fruitbody 2–9 cm tall, 2–7 cm broad, pear-shaped, top-shaped or subspherical with a broad umbo and short stalk, white at first then ochraceous brown, when young covered with large, white or creamy, cottony velar remnants, which tend to persist and form a ring-like zone around the base of the head; outer wall thickly and closely beset with relatively short spines; inner wall pinkish white then brownish, shining. Subgleba brownish olive or brownish violet. Spores 4–5 diam., finely warted or shortly echinulate, chocolate-brown in mass, mixed with pedicel remnants. Capillitium threads dark brown, 6–7 thick, with few pores or septa. On chalky soil amongst leaf mould, mostly in beech woods, uncommon.

Lycoperdon marginatum Vitt.
Fruitbody 1–5 cm diam., almost spherical, rarely top-shaped, white when young, brownish when old; outer wall covered at first with white, pyramid-forming and so appearing thick spines which drop off readily leaving a uniform layer of brown sphaerocysts on the downy inner wall. Subgleba brown with medium-sized cavities, sometimes separated from gleba by a very thin diaphragm. Spores 4–5 diam., finely warted, with pedicel not more than 2–3 long. Capillitium threads yellowish brown, elastic, 5–7 thick, little branched, pores sometimes very large. In sandy areas, dry grassland and open pine woods. British records uncertain.

Lycoperdon molle Pers. (Fig. 510)
Fruitbody 5–9 cm tall, 2–6 cm broad, subspherical, pear- or top-shaped, with thick stalk, often milky-coffee coloured but very variable; outer wall with short, simple, soft, pale greyish brown spines some of which lie together with their tips convergent, intermediate ones granular furfuraceous; inner wall hardly visible, cream to yellowish brown. When spines fall off the whole fruitbody becomes smooth. Gleba olivaceous brown with fairly distinct columella. Subgleba white at first then brownish violet, with fairly large cavities. Spores spherical, 4–5.5 diam., reddish or chocolate brown in mass, coarsely verrucose, mixed with detached pedicel remnants

10–20 long. Capillitium threads flexuous, brown, 5–10 diam., with distinctly irregular pores. On soil in deciduous and conifer woods, usually gregarious, quite common.

Lycoperdon pedicellatum Peck (Fig. 511)
Fruitbody top-shaped, pear-shaped or capitate, usually clearly but shortly stalked, up to 6 cm tall and 3.5 cm broad, white or cream, brown when old; outer wall with deciduous, rather stiff, conical spines up to 2 mm long, stalk covered with granules; inner wall greyish brown or ochraceous, parchment-like, smooth or sometimes areolate. Gleba cinnamon to olivaceous brown. Subgleba brownish violet with medium-sized cavities, occupying whole of stalk. Spores spherical to ovoid, 4–5 × 3.5–4.5, brown, finely warted, with pedicels 10–35 long. Capillitium threads brown, 4–9 thick, little branched, with pores but few septa. In damp grassy places, boggy woods etc.

Lycoperdon perlatum Pers. (Fig. 512)
Fruitbody 2–9 cm tall, 2–7 cm broad, white at first, becoming yellowish brown, pear-shaped or capitate, head subspherical, somewhat flattened, puckered below; outer wall of head with conical, pale, soft, deciduous spines 1–2 mm long, encircled by shorter, pale or darker warts or granules; inner wall matt, cream-coloured, tough, sometimes becoming indistinctly areolate when spines drop off; stem at first warted then smooth. Gleba olivaceous brown at maturity. Subgleba white, then brownish, with fairly large cavities. Spores spherical, olivaceous brown, minutely warted, 3–4.5 diam., mixed with pedicel remnants. Capillitium threads yellowish brown, 3–7 thick, with pores but few septa. In conifer and mixed woods and also quite often in open grassy places, very common.

Lycoperdon pyriforme Schaeff. ex Pers. (Fig. 513)
Fruitbody pear-shaped to capitate, white at first then ochraceous to brown, 2–7 cm tall, 1–5 cm broad, with white rhizomorphs; outer wall finely warted or scurfy; inner wall cream-coloured, papery, with fairly large mouth. Gleba olivaceous brown. Subgleba remaining white. Spores spherical, smooth, 3–4.5 diam. Capillitium threads brownish, 3–6 thick, without pores. Gregarious, always on dead wood, mainly on stumps.

Lycoperdon umbrinum Pers. (Fig. 514)
Fruitbody pear-shaped, top-shaped or subspherical with short, partly buried stalk, attached to white mycelium, 2–5 cm tall, 1–4 cm broad, at first pale brown then reddish to blackish brown; outer wall with slender, per-sistent spines up to 1 mm long, without granules or coarse warts amongst them; inner wall papery, shiny, visible between spines. Gleba with distinct columella, yellowish brown. Subgleba white to greyish brown. Spores spherical, 3.5–5.5. diam., finely warted, with pedicel 0.5–1.5 long. Capil-litium threads 4.5–10 thick, with large pores. Mostly in conifer woods on sandy soils.

Lysurus cruciatus (Lepr. and Mont.) Lloyd (Fig. 515)
Fruitbody when young spherical, white, 2–3 cm diam., when mature composed of a whitish, cylindrical, hollow, cellular stalk, 6–12 × 2 cm, sheathed by a volva at the base, divided at the apex into dark reddish brown, longitudinally grooved and transversely ribbed, commonly incurved arms or lobes, 1–3 cm long, coated on the inside by mucilage containing spores. Spores reddish brown in mass, oblong-ellipsoid, 3–5 × 1.5–2. In meadows, gardens and greenhouses, rare.

Melanogaster

Fruitbody wholly or partly embedded in soil, spherical or tuberous, lumpy, fleshy to gelatinous, wall brown, soft, not detachable from gleba. Hyphae with clamps. Gleba with roundish to somewhat angular, always dark or black, mucus-filled cavities, and pale trama plates which gelatinize; without true hymenium. Spores dark yellowish brown or blackish purple, flattened at base and with a pore, wall at least 0.5 thick.

KEY

Spores not more than 11 long .... *broomeianus*
Spores always more than 11 long 1
1. Spores 11–15 long, without a papilla .... *intermedius*
Spores 12–22 long, with a papilla .... *ambiguus*

Melanogaster ambiguus (Vitt.) Tul. (Fig. 516)
Fruitbody 1–4 cm diam., irregularly tuberous, lumpy, often fibrillose, olivaceous brown to dark brown, blackish brown when old. Gleba becoming blackish with white or cream trama plates. Spores 12–22 × 5–12, broadly spindle-shaped, lemon-shaped or ovoid with longish papilla at apex and frill from remainder of sterigma at base, dark brown, smooth. Solitary or gregarious in chalky or sandy loam, mostly under *Carpinus* and *Quercus* but occasionally with conifers.

Melanogaster broomeianus Berk. (Fig. 517)
Fruitbody 2–5 cm diam., irregularly tuberous, lumpy, finely hairy, at first ochraceous, then dark brown with a reddish sheen, finally blackish. Gleba when ripe reddish brown to blackish, with white, yellow or reddish trama plates. Spores cylindrical to narrowly ellipsoid, dark brown, smooth, 6–11 × 3–4.5 including attached, short, hyaline remains of sterigma. In sandy and calcareous soil, mainly under deciduous trees.

Melanogaster intermedius (Berk.) Zeller and Dodge
Fruitbody 2–5 cm diam., irregularly tuberous, dark reddish brown to blackish brown. Gleba with trama plates becoming reddish with age although

initially yellowish. Spores 11–15 × 8–10, ellipsoid clearly tapered at each end, without a papilla. Under broad-leaved trees.

Mutinus

Fruitbodies arising from thick, white, cord-like mycelium, when young, in the so-called witches' egg stage, ovoid or pear-shaped, splitting at the apex into several lobes. The mature fruitbody consists of a cylindrical, spongy, hollow stalk which terminates in a slender, tapered, sometimes curved head covered directly by the dark olivaceous, slimy gleba which often washes off to reveal the reddish or reddish orange colour of the head itself. The wall of the witches' egg remains at the base as a volva. Spores ellipsoid, pale yellowish, smooth.

KEY

Stalk and fertile part equal in length .... *bambusinus*
Stalk 3 times the length of the fertile part 1
1.   Stalk cylindrical, common in Britain .... *caninus*
Stalk tapered towards base, rare in Britain .... *ravenelii*

Mutinus bambusinus (Zoll.) E. Fischer (Fig. 518)
Fruitbody when young egg-shaped, about 3.5 × 2.5. cm, white to pale grey, when mature about 10–15 × 1 cm, the lower part pale salmon pink, thin-walled, the upper, fertile part equal to it in length, red or purplish red but usually covered, except for a sterile acutely pointed tip by the dark olive gleba. Spores 2–4 × 1. Foetid. Recorded under fruit trees, an imported, very rare species.

Mutinus caninus (Huds. ex Pers.) Fr. (Fig. 519)
Dog Stinkhorn
Fruitbody when young ellipsoid or ovoid, up to 4 × 2.5 cm, white or yellowish, when mature 7–15 × 1 cm, the lower three quarters cylindrical, whitish or pale pinkish buff, thin-walled, delicate, spongy, the upper quarter pointed at the tip, orange–red, covered at first for the most part by the dark olivaceous brown gleba, sheathing volva at base. Spores 4–5 × 1.5–2. Not foetid. Amongst fallen leaves, mostly in deciduous woods and especially common under *Fagus*, June–December.

Mutinus ravenelii (Berk. and Curt.) E. Fischer
Fruitbody when young egg-shaped, white, about 2 × 1.5 cm, when mature 6–9 × 1 cm, pale red, tapered at both ends, often perforated at apex, gleba olivaceous. Spores 3.5–5 × 2. In gardens parks and propagating houses, rare.

Mycenastrum corium (Guersent ex DC.) Desvaux (Fig. 520)

Fruitbody with lower part immersed in soil, spherical or pear-shaped, 5–12 cm diam., basal mycelial cord thick; outer wall 1-layered, its hyphae with conspicuous clamps, white, splitting to form broad scales and soon peeling off; inner wall brownish, tough, leathery to corky, 1-4 mm thick, smooth, dividing in an irregular stellate manner into about 10 lobes. Gleba at first white, then olive green, finally brownish, powdery floccose, without columella, easily separating from wall. No subgleba. Spores more or less spherical, 8-11 diam., brownish, with one guttule, wall reticulate. Capillitium threads yellowish, olive green, or pale reddish brown, elastic, little branched, with numerous sharply pointed, straight or curved spines; main stem up to 18 thick, often S-shaped. In sandy places and dry, open woods.

Mycocalia

Fruitbodies very small, not more than 2 mm diam., solitary or in groups, spherical or lenticular; wall thin, evanescent, made up of non-spiny, thin-walled hyphae with clamps. Peridioles lenticular, without attachment cords (funicles). Basidia old often with thick walls. Spores hyaline or yellowish brown.

KEY

	Spores yellowish brown, about 13 × 5.5 .... *sphagneti*
	Spores hyaline, smaller 1
1.	Peridioles 1 to each fruitbody, dark brick-red or yellow .... *minutissima*
	Peridioles usually more than one per fruitbody, not dark brick-red or yellow 2
2.	Peridioles yellowish brown or tan .... *denudata*
	Peridioles dark reddish brown to purplish black .... *duriaeana*

Mycocalia denudata (Fr.) Palmer (Fig. 521)
Fruitbodies spherical, up to 1.5 mm diam., often in groups and sometimes confluent, when fresh frequently covered with numerous droplets; wall thin, white to pale yellowish. Peridioles lying in a bed of jelly, flat disc-shaped, 300-400 × 70-140, yellowish brown to tan, becoming biconcave when dry. Spores hyaline, 6-10 × 5-7. On dead culms of grasses, sedges, *Juncus* etc., occasionally also on fallen twigs.

Mycocalia duriaeana (Tul.) Palmer (Fig. 522)
Fruitbodies similar to those of *M. denudata*. Peridioles dark reddish brown to blackish purple, about 300 × 150. Spores hyaline, about 7 × 5.5. On plant remains, e.g. of *Ammophila*, on dunes, also on pine wood and needles, moss etc.

Mycocalia minutissima (Palmer) Palmer (Fig. 523)
Fruitbodies solitary or in groups but not confluent, white, with only one peridiole developing in a drop of slime. Peridioles dark brick-red to yellow,

lenticular, biconcave when dry. Spores hyaline, 5–6 × 3.5–4. In very damp places, on rotting plant remains, e.g. of grasses, rushes and birch leaves.

Mycocalia sphagneti Palmer
Fruitbodies solitary or in groups but not confluent, each containing only one peridiole; wall white, at first woolly, then smooth. Peridioles lenticular, 450–700 × 100–300, at first white then dark red or blackish. Spores pale yellowish brown, almost smooth, mostly about 13 × 5.5. On the lower parts of *Juncus* growing in *Sphagnum*, June–July; recorded also on *Eriophorum* and *Nardus*.

Myriostoma *coliforme* (Dicks. ex Pers.) Corda (Fig. 524)
Fruitbody when young spherical, 4–8 cm diam., outer wall splitting at maturity into 4–12, usually 8, tapered lobes and then up to 25 cm across, rather pale brown. Spore sac round or somewhat flattened, up to 7 cm diam., pale silvery brown, supported by a number of 3–5 mm long stalks and with several to many small mouths. Gleba with several columellae. Spores spherical, brown, 4–6 diam., spiny. Capillitium threads spiny. On sandy soil beneath bushes, also in open woods, last seen in Britain in 1880.

Nidularia *farcta* (Roth. ex Pers.) Fr. (Fig. 525)
Fruitbodies solitary or gregarious, almost spherical, up to 2 cm tall and 1.5 cm broad, creamy, cinnamon or brown; wall one-layered, thick, felty, containing brown, spiny hyphae, downy on surface; when mature the whole upper portion breaks away forming a large mouth. Peridioles numerous, lenticular, 0.5–2 mm diam., chestnut brown, shining, embedded in slime, not attached to wall by funicles. Spores broadly ellipsoid, 6–10 × 4–7, hyaline. On partly rotted wood and sawdust.

Octaviania *asterosperma* Vitt. (Fig. 526)
Fruitbody hypogeous, containing watery latex, spherical or ellipsoid, 1–5 cm diam., with cushion-shaped sterile base, white at first, becoming greenish blue or red when rubbed or exposed to air, finally blackish brown; wall up to 0.5 mm thick. Gleba at first watery white or pale yellow, then blackish brown, fleshy, marbled with pale trama plates and small, irregular, darker cavities filled with spores. Spores spherical or subspherical, pale yellowish brown, thick-walled, mostly 10–13 diam., with some up to 18–20 diam., covered with large, conical spines, not amyloid. In soil under broad-leaved trees.

Phallus
Fruitbodies when young frequently called witches' eggs, spherical or ovoid, attached to mycelial cords; outer wall membranous, covering a thick gelatinous layer which in turn surrounds an olivaceous gleba, whitish glebal cavities and a compressed stalk. The egg eventually ruptures apically and

maturity is then attained in as little as 1.5 hours. Mature fruitbody consists of a long, fat, almost cylindrical, hollow, spongy stalk with the remains of the egg forming a volva at the base and a bell-shaped to conical head formed by a pendent, honeycomb-like structure attached only at the apex of the stalk and surrounding it like a thimble; this is coated with dark olivaceous slime containing the spores. Spores ellipsoid, smooth, pale olivaceous.

KEY

Fruitbody, especially egg and volva rosy or violaceous, smell sweet, pleasant .... *hadriani*
Fruitbody mostly without any rosy or violaceous colour, smell strong, foetid, unpleasant .... *impudicus*

Phallus hadriani Vent. ex Pers. (Fig. 527)
Dune Stinkhorn
Fruitbody when in witches' egg stage 4–6 × 3–4 cm, soon becoming rosy to violet with similarly coloured mycelial cords, when mature with stalk 10–18 × 3–4 cm, white sometimes tinted in part rose or violet, head covered with olivaceous slime, volva elongated and retaining its distinctive colour though browning somewhat when old. Spores 3–4 × 1.5–2. Sweet-smelling like hyacinths or violets. Mostly on sand-dunes amongst marram grass but occasionally in other sandy places, rarely in woods and then only near the sea.

Phallus impudicus L. ex Pers. (Fig. 528)
Stinkhorn or Wood Witch
Fruitbody in witches' egg stage white or yellowish, 4–6 × 3–5 cm, with pure white mycelial cords, when mature with stalk 10–25 × 4–5 cm, white, head about a quarter to one fifth of the total length, covered with dark olivaceous slime, volva white or pale yellowish. Spores 3.5–5 × 1.5–2. Stinking; spores dispersed by flies which are attracted in large numbers by the smell. Very common in woods.

Pisolithus *arhizus* (Pers.) Rauschert (Fig. 529)
Fruitbody irregularly club-shaped or capitate, up to 30 cm tall, 6–12 cm broad, ochraceous or rather olivaceous brown, the lower, thick stalk or root-like part immersed in soil often with chrome yellow markings and sand-grain incrustation; wall thin, becoming brittle, disintegrates at maturity to expose the gleba. Gleba composed at first of numerous yellow to brown, pea-shaped peridioles which break down to form a powdery mass of spores when ripe. There is no true capillitium, just some septate, hyaline hyphae. Spores spherical, 8–12 diam., ochraceous or cinnamon brown, warted to spiny, each with one guttule. On sandy or gravelly soil by

roadsides and in woods, with pine and birch. They sometimes look rather like horse-droppings.

Queletia mirabilis Fr. (Fig. 530)
Fruitbody whitish to brownish, head more or less spherical, flattened above, 2–4 cm diam., stalk 4–15 × 2–4 cm, fibrously grooved and coarsely scaly; wall thin, breaking up and flaking away to expose the gleba. Gleba at first whitish, finally orange–brown. Spores spherical, 5–6 diam., often with short pedicel, warted, yellowish to rusty brown. Capillitium threads twisted, simple or branched, thin-walled, pale yellowish brown, 7–10 thick. Mostly on old wood piles, partly buried amongst rotting leaves, rare.

Rhizopogon
Fruitbodies either totally immersed in soil or partly free, spherical or irregular in shape, sessile, whitish when young, then ochraceous or brownish, some species reddening on exposure to air, the surface usually covered with mycelial strands; wall with one or two layers. Gleba not easily separated from wall, softly fleshy, not cartilaginous, at first white then mostly olive, with numerous small, narrow cavities lined by hymenium, deliquescing when ripe. No columella, no subgleba. Spores almost hyaline or yellowish to olivaceous, cylindrical, narrowly ellipsoid or spindle-shaped, smooth. Mostly under conifers, mycorrhizal.

KEY

Fruitbody never red or reddening when bruised	1
Fruitbody reddening at least when bruised or exposed to the air	2
1. Fruitbody yellow–ochre to olivaceous brown	*luteolus*
Fruitbody whitish, becoming black when bruised	*reticulatus*
2. Spores 8–12 long	*roseolus*
Spores 5–8 long	3
3. Fruitbody 1–2 cm diam., only under *Pseudotsuga*	*hawkeri*
Fruitbody 2–5 cm diam., mostly under *Pinus*	*vulgaris*

Rhizopogon hawkeri A.H. Smith
Fruitbody 1–2 cm diam., spherical to ovoid, with thin mycelial strands or rhizoids only at the base, at first greyish white, then with brown spots, later red, when old ochre with reddish sheen. Gleba white, then ochre. Spores hyaline, cylindrical to ellipsoid, without pedicel, 5–8 × 2–2.5. Under *Pseudotsuga*.

Rhizopogon luteolus Fr. (Fig. 531)
Fruitbody 2–5 cm diam., spherical or tuberous and knobby, pale to dark yellow–ochre, olivaceous brown when old, never reddening, covered with numerous dark tawny fibrils; wall thick and tough. Gleba whitish yellow to

dark olive, with very small, round or irregular cavities which become filled with spores. Trama plates white or yellowish. Spores $6-9 \times 2.5-3.5$, narrowly ellipsoid, olivaceous, with two guttules. Taste mild, smell unpleasant. In sandy places, in pine woods under moss, not uncommon in Scotland.

Rhizopogon reticulatus Hawker
Fruitbody 2 cm diam., spherical to somewhat flattened, soft, white when young, blackening when bruised, covered with a network of fine, black mycelial strands; wall thin, fragile. Gleba when young white, turning yellowish olive to greenish brown, finally blackish. Spores $6.5-7.5 \times 1.5-3$. Mostly under spruce.

Rhizopogon roseolus (Corda) T.M. Fries (Fig. 532)
Fruitbody 1-5 cm diam., tuberous, knobby, often with white mycelial cord at base, white or yellowish, turning rose on exposure to air, finally reddish brown, thin-walled, covered with adpressed, white, yellowish, reddish or dark fibres. Gleba at first bright yellow, then pale olive to yellowish brown, with numerous narrow cavities. Spores $8-12 \times 3-5$, ellipsoid to fusiform, pale yellowish. Smell unpleasant. In woods.

Rhizopogon vulgaris (Vitt.) M. Lange
Fruitbody 2-5 cm diam., whitish or yellowish, becoming brown or blackish, reddening on exposure to air or when bruised, surface covered with dark but not always very distinct fibrils; wall fairly thick, when freshly cut brick red or blood red. Gleba at first whitish then dark yellowish brown, often with a brick-red sheen, with numerous narrow cavities. Spores $5-8 \times 2-3$, cylindrical to narrowly ellipsoid, almost hyaline. Commonly under pines and frequently coming up to the surface.

Scleroderma

Fruitbodies growing mostly above ground, spherical, pear-shaped or somewhat irregular, in some species prolonged downwards into a stem-like base, often with well developed mycelial cords; wall tough and leathery or more fragile, smooth, warted or scaly, mostly yellow or brownish, splitting open at apex irregularly or in a stellate manner when mature, Gleba when ripe brown or blackish brown, with whitish trama plates. Cavities and hymenium poorly developed and usually no capillitium threads. Spores spherical, brown, mostly 8-15 diam., with longish spines or reticulate ornamentation.

KEY

Spores with isolated spines only 1
Spores always with a reticulum 3

1. Fruitbody not clearly tapered into a stem-like base .... *cepa*
    Fruitbody clearly tapered into a stem-like base 2
2. Small scales on fruitbody always surrounded by areolae or rings
    .... *areolatum*
    Scales never surrounded by areolae or rings .... *verrucosum*
3. Reticulum shallow, always less than 1 high .... *polyrhizum*
    Reticulum deep, 1.5 or more high 4
4. Reticulum of even height, without projecting spines .... *bovista*
    Spore ornamentation consisting of spines joined at their bases to form
    a reticulum .... *citrinum*

Scleroderma areolatum Ehrenb. (Fig. 533)
Fruitbody 2–4 cm diam., subspherical, tapered into a stem-like base about
1.5 cm long; wall yellowish brown, covered with small, dark, adpressed
scales each surrounded by an areola or ring, which persists when the scales
become detached. Gleba dark purplish brown. Spores spherical, dark
brown, 9–14 diam., without their ornamentation, spines 1.5 long. In
humus-rich soil in and along the borders of deciduous woods.

Scleroderma bovista Fr. (Fig. 534)
Fruitbody almost spherical, 2–6 cm diam., sessile or almost so, yellowish or
slightly orange, sometimes reddish violet at base; wall thin, tough, leathery,
smooth or, when old, with adpressed reddish brown scales, rupturing irregularly at apex. Hyphae with clamps. Gleba brown, with yellowish tramal
plates, finally almost black and powdery. Spores spherical, 9–13 diam.,
without their ornamentation which consists of a hyaline reticulum about
1.5 high without any clearly projecting spines, mesh wide. On sandy soil in
woods, on heaths and sometimes dunes.

Scleroderma cepa Pers.
Fruitbody almost spherical, 3–5 cm diam., yellowish or brownish orange,
smooth or coarsely fissured; wall thick, clamps rarely seen, rupturing
irregularly at apex. Spores 9–15 diam., without their ornamentation, spines
about 1.5 long. In sandy or humus-rich soil in woods.

Scleroderma citrinum Pers. (Fig. 535)
Common Earth-Ball
Fruitbodies often gregarious, subspherical, almost sessile, 3–12 cm diam.,
usually pale chrome or lemon yellow, sometimes ochraceous or somewhat
orange, coarsely scaly; wall tough and very thick (4 mm), eventually breaking open irregularly at apex. Hyphae with clamps. Gleba purplish black,
marbled by white tramal plates. Spores spherical, brown, 9–12 diam., without their ornamentation which consists of spines up to 1.5 high, joined at
their bases to form a network. Very common on sandy soil, often with
mosses, in and along the edges of woods and also on heaths.

Scleroderma polyrhizum T.F. Gmel. ex Pers.
Fruitbody 4–13 cm diam., subspherical, greyish to pale yellowish, rather smooth; wall very thick, splitting open at the apex in a stellate manner, hyphae with clamps. Gleba dark brown. Spores 7–10 diam., very shallowly and irregularly reticulate. Uncommon in Britain.

Scleroderma verrucosum Bull. ex Pers. (Fig. 536)
Fruitbody 3–7 cm diam., tapered below into a thick, grooved, stem-like base about 5 cm long, rather reddish brown when young, becoming brownish yellow; wall rather thin and brittle, at first smooth, then cracked except at the apex to form irregular scales sometimes with raised edges but never surrounded by areolae or rings. Hyphae almost always without clamps. Gleba olivaceous brown. Spores spherical, dark brown, 8–11 diam., without their ornamentation, spines separate, 1–1.5 long. Usually on humus-rich soil in woods and parks.

Sclerogaster

Fruitbodies hypogeous, not more than 2 cm and usually less than 1 cm diam., spherical, sessile, white or whitish, with a dense mycelial network at the base; wall soft, woolly or cottony, free or adhering to gleba. Gleba compact, when fresh jelly-like, very hard when dry, white at first then coloured, with small cavities, without columella and hyphae without clamps. Spores small, not more than 7 diam., spherical, hyaline to yellowish, smooth or finely spiny or warted, not amyloid.

KEY

Fruitbodies about 5 × 3 mm, growing between grass roots
.... *broomeianus*
Fruitbodies larger, not growing between grass roots 1
1. Spores 3–6 diam., almost smooth or minutely spiny .... *lanatus*
Spores 5–7 diam., covered with small truncate warts .... *compactus*

Sclerogaster broomeianus Zeller and Dodge
Fruitbody about 5 × 3 mm diam., white; wall one-layered, becoming free from gleba. Spores 5–7 diam., finely spiny. Growing between grass roots.

Sclerogaster compactus (Tul.) Sacc. (Fig. 537)
Fruitbody 0.5–2 cm diam., at first white then pale yellowish. Gleba at first white, then yellow and finally orange–yellow. Spores 5–7 diam., yellowish, minutely verruculose, warts crowded, low, truncate. In woods.

Sclerogaster lanatus Hesse
Fruitbody up to 1 cm diam., white; wall 2-layered, woolly or felted, with hyphae of outer layer thickly coated by crystals. Gleba becoming yellowish.

Spores 3–6 diam., almost smooth or minutely spiny, yellow in mass. In soil under *Picea.*

Sphaerobolus *stellatus* Tode ex Pers. (Fig. 538)
Fruitbodies 1.5–2.5 mm diam., seated in a cobwebby mycelial mat, at first spherical or ovoid, white or yellowish, splitting at the apex in a stellate manner to form 5–9 acutely pointed lobes surrounding the single ellipsoid to spherical, orange to reddish brown peridiole which contains the glebal mass of spores embedded in slime. Wall with 4–5 layers, the two inner ones forming a sort of catapult which projects the peridiole a distance of up to 3 m. Spores broadly ellipsoid, 6–10 × 4–6, hyaline, smooth. The basidia bear 4–8 spores. Mostly on fallen debris including rotten wood; rarely on dung. A common species but, because of its size, easily overlooked.

Stephanospora *caroticolor* (Berk.) Pat. (Fig. 539)
Fruitbody hypogeous, but pushing up just to the surface when quite mature, spherical or tuberous, lumpy, 3–5 cm diam., sessile, orange or carrot colour; wall thin, tomentose or velvety. Hyphae without clamps. Gleba orange to brick red, with white marbling, cavities small, irregular, winding, no columella. Spores 9–16 × 7–12, oblong-ellipsoid, yellowish, with broad spines and a distinctive flared collar at the base. In rich, loamy soil of deciduous and mixed woods.

Tulostoma
Fruitbodies composed of a slender, tough, hard stalk surmounted by a spherical or subspherical fertile head, the tissue of the two parts being anatomically quite different. Outer wall of head brittle, mostly soon disappearing; inner wall papery or parchment-like, with a small apical pore or mouth surrounded by a flat or projecting and tubular peristome. Base of stalk often bulbous and in some species with a volva. Subgleba none. Spores brown, with rather thick walls, smooth, warted or spiny. Capillitium threads little branched, thick-walled, swollen at septa. When young the fruitbody is buried in the soil.

KEY

    Stalk markedly scaly, scales spreading .... *squamosum*
    Stalk smooth or with closely adpressed scales 1
1.   Peristome rather flat, irregularly fibrillose or split .... *fimbriatum*
    Peristome raised, cylindrical, smooth, encircled at its base by a dark brown halo 2
2.   Stalk dark brown .... *melanocyclum*
    Stalk whitish, creamy or pale brown .... *brumale*

Tulostoma brumale Pers. (Fig. 540)
Fruitbody with subspherical, flattened head 0.5–2 cm diam., outer wall disappearing; inner wall smooth, parchment-like, creamy or pale ochraceous, mouth 1–1.5 mm wide, with projecting cylindrical peristome encircled at its base by a dark brown halo. Stalk 2–7 cm × 1–4 mm, often bulbous at base, whitish, cream or pale brown, not scaly or with some very small scales, always bare when old. Spores in mass reddish brown, spherical, 3.5–5.5 diam., singly pale yellowish, finely but distinctly warted. Capillitium threads 2.5–6.5 thick, almost hyaline, swollen at septa, some encrusted with crystals. On dunes and light chalky soil; also on mortar of old stone walls.

Tulostoma fimbriatum Fr.
Fruitbody with more or less spherical head flattened below, 0.5–1.5 cm diam.; outer wall brown, thin, sometimes persisting and sand-encrusted on lower part; inner wall dirty white to pale grey, papery, mouth with flat or very slightly raised, irregularly fibrillose or split peristome. Stalk 2–6 cm × 2–4 mm, brown, clearly furrowed, with adpressed scales, often bulbous at base. Spores 5–8 (10) diam., distinctly warted. Capillitium threads 4–6 diam. On dry turf, rare in Britain.

Tulostoma melanocyclum Bres.
Fruitbody with spherical head 6–12 mm diam., smooth, pale brownish or reddish brown, seldom white, mouth less than 1.5 mm diam., with short, smooth, cylindrical peristome encircled at its base by a dark brown halo. Stalk dark brown, 2–7 cm × 1.5–4 mm, with adpressed scales up to 2.5 mm long. Spores subspherical, 5–7 diam., densely spinulose. Capillitium threads up to 10 thick, without crystals, scarcely swollen at septa. On calcareous dunes and other sandy soils, uncommon in Britain.

Tulostoma squamosum Gmel. ex Pers.
Fruitbody with spherical to elongated or flattened head, 7–12 mm diam.; outer wall fragile, membranous, whitish; inner wall thin, papery, whitish to yellowish brown, mouth less than 1.5 mm diam., peristome cylindrical, smooth, without dark brown halo around its base. Stalk dark brown, 3–6 cm × 2–4 mm, distinctly scaly with scales spreading. Spores spherical, 5–6.5 diam., yellowish, with distinct, rather long spines. Capillitium threads without crystals. On dry calcareous soil.

Vascellum pratense (Pers.) Kreisel (Fig. 541)
Fruitbody 2–5 cm diam., top-shaped or subspherical, somewhat flattened, with short, fat false stem; outer wall with sphaerocysts, at first white or creamy white, scurfy, with some spines 1–1.5 mm long, which may be washed off by rain, darkening with age; inner wall papery, thin, shiny, greyish brown at first with a small mouth at the apex but later most of the upper wall comes away leaving a very large opening. Gleba brown at maturity, separated from subgleba by a parchment-like diaphragm. No columella.

Subgleba spongy, brown. Spores spherical, thin walled, 3–4.5 diam., minutely warted, with or without a short pedicel. Capillitium threads replaced by mostly unbranched, hyaline or pale yellow, thin-walled, 3–8 wide hyphae. On lawns, golf courses, pastures etc., summer–autumn, common.

Wakefieldia *macrospora* (Hawker) Hawker (Fig. 542)
Fruitbody hypogeous, 0.5–2.5 cm diam., spherical or irregular, without stalk or sterile base, with thin mycelial cord, white when young, turning ochraceous; wall thin, often splitting. Hyphae without clamps. Gleba at first rose, then purple to purplish brown, with small cavities and no columella, trama plates thin. Spores when mature spherical, golden brown, 12–18 diam., thick-walled, with irregular, truncate warts up to 1.5 tall and 2–3 broad, with hyaline apiculus 2–3 long. On chalky soil, under beech, rare.

Zelleromyces *stephensii* (Berk.) A.H. Smith (Fig. 543)
Fruitbody hypogeous, 1–4 cm diam., spherical or tuberous, sometimes two-lobed, hard when dry, at first yellow-flecked or entirely yellow, finally orange or reddish brown, when cut yielding white latex from trama plates; wall smooth, grooved or ribbed, fragile, enclosing gleba but often perforated at base. Gleba at first pale reddish yellow, then ochraceous, with columella, at least when young, no clamps or sphaerocysts. Spores subspherical or broadly ellipsoid, 10–16 × 10–12, yellowish or pale ochraceous, surface ornamented with rods or spines up to about 1 tall, amyloid, pedicel 3–4 long. On chalky, loamy soil in woods.

HOST INDEX

Abies
Aleurodiscus amorphus
Amylostereum chailletii
Calocera furcata
Ganoderma carnosum
Hericium coralloides
Hymenochaete mougeotii
Pseudohydnum gelatinosum
Acer
Antrodia albida
Calocera glossoides
Dendrothele acerina
Spongipellis pachyodon
Tremella moriformis
Typhula erythropus
Aesculus
Aurantioporus fissilis
Agropyron
Cellypha goldbachii
Algae
Athelia arachnoidea
Alnus
Achroomyces effusus
A. vestitus
Antrodiella semisupina
Daedaleopsis confragosa
Dichomitus campestris
Exidia repanda
Flagelloscypha faginea
Inonotus radiatus
Megalocystidium leucoxanthum
Myxarium crystallinum
Peniophora erikssonii
Phellinus trivialis
Plicaturopsis crispa
Polyporus badius
Protodontia subgelatinosa
Stereum subtomentosum
Trametes pubescens
Typhula corallina
T. erythropus
T. phacorrhiza
Ammophila
Mycocalia duriaeana
Anthriscus
Calyptella capula

Arrhenatherum
Cellypha goldbachii
Athelia
Tulasnella inclusa
Beta
Calyptella campanula
Thanatephorus cucumeris
Betula
Antrodia albida
Antrodiella semisupina
Cerrena unicolor
Creolophus cirrhatus
Daedaleopsis confragosa
Exidia cartilaginea
E. repanda
Flagelloscypha faginea
Fomes fomentarius
Gloeoporus dichrous
Inonotus obliquus
Lenzites betulina
Merismodes confusus
Mycocalia minutissima
Myxarium crystallinum
Phellinus igniarius
P. laevigatus
P. nigricans
Piptoporus betulinus
Protodontia subgelatinosa
Tomentella albomarginata
Trametes pubescens
Tulasnella violea
Blechnum
Flagelloscypha morlichensis
Bonfire sites
Aureoboletus cramesinus
Cotylidia undulata
Cyathus stercoreus
Geopetalum carbonarium
Botryohypochnus
Achroomyces peniophorae
Botryosphaeria
Achroomyces sebaceus
Brassica
Thanatephorus cucumeris
Bryophyta
see Mosses

GLOSSARY

Acicular having the form of a needle
Acuminate narrowed to a sharp point
Adpressed closely flattened down
Agglutinated stuck together
Allantoid sausage-shaped and slightly curved
Alveolae small hollows or pits in a surface
Alveolate pitted
Amphigenous occurring on both opposed surfaces
Ampulliform flask-shaped with the swollen part at the base
Amygdaliform almond-shaped
Amyloid stained blue–black by Melzer's iodine solution
Anastomosing joining together, often making a network
Apical at the tip
Apiculate having an apiculus
Apiculus a short, often pointed projection which attaches a basidiospore
  to its sterigma
Apophysis a small swelling at the base of a spore-sac where it joins the stalk
  in some *Geastrum* species
Arachnoid formed of delicate fibres, resembling a cobweb
Areolae small areas limited by cracks or lines
Areolate divided by cracks or lines into small areas
Asterosetae setae with branches radiating in a star-like manner
Athelioid cobwebby or thinly membranous, like *Athelia*
Avellaneous hazel-nut brown
Basidia organs which bear most commonly four, although sometimes
  fewer or more, basidiospores on rather slender sterigmata
Basidia (pleuro-) broad-based basidia which develop laterally on hyphae
  and appear to have roots or basal projections
Bifurcate forked or divided into two parts
Binding hyphae thick-walled, much branched, narrow hyphae without
  septa
Byssoid made up of slender threads
C+ stained deeply by cotton blue in lactic acid or lacto-phenol
C− not stained deeply by cotton blue
Caespitose in groups or tufts
Campanulate bell-shaped
Capillitium a mass of sterile threads amongst spores
Capitate having a well-defined head

Carminophilic granules granules which stain black when treated with aceto-carmine and heated

Cartilaginous firm and tough but easily bent

Catenate developing in chains

Cerebriform convoluted like a brain

Chlamydospores thick-walled non-deciduous spores

Ciliate with short, fine, hair-like outgrowths

Clamps short, narrow, loop-like tubes connecting two adjacent cells of a hypha

Clavarioid *Clavaria*-like

Clavate club-shaped

Columella a sterile central axis inside a fruitbody

Concolorous having the same colour

Concrescent growing together so as to become one structure

Confluent running together, meeting and uniting into one

Conidia asexual spores which, when mature, are liberated from a conidiophore

Conidiophores hyphae which bear one conidium or several conidia

Context inner flesh of a fruitbody

Coralloid having a branching structure like coral

Coremia loosely bound together fascicles

Cortex rind or outer layer

Corticate having a rind or cortex

Crenate having the edge indented, with resulting rounded teeth

Crisped crinkled or finely wavy

Cruciate in the form of a cross

Crustose crust-like

Cupulate cup-shaped

Cuspidate rounded at the end but with a protruding, small, sharp point in the middle

Cyanophilous staining readily with cotton blue

Cyathiform like a cup which is wider at the top than at the bottom

Cystidia sterile, distinctively-shaped terminal parts of hyphae

Cystidia (g) thin-walled cystidia usually with yellowish or refractive oily contents

Cystidia (m) encrusted cystidia with the upper part thick-walled at maturity

Cystidioles projecting sterile hymenial cells about the same width as basidia

Daedaleoid radially elongated, wavy and maze-like

Decurrent running down onto either a stalk or a substrate

Deflexed bent over abruptly

Deliquescent liquefying

Dendrohyphidia thin-walled, only slightly modified ends of hyphae which tend to be much branched

Dendroid tree-like

Dentate toothed

Denticulate having small teeth

Depressed slightly concave

Dextrinoid deeply stained yellowish to reddish brown by Melzer's iodine solution

Diaphragm a membrane separating the gleba from the subgleba

Dichotomous branching, often successively, to form two more or less equal arms

Dimidiate with one half much smaller than the other

Discoid flat and circular, disc-like

Discrete separate

Dissepiments partitions, usually separating tubes or pores

Doliiform barrel-shaped

Echinulate spiny

Effused spread out

Effuso-reflexed partly spread out over the surface of the substrate and partly turned up to form narrow brackets

Elaters narrow elastic threads which aid spore-dispersal

Ellipsoid solid bodies like spores elliptic in optical section

Epispore outer spore wall

Erumpent bursting through to the surface

Evanescent fugacious, having a short existence

Excentric to one side of the centre, not central

Falcate sickle-shaped

Farinose as if dusted with flour

Fasciculate arranged in bundles

FeSO$_4$ Ferrous sulphate in distilled water with 1–2 drops of dilute sulphuric acid

Fibrillose covered with silk-like fibres

Filiform slender and thread-like

Fimbriate fringed

Flabelliform fan-shaped

Floccose cottony

Fugacious soon disappearing

Fuliginous sooty

Fulvous tawny

Funicles hyphal cords in bird's-nest fungi which temporarily attach peridioles to the inner wall of the peridium

Furfuraceous scurfy

Fusiform spindle-shaped, tapered at each end

Generative hyphae thin-walled, branched, usually septate hyphae

Geniculate bent like a knee

Gleba sporing tissue in gasteromycetes

Gloeocystidia thin-walled cystidia usually with yellowish or refractive oily contents

Granulate covered with very small granules

Guttules oil-like droplets or vacuoles seen most clearly in water mounts

Hamate hooked

Haustoria hyphal branches which penetrate host cells

Helically spirally twisted in the form of a corkscrew

Hyaline transparent or almost so

Hydnoid distinctly toothed

Hygrophanous appearing water-soaked when wet

Hymenium the spore-bearing layer of a fruitbody

Hyphae fungus threads or filaments

Hyphidia thin-walled, only slightly modified ends of hyphae

Hypochnoid resupinate with effused, rather loosely interwoven hyphae

Hypogeous growing in the soil below the surface

Imbricate covering one another partly like tiles on a roof

Incised slit or cut

Indusium a cover or net-like pendent veil

Infundibuliform funnel-shaped

Involute rolled in

Irpicoid with flattened teeth sometimes fused at base

Isabelline tawny yellow

Isodiametric as long as wide

J+ amyloid, stained blue–black by Melzer's iodine solution

J− not amyloid, not stained blue–black by Melzer's iodine solution

KOH potassium hydroxide, mostly used as a 2–10% solution in distilled water

Labyrinthine resembling a labyrinth or maze-like network of chambers

Lacerate roughly cut or torn

Lachrymoid tear-shaped

Laciniate as if torn into strips

Lacunose having hollows

Lageniform flask-shaped, swollen at base

Lamellae thin plates

Lamellate made up of thin plates

Lanceolate spear-shaped

Lenticular shaped like a biconvex lens

Limoniform lemon-shaped

Locules cavities

Melzer's iodine solution a reagent containing iodine, potassium iodide, chloral hydrate and distilled water

Membranous like parchment or a thin skin

Merulioid folded hymenium, the folds anastomosing to form a semi-poroid network all parts of which may be fertile including the tops of the folds

Metachromatic taking a colour different from that of the stain solution

Metuloids encrusted cystidia, thick-walled at least in the upper part when mature

Moniliform swollen at intervals like a string of beads

Mucronate ending in a short sharp point

Muriform longitudinally and transversely septate

Mycelium a mass or group of hyphae making up the thallus of a fungus

Navicular boat-shaped

Obclavate the shape of a club upside down, thickened towards the base

Obconical inversely cone-shaped

Obovoid egg-shaped with the broad end at the base

Obpyriform pear-shaped with the broad end at the base

Obtuse rounded or blunt

Odontoid toothed

Operculate opening by a little lid

Orbicular round

Ovoid egg-shaped with the broad end at the top

Palisade cells or hyphae elongated vertically and lying parallel to one another in rows

Pectinate comb-like

Pedicels short stalks

Pellicular like a thin skin or membrane

Pendent hanging down

Penicillate splayed out in the form of a brush

Peridioles small oval, lenticular or spherical bodies each with a thick wall enclosing a portion of gleba

Periderm wall of a fruitbody

Perispore a layer of the spore wall, often fugacious, enveloping the whole spore

Peristome that part of the wall of a spore-sac which surrounds the mouth

Peronate sheathing

Phlebioid folded hymenium with folds irregularly arranged or more or less parallel but rarely anastomosing

Pileate having a distinct cap or pileus

Pilei caps

Pilose covered densely with soft hairs

Polygonal many-angled

Poroid having pores or openings at the ends of tubes

Probasidia first stages in the formation of basidia, differing in shape from the mature ones, often round and somewhat thick–walled

Pruinose with a frost-like or floury covering

Pseudocolumella the upper part of a stalk, sterile and often with cavities, which is surrounded by gleba

Pseudodiaphragm a false diaphragm, not really a membrane

Pseudoparenchymatous appearing cellular like the parenchyma of flowering plants but made up of hyphal elements

Pseudostromata stromata formed partly of fungus thallus tissue and partly of host tissue

Puberulent minutely downy

Pubescent softly hairy or downy

Pulvinate cushion-shaped

Punctate marked with very small spots or hollows

Pustulate blister-like

Pyriform pear-shaped

Reniform kidney-shaped

Resupinate an adjective used to describe fruitbodies which lie flat on the substrate with the hymenium on the outside

Reticulate net-like

Rhizoids thin, root-like outgrowths

Rhizomorphs cord-like structures made up of hyphae

Rimose cracked

Rostrate beaked or strongly attenuated at the apex

Rugose coarsely wrinkled

Rugulose finely wrinkled

Saccate like a bag

Sclerotia masses of aggregated hyphae, often with a rind of thick-walled cells

Scurfy covered with bran-like flakes

Septa cell walls dividing up a hypha

Sessile without a stalk

Setae bristle–like, often erect and thick-walled hyphae

Setose bearing setae

Sinuate wavy

Skeletal hyphae thick-walled, mostly non–septate hyphae

Spathulate shaped like a spoon

Spatulate an accepted alternative spelling of spathulate

Sphaerocysts inflated, often spherical, cells in fungus tissue which separate readily making the flesh fragile

Spheropedunculate with spherical head and narrow stalk

Spicules needle-like outgrowths

Spinulose delicately spiny

Spores reproductive cells in fungi and other lower plants

Squamules small scales

Squamulose having small scales

Stellate star-shaped

Stephanocysts almost spherical, one-septate cystidia with small teeth forming a girdle around the middle

Sterigmata slender, cylindrical or subulate structures by which spores are attached to basidia

Stipitate stalked
Striae delicate lines, grooves or ridges
Striate marked with striae
Stroma an often cushion-like mass of fungal cells or closely interwoven hyphae
Subgleba sterile tissue below and supporting the gleba
Subhymenium tissue immediately below the hymenium
Subicular belonging to a subiculum
Subiculum a mycelial mat mostly lying under fruitbodies
Substipitate with a short or poorly defined stalk
Substrate the host plant or other material on which a fungus grows
Subulate rather slender and tapered to a point like an awl
Suburniform almost urn-shaped
Sulcate grooved
Terrestrial growing on land, on the ground
Tessellated made up of small cubes like a mosaic
Tetraradiate four-armed
Tomentose downy
Tomentum a dense covering of fine hairs
Torulose swollen at intervals like a chain of beads
Trama a layer of hyphae in the centre of a lamella, a tooth or the dissepiment between tubes
Tramal belonging to or part of the trama
Tremelloid jelly-like
Trichotomous branching to form three more or less equal arms
Triquetrous three-edged, three-cornered
Truncate ending abruptly as if cut straight across
Tubercle a wart-like projection
Tuberculate covered with wart-like projections
Tuberous resembling tubers or swollen roots
Turbinate top-shaped
Umbilicate with a small hollow
Umbonate having a central swelling like a boss on a shield
Uncinate hooked
Ungulate hoof-shaped
Urniform urn-like
Utriform bag-like
Velutinous velvety
Ventricose inflated, very broadly fusiform
Verrucose warted
Verruculose covered with very small warts
Verticil a whorl
Verticillate arranged in whorls
Vesiculose bladder-like

Vinaceous wine-coloured, purplish red
Volva the remains of a peridium or wall which persists as a cup surrounding the base of a stalk or fruitbody
Zonate zoned, banded

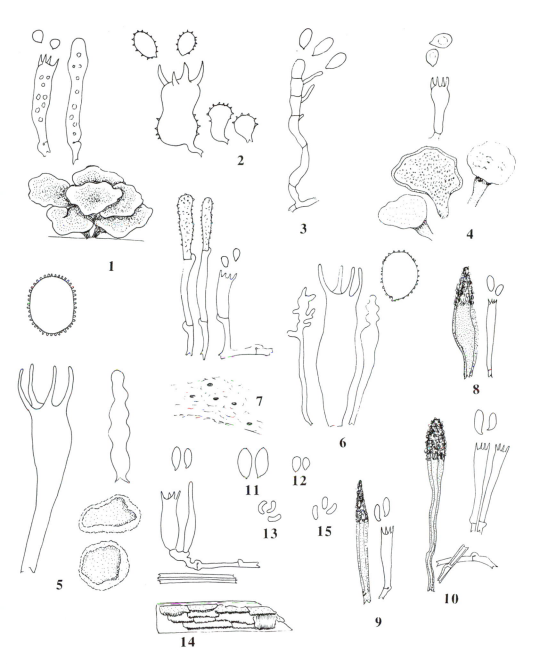

Plate 1 Hymenomycetes. 1 *Abortiporus biennis*; 2 *Acanthobasidium delicatum*; 3 *Achroomyces fimetarius*; 4 *Albatrellus cristatus*; 5 *Aleurodiscus amorphus*; 6 *A.aurantius*; 7 *Amphinema byssoides*; 8 *Amylostereum areolatum*; 9 *A.chailletii*; 10 *A.laevigatum*; 11 *Antrodia albida*; 12 *A.gossypium*; 13 *A.lenis*; 14 *A.serialis*; 15 *A.xantha*.

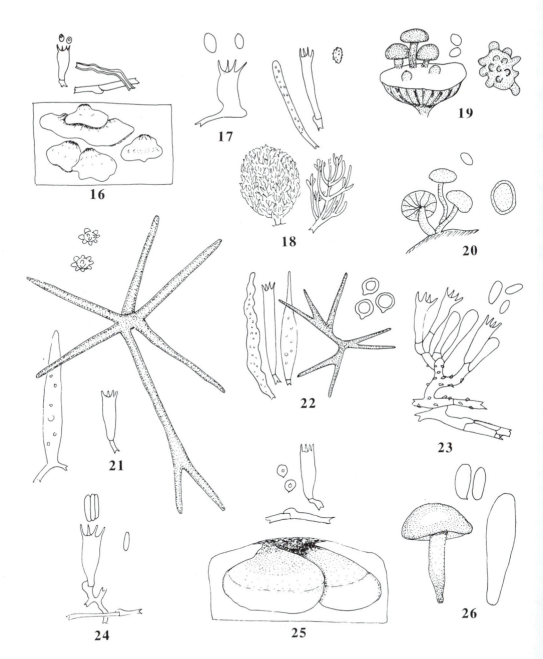

Plate 2 Hymenomycetes. 16 *Antrodiella semisupina*; 17 *Aphanobasidium filicinum*; 18 *Artomyces pyxidatus;* 19 *Asterophora lycoperdoides*; 20 *A.parasitica*; 21 *Asterostroma cervicolor*; 22 *A.laxum*; 23 *Athelia epiphylla*; 24 *Athelopsis glaucina*; 25 *Aurantioporus fissilis*; 26 *Aureoboletus cramesinus.*

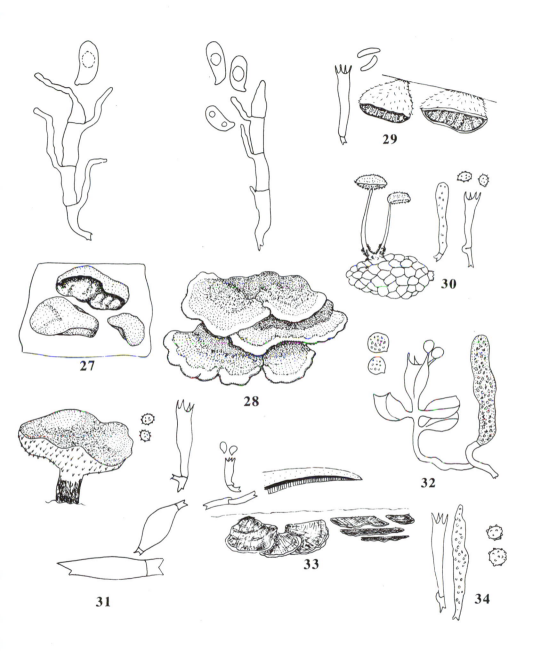

Plate 3 Hymenomycetes. 27 *Auricularia auricula-judae*; 28 *A.mesenterica*; 29 *Auriculariopsis ampla*; 30 *Auriscalpium vulgare*; 31 *Bankera fuligineo-alba*; 32 *Basidiodendron caesiocinereum*; 33 *Bjerkandera adusta*; 34 *Boidinia furfuracea*.

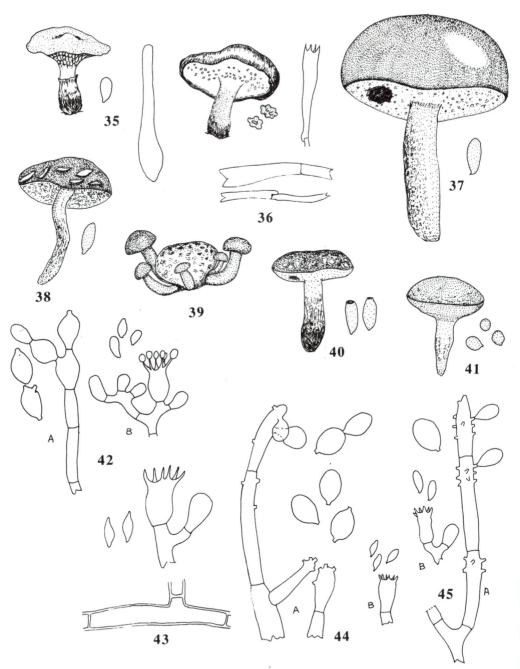

Plate 4 Hymenomycetes. 35 *Boletinus cavipes*; 36 *Boletopsis leucomelaena*; 37 *Boletus badius*; 38 *B.chrysenteron*; 39 *B.parasiticus*; 40 *B.porosporus*; 41 *B.rubinus*; 42 *Botryobasidium aureum*, A conidial, B basidial; 43 *B.botryosum*; 44 *B.candicans*, A conidial, B basidial; 45 *B.conspersum*, A conidial, B basidial.

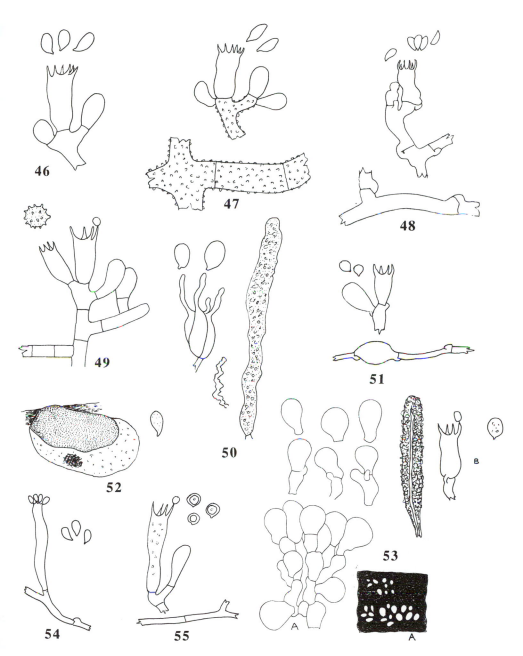

Plate 5 Hymenomycetes. 46 *Botryobasidium obtusisporum*; 47 *B.pruinatum*; 48 *B.subcoronatum*; 49 *Botryohypochnus isabellinus*; 50 *Bourdotia galzinii*; 51 *Brevicellicium olivascens*; 52 *Buglossoporus pulvinus*; 53 *Bulbillomyces farinosus*, A conidial, B basidial; 54 *Butlerelfia eustacei*; 55 *Byssocorticium pulchrum*.

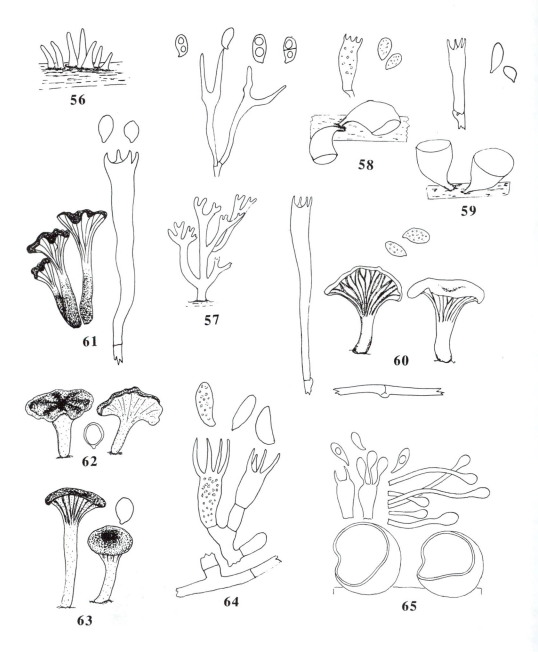

Plate 6 Hymenomycetes. 56 *Calocera cornea*; 57 *C.viscosa*; 58 *Calyptella campanula*; 59 *C.capula*; 60 *Cantharellus cibarius*; 61 *C.cinereus*; 62 *C.lutescens*; 63 *C.tubaeformis*; 64 *Cejpomyces terrigenus*; 65 *Cellypha goldbachii.*

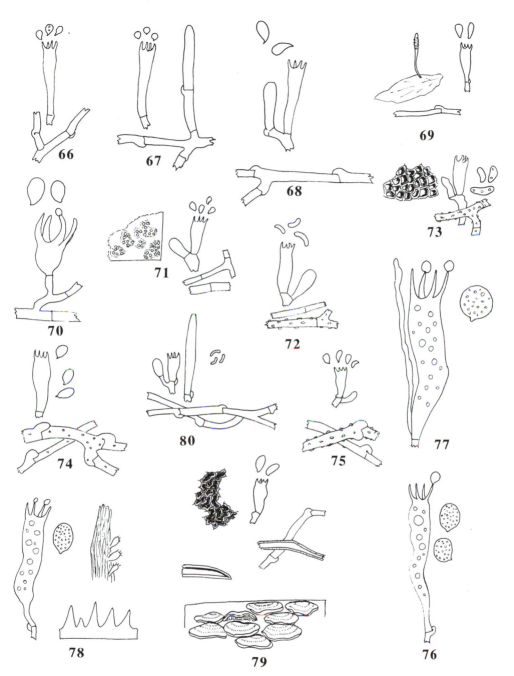

Plate 7 Hymenomycetes. 66 *Ceraceomyces serpens*; 67 *C.sublaevis*; 68 *C.tessulatus*; 69 *Ceratellopsis aculeata*; 70 *Ceratobasidium cornigerum*; 71 *Ceriporia excelsa*; 72 *C.purpurea*; 73 *C.reticulata*; 74 *Ceriporiopsis aneirinus*; 75 *C.gilvescens*; 76 *Cerocorticium confluens*; 77 *C.hiemale*; 78 *C.molare*; 79 *Cerrena unicolor*; 80 *Chaetoporellus latitans*.

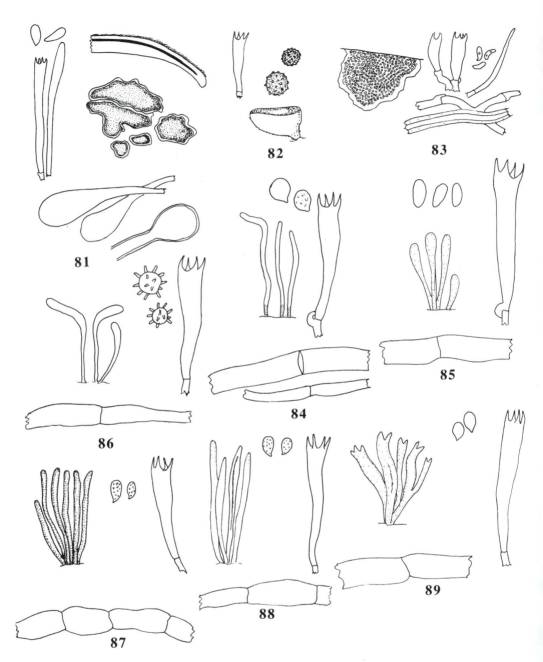

Plate 8 Hymenomycetes. 81 *Chondrostereum purpureum*; 82 *Chromocyphella muscicola*; 83 *Cinereomyces lindbladii*; 84 *Clavaria acuta*; 85 *C.argillacea*; 86 *C.asterospora*; 87 *C.fumosa*; 88 *C.vermicularis*; 89 *C.zollingeri*.

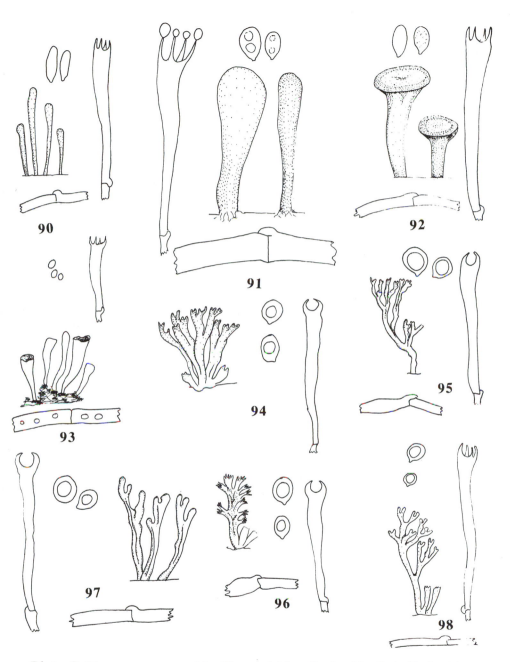

Plate 9 Hymenomycetes. 90 *Clavariadelphus ligula*; 91 *C.pistillaris*; 92 *C.truncatus*; 93 *Clavicorona taxophila*; 94 *Clavulina amethystina*; 95 *C.cinerea*; 96 *C.cristata*; 97 *C.rugosa*; 98 *Clavulinopsis cinereoides*.

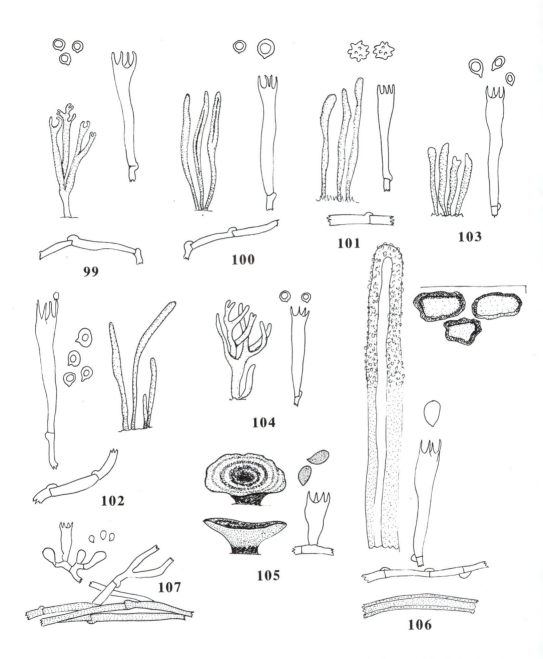

Plate 10 Hymenomycetes. 99 *Clavulinopsis corniculata*; 100 *C.fusiformis*; 101 *C.helvola*; 102 *C.laeticolor*; 103 *C.luteo-alba*; 104 *C.subtilis*; 105 *Coltricia perennis*; 106 *Columnocystis abietina*; 107 *Confertobasidium olivaceo-album*.

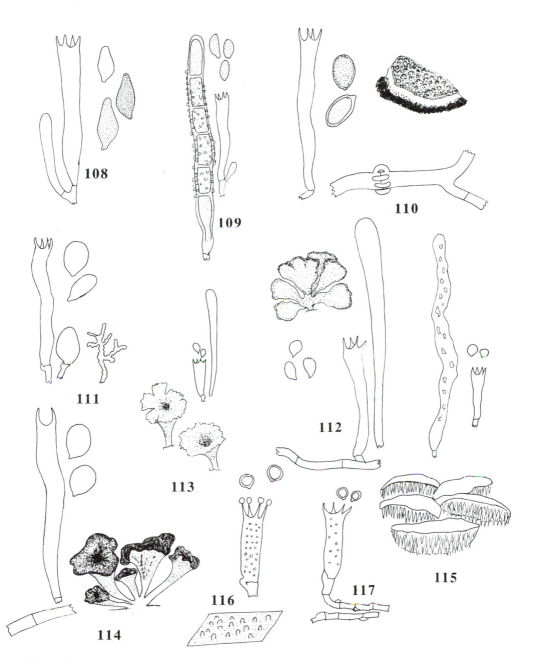

Plate 11 Hymenomycetes. 108 *Coniophora fusispora*; 109 *C.olivacea*; 110 *C.puteana*; 111 *Corticium roseum*; 112 *Cotylidia pannosa*; 113 *C.undulata*; 114 *Craterellus cornucopioides*; 115 *Creolophus cirrhatus*; 116 *Cristinia gallica*; 117 *C.helvetica*.

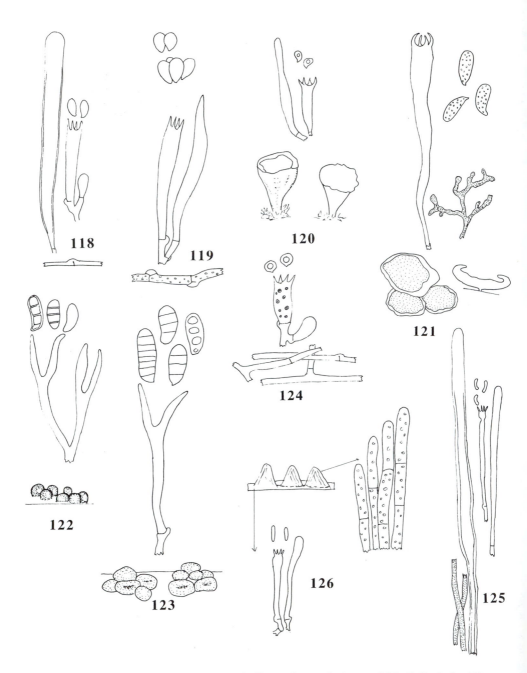

Plate 12 Hymenomycetes. 118 *Crustoderma dryinum*; 119 *Cylindrobasidium evolvens*; 120 *Cyphellostereum laeve*; 121 *Cytidia salicina*; 122 *Dacrymyces stillatus*; 123 *D.variisporus*; 124 *Dacryobasidium coprophilum*; 125 *Dacryobolus karstenii*; 126 *D.sudans*.

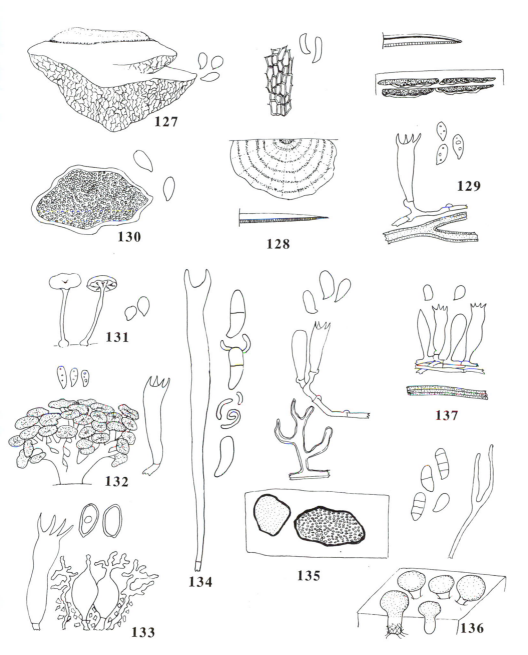

Plate 13 Hymenomycetes. 127 *Daedalea quercina*; 128 *Daedaleopsis confragosa*; 129 *Datronia mollis*; 130 *D.stereoides*; 131 *Delicatula integrella*; 132 *Dendropolyporus umbellatus*; 133 *Dendrothele acerina*; 134 *Dicellomyces scirpi*; 135 *Dichomitus campestris*; 136 *Ditiola radicata*; 137 *Donkioporia expansa*.

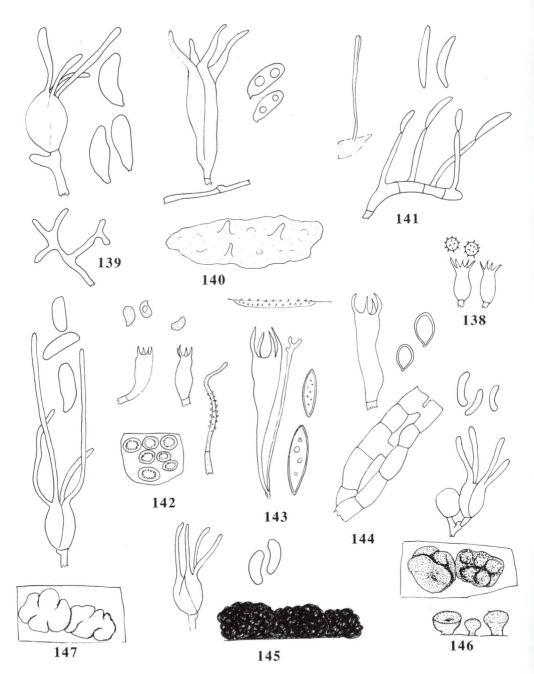

Plate 14 Hymenomycetes. 138 *Echinotrema clanculare*; 139 *Efibulobasidium albescens*; 140 *Eichleriella deglubens*; 141 *Eocronartium muscicola*; 142 *Episphaeria fraxinicola*; 143 *Epithele typhae*; 144 *Erythricium laetum*; 145 *Exidia glandulosa*; 146 *E.recisa*; 147 *E.thuretiana*.

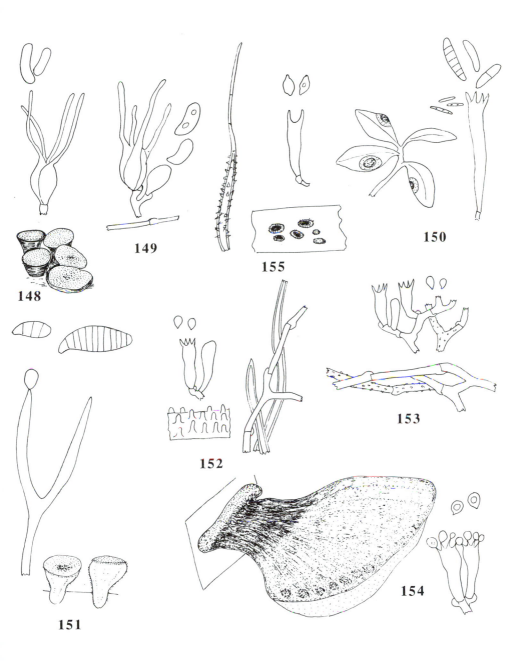

Plate 15 Hymenomycetes. 148 *Exidia truncata*; 149 *Exidiopsis effusa*; 150 *Exobasidium vaccinii*; 151 *Femsjonia pezizaeformis*; 152 *Fibrodontia gossypina*; 153 *Fibulomyces mutabilis*; 154 *Fistulina hepatica*; 155 *Flagelloscypha minutissima*.

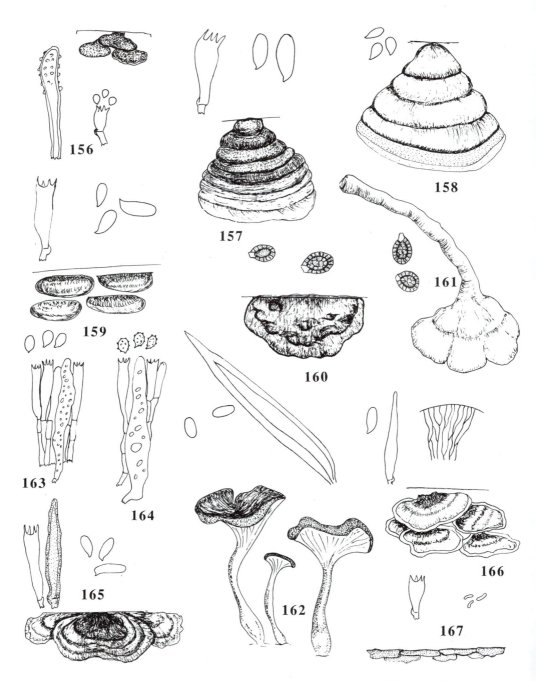

Plate 16 Hymenomycetes. 156 *Flaviporus brownei*; 157 *Fomes fomentarius*; 158 *Fomitopsis pinicola*; 159 *Funalia gallica*; 160 *Ganoderma adspersum*; 161 *G.lucidum*; 162 *Geopetalum carbonarium*; 163 *Gloeocystidiellum ochraceum*; 164 *G.porosum*; 165 *Gloeophyllum abietinum*; 166 *G.sepiarium*; 167 *Gloeoporus dichrous*.

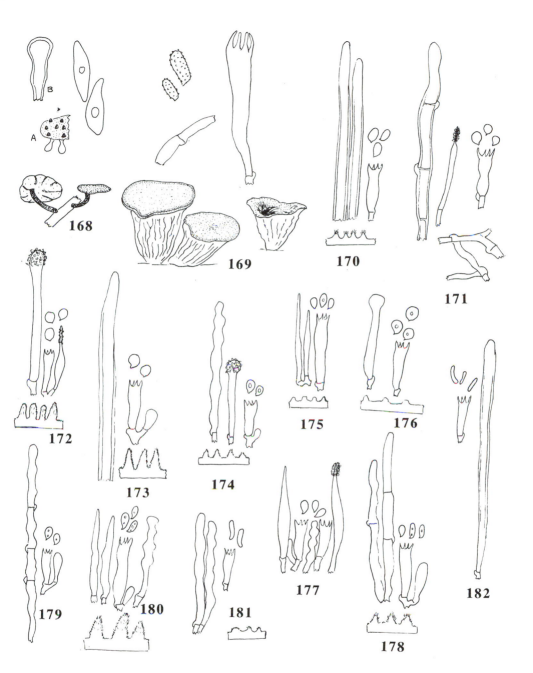

Plate 17 Hymenomycetes. 168 *Gloiocephala menieri*, A cap surface, B cystidium; 169 *Gomphus clavatus*; 170 *Grandinia abieticola*; 171 *G.alutaria*; 172 *G.arguta*; 173 *G.barba-jovis*; 174 *G.breviseta*; 175 *G.crustosa*; 176 *G.granulosa*; 177 *G.hastata*; 178 *G.nespori*; 179 *G.pallidula*; 180 *G.quercina*; 181 *G.stenospora*; 182 *G.subalutacea*.

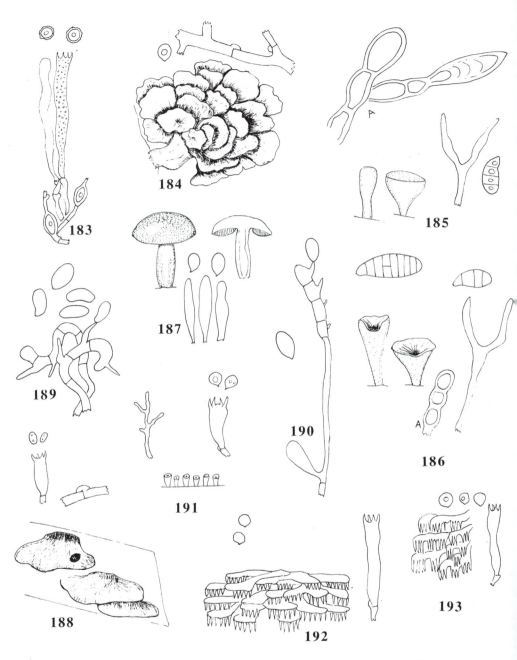

Plate 18 Hymenomycetes. 183 *Granulobasidium vellereum*; 184 *Grifola frondosa*; 185 *Guepiniopsis alpina*, A palisade elements; 186 *G.chrysocoma*; 187 *Gyroporus castaneus*; 188 *Hapalopilus rutilans*; 189 *Helicobasidium brebissonii*; 190 *Helicogloea lagerheimii*; 191 *Henningsomyces candidus*; 192 *Hericium clathroides*; 193 *H.coralloides*.

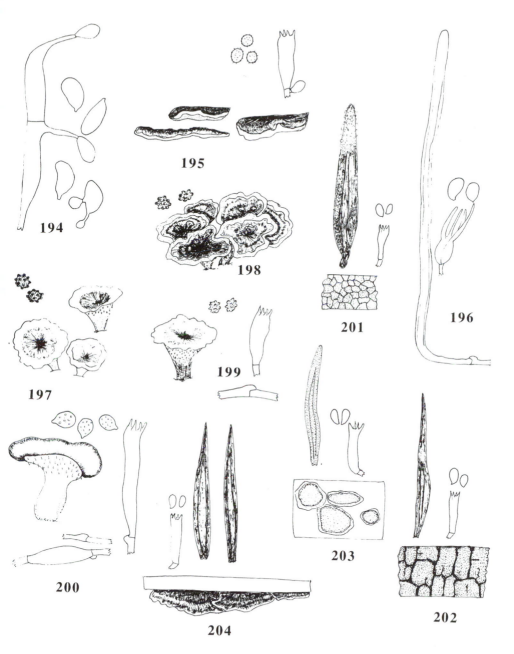

Plate 19 Hymenomycetes. 194 *Herpobasidium filicinum*; 195 *Heterobasidion annosum*; 196 *Heterochaetella dubia*; 197 *Hydnellum aurantiacum*; 198 *H.concrescens*; 199 *H.peckii*; 200 *Hydnum repandum*; 201 *Hymenochaete corrugata*; 202 *H.fuliginosa*; 203 *H.mougeotii*; 204 *H.rubiginosa*.

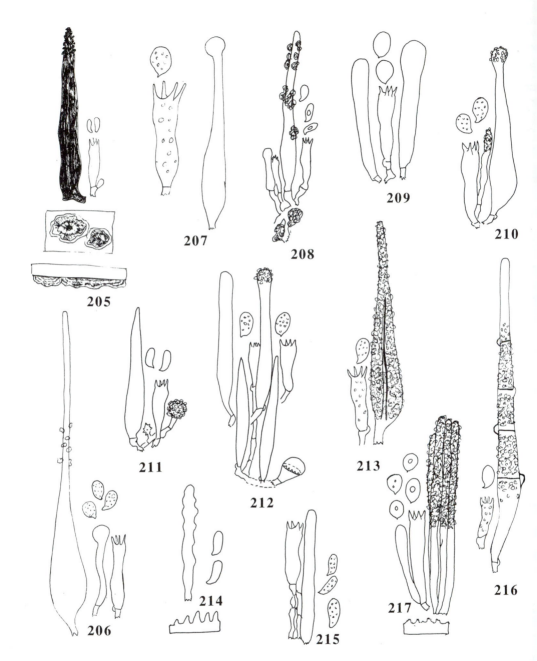

Plate 20 Hymenomycetes. 205 *Hymenochaete tabacina*; 206 *Hyphoderma argillaceum*; 207 *H.capitatum*; 208 *H.macedonicum*; 209 *H.obtusum*; 210 *H.orphanellum*; 211 *H.pallidum*; 212 *H.praetermissum*; 213 *H.puberum*; 214 *H.radula*; 215 *H.roseocremeum*; 216 *H.setigerum*; 217 *Hyphodermella corrugata*.

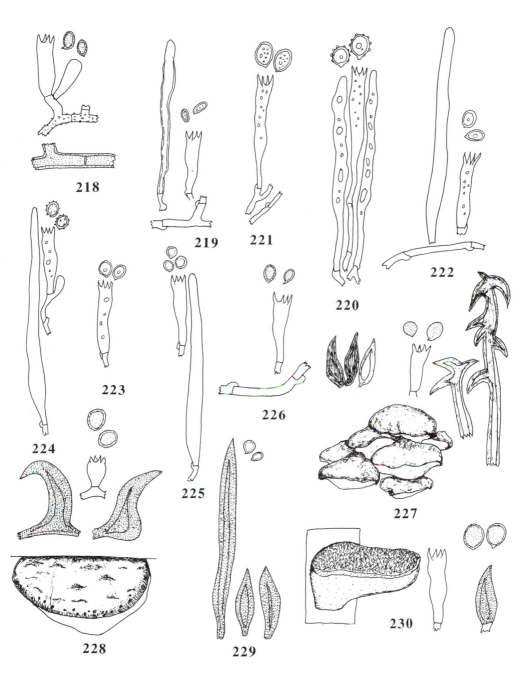

Plate 21 Hymenomycetes. 218 *Hypochnella violacea*; 219 *Hypochniciellum molle*; 220 *Hypochnicium analogum*; 221 *H.bombycinum*; 222 *H.geogenium*; 223 *H.lundellii*; 224 *H.punctulatum*; 225 *H.sphaerosporum*; 226 *Hypochnopsis mustialaensis*; 227 *Inonotus cuticularis*; 228 *I.dryadeus*; 229 *I.hastifer*; 230 *I.hispidus*.

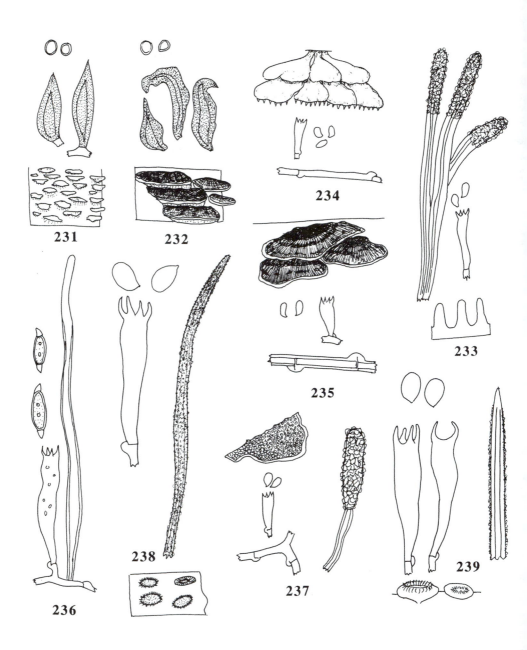

Plate 22 Hymenomycetes. 231 *Inonotus nodulosus*; 232 *I.radiatus*; 233 *Irpex lacteus*; 234 *Irpicodon pendulus*; 235 *Ischnoderma benzoinum*; 236 *Jaapia argillacea*; 237 *Junghuhnia nitida*; 238 *Lachnella alboviolascens*; 239 *L.villosa*.

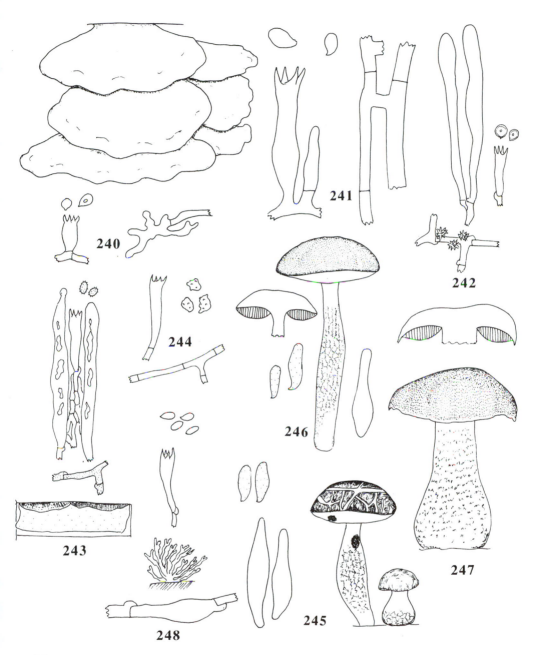

Plate 23 Hymenomycetes. 240 *Laetiporus sulphureus*; 241 *Laetisaria fuciformis*; 242 *Lagarobasidium dendriticum*; 243 *Laxitextum bicolor*; 244 *Lazulinospora cyanea*; 245 *Leccinum carpini*; 246 *L.scabrum*; 247 *L.versipelle*; 248 *Lentaria delicata*.

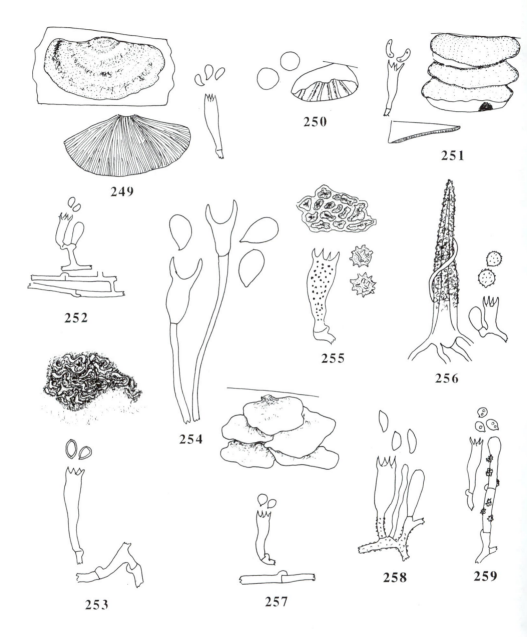

Plate 24 Hymenomycetes. 249 *Lenzites betulina*; 250 *Leptoglossum retirugum*; 251 *Leptoporus mollis*; 252 *Leptosporomyces galzinii*; 253 *Leucogyrophana mollusca*; 254 *Limonomyces culmigena*; 255 *Lindtneria trachyspora*; 256 *Litschauerella clematidis*; 257 *Loweomyces wynnei*; 258 *Luellia recondita*; 259 *Lyomyces sambuci*.

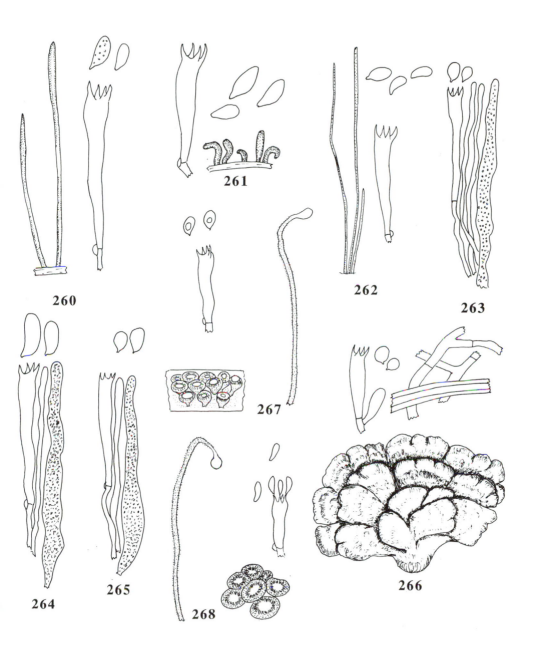

Plate 25 Hymenomycetes. 260 *Macrotyphula fistulosa*; 261 *M.fistulosa* var.*contorta*; 262 *M.juncea*; 263 *Megalocystidium lactescens*; 264 *M.leucoxanthum*; 265 *M.luridum*; 266 *Meripilus giganteus*; 267 *Merismodes anomalus*; 268 *M.confusus.*

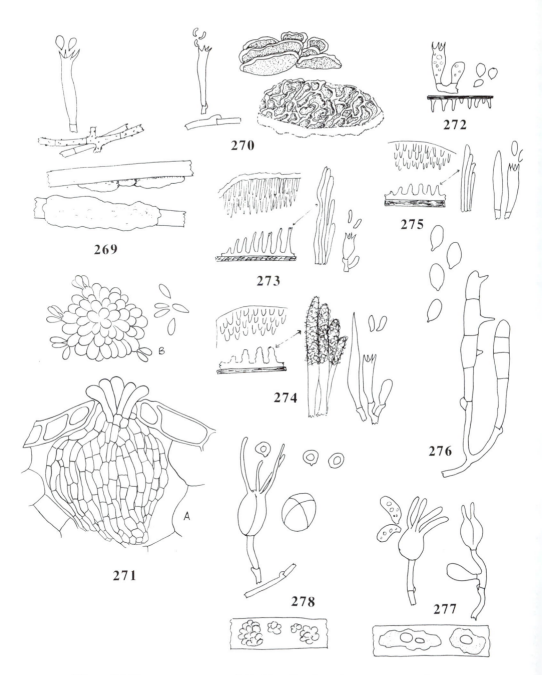

Plate 26 Hymenomycetes. 269 *Meruliopsis corium*; 270 *Merulius tremellosus*; 271 *Microstroma juglandis*, A leaf section, B surface view; 272 *Mucronella calva* var.*aggregata*; 273 *Mycoacia aurea*; 274 *M.fuscoatra*; 275 *M.uda*; 276 *Mycogloea macrospora*; 277 *Myxarium nucleatum*; 278 *M.sphaerosporum*.

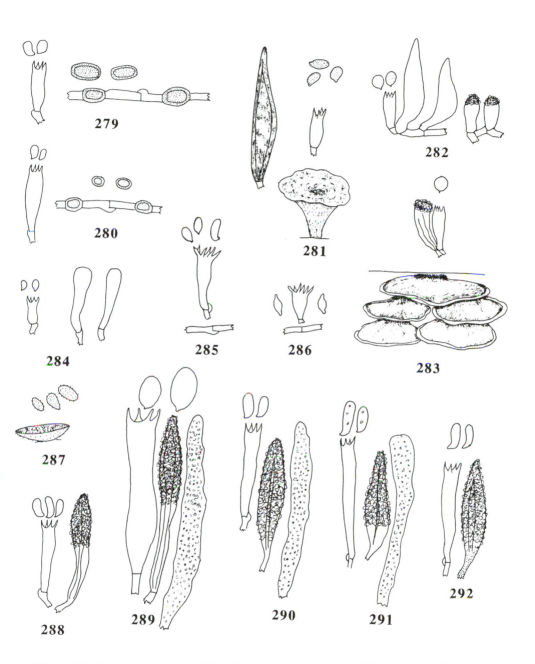

Plate 27 Hymenomycetes. 279 *Oligoporus ptychogaster*; 280 *O.rennyi*; 281 *Onnia tomentosa*; 282 *Oxyporus corticola*; 283 *O.populinus*; 284 *Parvobasidium cretatum*; 285 *Paullicorticium niveo-cremeum*; 286 *P.pearsonii*; 287 *Pellidiscus pallidus*; 288 *Peniophora cinerea*; 289 *P.erikssonii*; 290 *P.incarnata*; 291 *P.laeta*; 292 *P.limitata*.

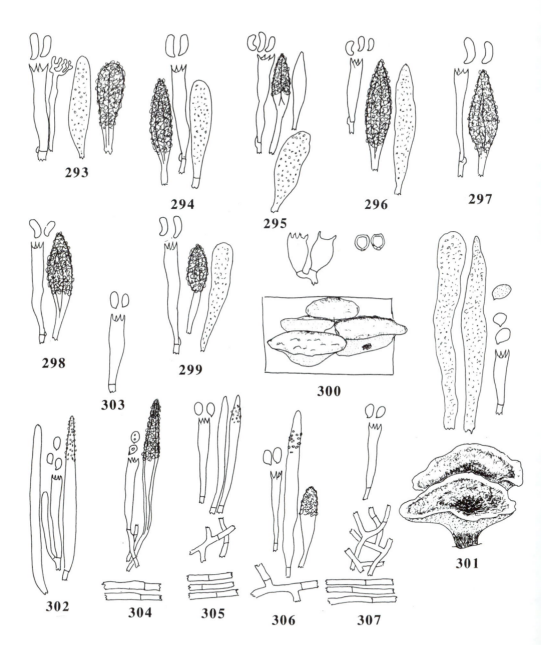

Plate 28 Hymenomycetes. 293 *Peniophora lycii*; 294 *P.nuda*; 295 *P.pini*; 296 *P.pithya*; 297 *P.quercina*; 298 *P.rufomarginata*; 299 *P.violacea-livida*; 300 *Perenniporia fraxinea*; 301 *Phaeolus schweinitzii*; 302 *Phanerochaete affinis*; 303 *P.avellanea*; 304 *P.filamentosa*; 305 *P.sanguinea*; 306 *P.sordida*; 307 *P.tuberculata.*

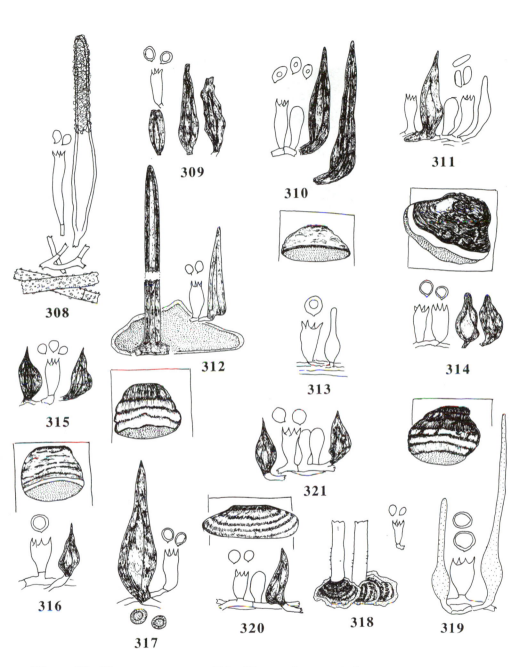

Plate 29 Hymenomycetes. 308 *Phanerochaete velutina*; 309 *Phellinus conchatus*; 310 *P.contiguus*; 311 *P.ferreus*; 312 *P.ferruginosus*; 313 *P.hippophäeicola*; 314 *P.igniarius*; 315 *P.laevigatus*; 316 *P.nigricans*; 317 *P.pini*; 318 *P.ribis*; 319 *P.robustus*; 320 *P.trivialis*; 321 *P.tuberculosus*.

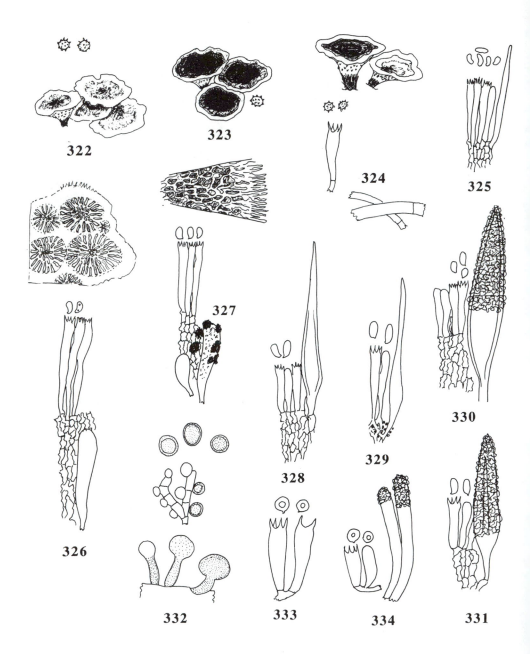

Plate 30 Hymenomycetes. 322 *Phellodon melaleucus*; 323 *P.niger*; 324 *P.tomentosus*; 325 *Phlebia livida*; 326 *P.radiata*; 327 *P.rufa*; 328 *P.segregata*; 329 *P.subochracea*; 330 *Phlebiopsis gigantea*; 331 *P.roumeguerii*; 332 *Phleogena faginea*; 333 *Physisporinus sanguinolentus*; 334 *P.vitreus*.

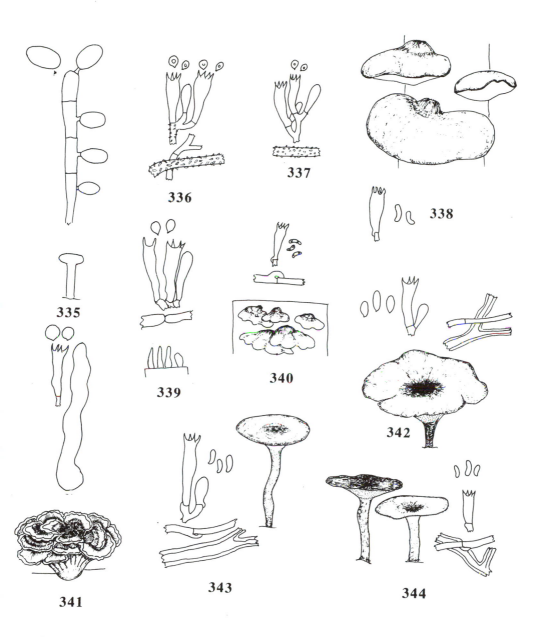

Plate 31 Hymenomycetes. 335 *Pilacrella solani*; 336 *Piloderma byssinum*;
337 *P.croceum*; 338 *Piptoporus betulinus*; 339 *Pistillaria pusilla*; 340
Plicaturopsis crispa; 341 *Podoscypha multizonata*; 342 *Polyporus badius*; 343
P.brumalis; 344 *P.ciliatus.*

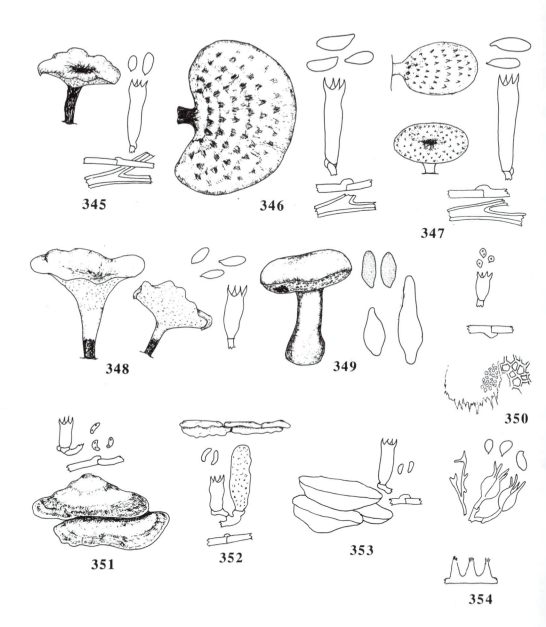

Plate 32 Hymenomycetes. 345 *Polyporus melanopus*; 346 *P.squamosus*; 347 *P.tuberaster*; 348 *P.varius*; 349 *Porphyrellus pseudoscaber*; 350 *Porpomyces mucidus*; 351 *Postia caesia*; 352 *P.leucomallela*; 353 *P.stiptica*; 354 *Protodontia subgelatinosa*.

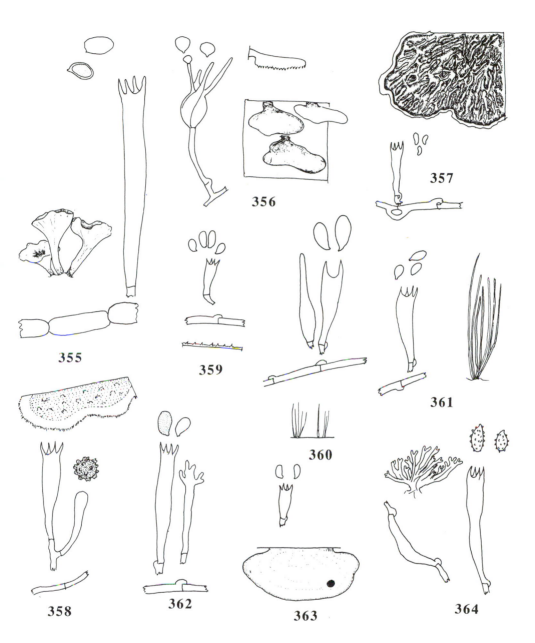

Plate 33 Hymenomycetes. 355 *Pseudocraterellus sinuosus*; 356 *Pseudohydnum gelatinosum*; 357 *Pseudomerulius aureus*; 358 *Pseudotomentella mucidula*; 359 *Pteridomyces galzinii*; 360 *Pterula gracilis*; 361 *P.multifida*; 362 *Pulcherricium caeruleum*; 363 *Pycnoporus cinnabarinus*; 364 *Ramaria abietina*.

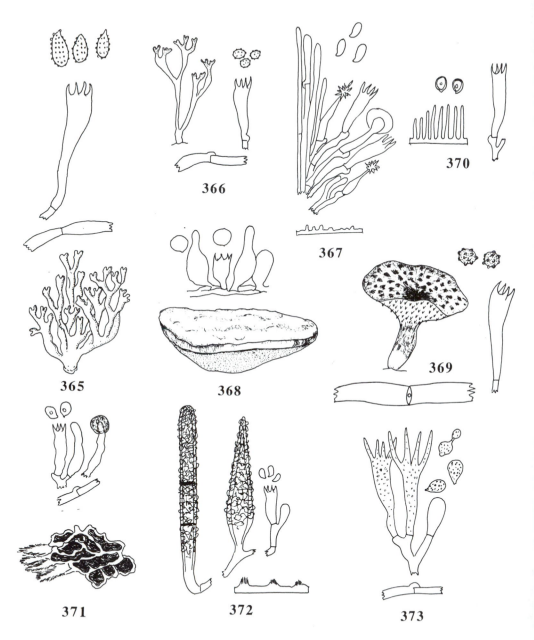

Plate 34 Hymenomycetes. 365 *Ramaria aurea*; 366 *Ramariopsis pulchella*; 367 *Resinicium bicolor*; 368 *Rigidoporus ulmarius*; 369 *Sarcodon scabrosus*; 370 *Sarcodontia setosa*; 371 *Schizopora paradoxa*; 372 *Scopuloides rimosa*; 373 *Scotomyces subviolaceus*.

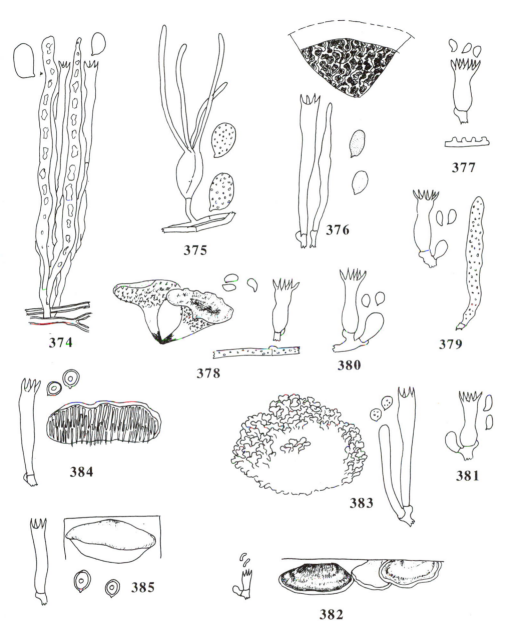

Plate 35 Hymenomycetes. 374 *Scytinostroma ochroleucum*; 375 *Sebacina incrustans*; 376 *Serpula lacrimans*; 377 *Sistotrema brinkmannii*; 378 *S.confluens*; 379 *S.coroniferum*; 380 *S.diademiferum*; 381 *S.oblongisporum*; 382 *Skeletocutis nivea*; 383 *Sparassis crispa*; 384 *Spongipellis pachyodon*; 385 *S.spumeus*.

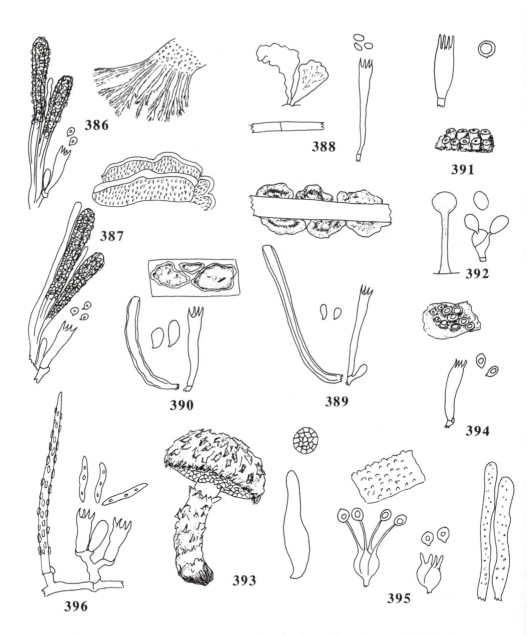

Plate 36 Hymenomycetes. 386 *Steccherinum fimbriatum*; 387 *S.ochraceum*; 388 *Stereopsis vitellina*; 389 *Stereum hirsutum*; 390 *S.rugosum*; 391 *Stigmatolemma poriaeforme*; 392 *Stilbum vulgare*; 393 *Strobilomyces floccopus*; 394 *Stromatoscypha fimbriata*; 395 *Stypella vermiformis*; 396 *Subulicystidium longisporum*.

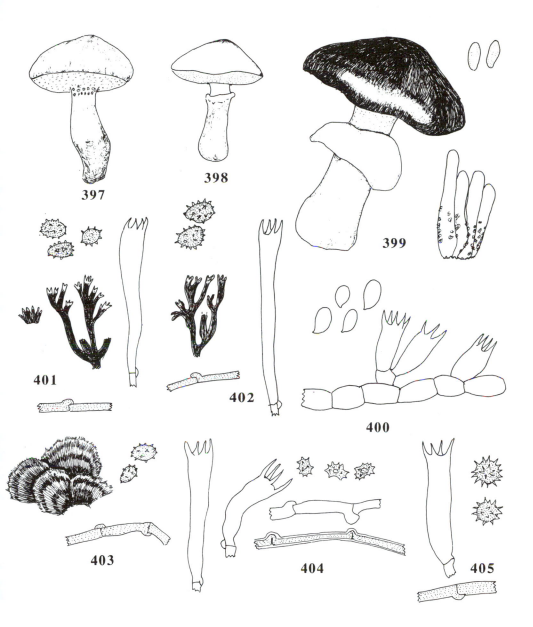

Plate 37 Hymenomycetes. 397 *Suillus granulatus*; 398 *S.grevillei*; 399 *S.luteus*; 400 *Thanatephorus cucumeris*; 401 *Thelephora anthocephala*; 402 *T.palmata*; 403 *T.terrestris*; 404 *Tomentella avellanea*; 405 *T.bryophila*.

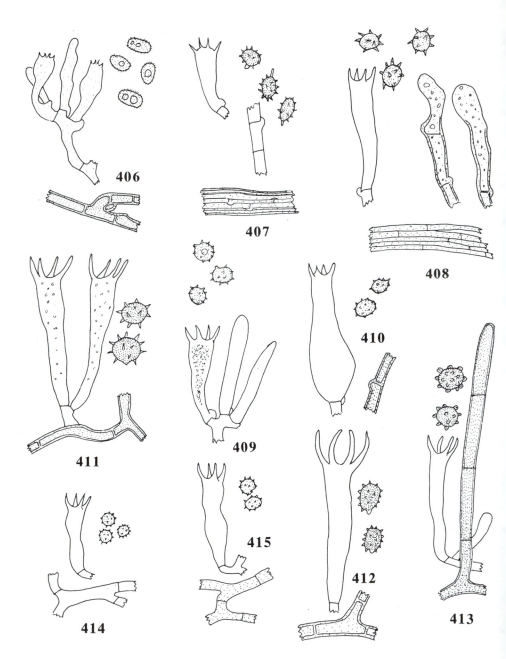

Plate 38 Hymenomycetes. 406 *Tomentella cladii*; 407 *T.ferruginea*; 408 *T.pilosa*; 409 *T.subtestacea*; 410 *T.terrestris*; 411 *Tomentellastrum alutaceo-umbrinum*; 412 *T.fuscocinereum*; 413 *Tomentellina fibrosa*; 414 *Tomentellopsis echinospora*; 415 *T.zygodesmoides*.

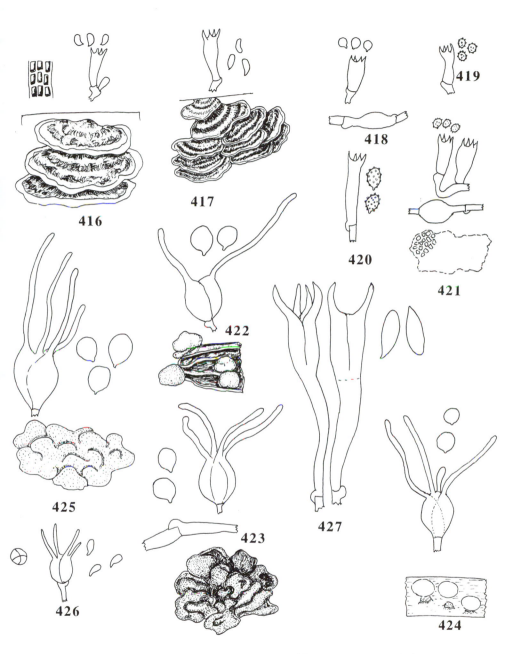

Plate 39 Hymenomycetes. 416 *Trametes gibbosa*; 417 *T.versicolor*; 418 *Trechispora cohaerens*; 419 *T.farinacea*; 420 *T.fastidiosa*; 421 *T.mollusca*; 422 *Tremella encephala*; 423 *T.foliacea*; 424 *T.globospora*; 425 *T.mesenterica*; 426 *T.translucens*; 427 *Tremellodendropsis tuberosa*.

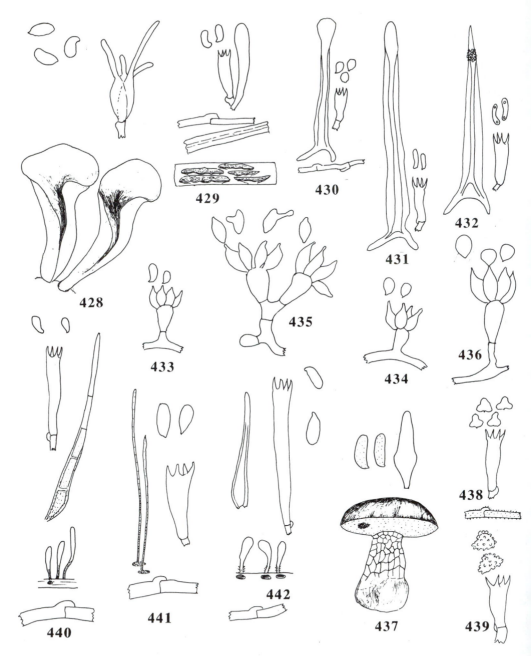

Plate 40 Hymenomycetes. 428 *Tremiscus helvelloides*; 429 *Trichaptum abietinum*; 430 *Tubulicrinis accedens*; 431 *T.glebulosus*; 432 *T.subulatus*; 433 *Tulasnella allantospora*; 434 *T.pruinosa*; 435 *T.violacea*; 436 *T.violea*; 437 *Tylopilus felleus*; 438 *Tylospora asterophora*; 439 *T.fibrillosa*; 440 *Typhula erythropus*; 441 *T.phacorrhiza*; 442 *T.quisquiliaris*.

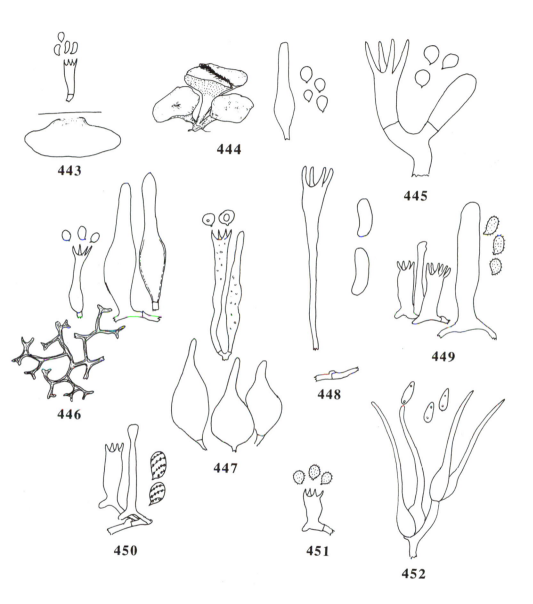

Plate 41 Hymenomycetes. 443 *Tyromyces chioneus*; 444 *Uloporus lividus*; 445 *Uthatobasidium ochraceum*; 446 *Vararia ochroleuca*; 447 *Vesiculomyces citrinus*; 448 *Vuilleminia comedens*; 449 *Xenasma pruinosum*; 450 *X.pulverulentum*; 451 *Xenasmatella tulasnelloidea*; 452 *Xenolachne longicornis*.

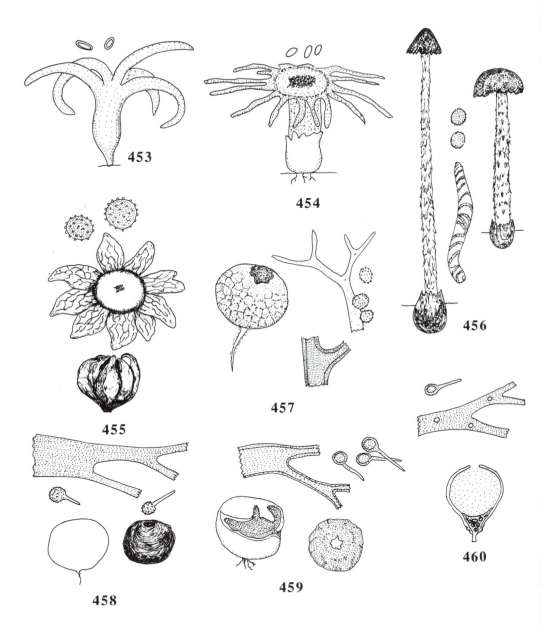

Plate 42 Gasteromycetes. 453 *Anthurus archeri*; 454 *Aseroe rubra*; 455 *Astraeus hygrometricus*; 456 *Battarraea phalloides*; 457 *Bovista aestivalis*; 458 *B.nigrescens*; 459 *B.plumbea*; 460 *Bovistella radicata*.

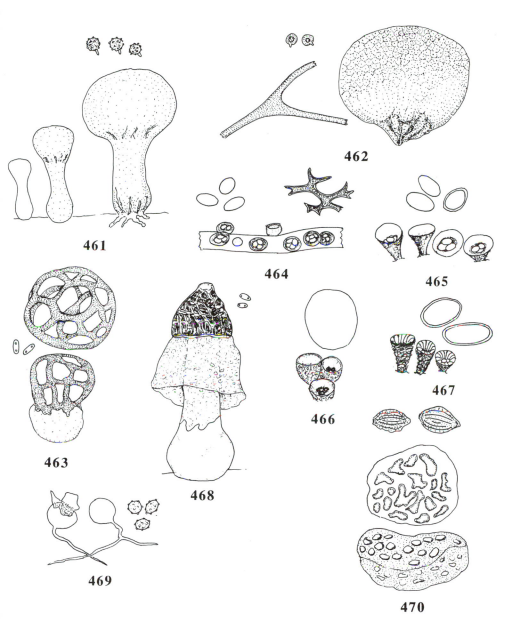

Plate 43 Gasteromycetes. 461 *Calvatia excipuliformis*; 462 *C.utriformis*; 463 *Clathrus ruber*; 464 *Crucibulum laeve*; 465 *Cyathus olla*; 466 *C.stercoreus*; 467 *C.striatus*; 468 *Dictyophora duplicata*; 469 *Gastrosporium simplex*; 470 *Gautieria morchellaeformis*.

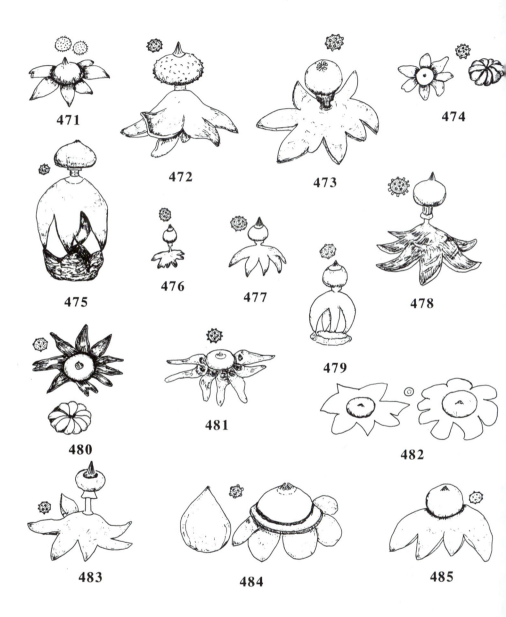

Plate 44 Gasteromycetes. 471 *Geastrum badium*; 472 *G.berkeleyi*; 473 *G.coronatum*; 474 *G.floriforme*; 475 *G.fornicatum*; 476 *G.minimum*; 477 *G.nanum*; 478 *G.pectinatum*; 479 *G.quadrifidum*; 480 *G.recolligens*; 481 *G.saccatum*; 482 *G.sessile*; 483 *G.striatum*; 484 *G.triplex*; 485 *G.vulgatum*.

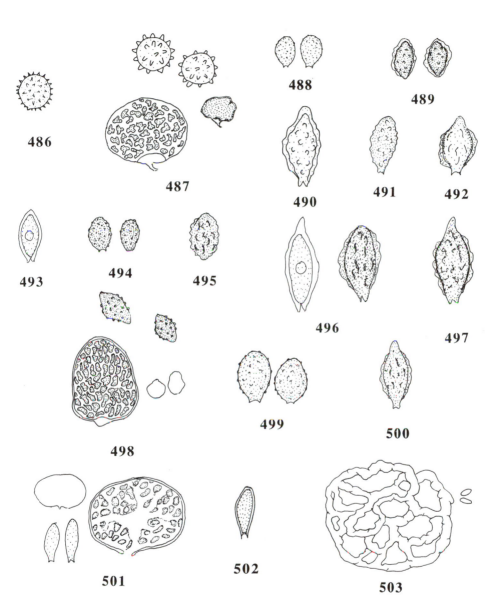

Plate 45 Gasteromycetes. 486 *Gymnomyces xanthosporus*; 487 *Hydnangium carneum*; 488 *Hymenogaster albus*; 489 *H.arenarius*; 490 *H.citrinus*; 491 *H.griseus*; 492 *H.hessei*; 493 *H.luteus*; 494 *H.lycoperdineus*; 495 *H.muticus*; 496 *H.olivaceus*; 497 *H.sulcatus*; 498 *H.tener*; 499 *H.thwaitesii*; 500 *H.vulgaris*; 501 *Hysterangium nephriticum*; 502 *H.thwaitesii*; 503 *Ileodictyon cibarius*.

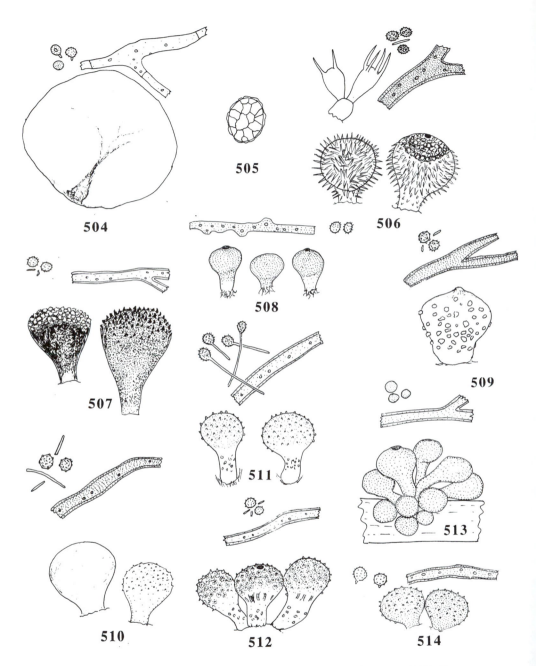

Plate 46 Gasteromycetes. 504 *Langermannia gigantea*; 505 *Leucogaster floccosus*; 506 *Lycoperdon echinatum*; 507 *L.foetidum*; 508 *L.lividum*; 509 *L.mammaeforme*; 510 *L.molle*; 511 *L.pedicellatum*; 512 *L.perlatum*; 513 *L.pyriforme*; 514 *L.umbrinum*.

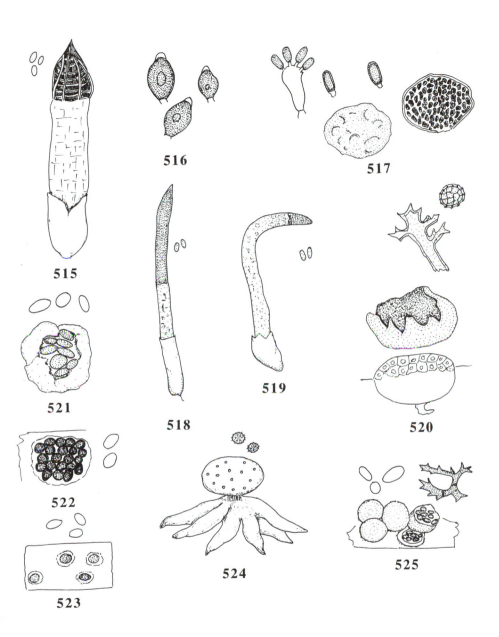

516

517

515

521

518

519

520

522

523

524

525

Plate 47 Gasteromycetes. 515 *Lysurus cruciatus*; 516 *Melanogaster ambiguus*; 517 *M.broomeianus*; 518 *Mutinus bambusinus*; 519 *M.caninus*; 520 *Mycenastrum corium*; 521 *Mycocalia denudata*; 522 *M.duriaeana*; 523 *M.minutissima*; 524 *Myriostoma coliforme*; 525 *Nidularia farcta*.

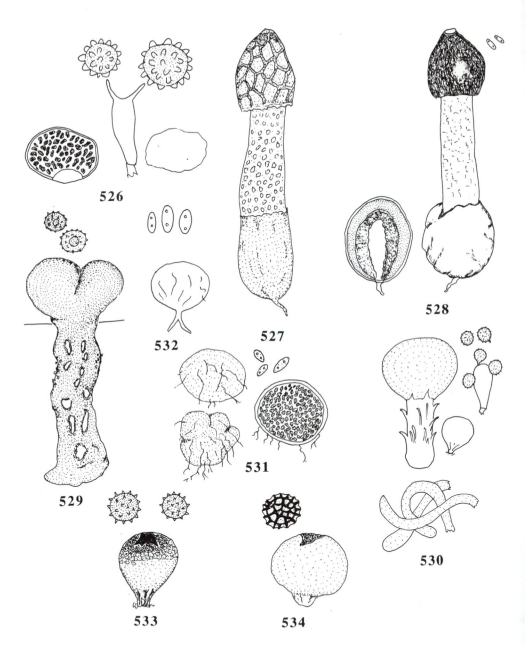

Plate 48 Gasteromycetes. 526 *Octaviania asterosperma*; 527 *Phallus hadriani*; 528 *P.impudicus*; 529 *Pisolithus arhizus*; 530 *Queletia mirabilis*; 531 *Rhizopogon luteolus*; 532 *R.roseolus*; 533 *Scleroderma areolatum*; 534 *S.bovista*.

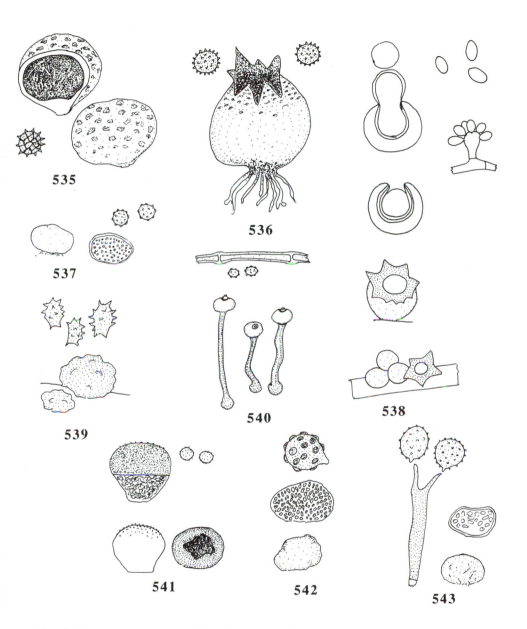

Plate 49 Gasteromycetes. 535 *Scleroderma citrinum*; 536 *S.verrucosum*; 537 *Sclerogaster compactus*; 538 *Sphaerobolus stellatus*; 539 *Stephanospora caroticolor*; 540 *Tulostoma brumale*; 541 *Vascellum pratense*; 542 *Wakefieldia macrospora*; 543 *Zelleromyces stephensii*.

INDEX

DATE DUE

~~MAY 1 4 1997~~			